Explosion Safety for Industries

Industrial explosions can occur anywhere where combustible powders and/or flammable vapours are handled, such as facilities dealing with wood, flour, starch, sugar, rubber, plastics, pharmaceuticals, oil, gas, and other industries. Explosions can lead to loss of lives and colossal costs to companies found culpable for failing to follow safety protocols.

Explosion Safety for Industries presents a set of preventative measures, rules, and standards to mitigate the risk of explosions by preventing potential explosive atmosphere formation. Coverage includes risks and hazards, equipment selection, earthing systems, static electricity, and lightning protection, among others. This book aims to help those responsible for the design, installation, operation and maintenance of process plants comply with the provisions of international safety legislation. Readers will develop a thorough understanding of the science behind explosions and the methods and measures needed to prevent accidents irrespective of their workplace or industry.

This book is a comprehensive guide for industrial explosion prevention and protection for safety engineers, specialists, and managers, as well as for physicists, chemists, mechanical engineers, and electrical engineers.

Explosion Safety for Industries
Prevention and Protection

Estellito Rangel Junior

CRC Press
Taylor & Francis Group
Boca Raton London New York

CRC Press is an imprint of the
Taylor & Francis Group, an **informa** business

Designed cover image: Getty Images

With thanks to Lucas Almeida for providing illustrations for this book.

First edition published 2026
by CRC Press
2385 NW Executive Center Drive, Suite 320, Boca Raton FL 33431

and by CRC Press
4 Park Square, Milton Park, Abingdon, Oxon, OX14 4RN

CRC Press is an imprint of Taylor & Francis Group, LLC

ISBN: 978-1-032-79707-6 (hbk)
ISBN: 978-1-032-81466-7 (pbk)
ISBN: 978-1-003-50000-1 (ebk)

DOI: 10.1201/9781003500001

Typeset in Times
by Apex CoVantage, LLC

Contents

Preface

Statistics show a large number of industrial fires and explosions happening in a wide range of industries such as oil, food, paints, insecticides, fertilizers, and chemicals. And with the construction of increasingly larger factories processing large volumes of flammable materials, this number is likely to increase in the coming years due to a variety of causes, from equipment malfunction to human error and inadequate procedures.

According to the National Fire Protection Association (NFPA), an average of 37,000 fires occurs annually at industrial and manufacturing properties in the United States. The number of fatalities is extremely limited—just one or two deaths per year on average—but the costs for loss of inventory and damages make up around US$ 1 billion in a year.

Among the various causes, and considering so different and complex industrial environments, some examples are worth highlighting:

- Combustible dust is often overlooked but highly deadly, and it is a major cause of fire in food manufacturing, woodworking, chemical, manufacturing, metalworking, and pharmaceutical segments.
- Hot work is often involved in fires due to the production of sparks, which can ignite nearby flammable materials.
- Industries that process flammable liquids and gases are the most involved with industrial fires and explosions, accounting for about 40% of all incidents, especially in plants with high pressures and temperatures.
- Equipment malfunctions are another common cause of industrial fires. Furnaces, boilers, and related heating equipment all have the potential to ignite when not maintained properly.
- Electrical fires constitute 14% of industrial incidents, including among their causes wiring that is not up to the standard, overloaded circuits, arc flash, lightning, and static discharge.

Therefore, to avoid explosions and fires in industrial environments, it is necessary to know the processes and the characteristics of the equipment and facilities, identify risk situations, be informed about events that have occurred in similar plants, and analyse their causes.

This book aims to present these situations and help in the preparation of a Fire and Explosion Risk Management Programme integrated into the companies' safety programme.

The 12 chapters cover everything from the physical and chemical characteristics of flammable gases and combustible dusts to the safety requirements of industrial electrical installations, including grounding, requirements for classified areas, static electricity, and lightning protection systems. They conclude by publishing recommendations for good practices for the execution of services and considering the impacts of human error on safety.

   Some of the differences between the laws of the countries for the prevention of explosions, particularly those of Brasil, Europe, and the United States, are also discussed. Some of the most catastrophic fire and explosion events are mentioned, as well as their causes and recommendations to prevent them from happening again. At the end of the chapters, a list of technical documents that were used as references is cited, which makes it possible to obtain additional information. The aim is to use simple and objective language, which helps the reader quickly understand the concepts involved. Technology and human knowledge are always evolving, and the author would like to thank you in advance for any suggestions for improving this book, which will help us to promote effective safety for facilities and workers in industrial processes.

   Enjoy your reading!

Foreword

The accidents involving Piper Alpha (North Sea, the UK, 1988, 167 deaths), the P-36 platform (Brasil, 2001, 11 deaths), and the FPSO Cidade de São Mateus (Brasil, 2015, 9 deaths) are examples of accidents where the presence of an explosive atmosphere led to explosions and/or fire scenarios resulting in death, multiple injuries, and financial losses due to damage to the asset and operational interruption.

Major accidents such as these require in-depth investigation into their causes. One of the objectives of the accident report is to share lessons learnt with the public. These lessons have the potential to not only improve the processes of the company where the disaster occurred but also provide support to technical engineering associations to update the requirements for standardizing their projects and processes.

In this book on explosion prevention in the industry, the author emphasizes the requirements for electrical installations in classified areas, encompassing design, installation, and maintenance. And when it comes to industry requirements and best practices, the author of this book is an authority. Estellito has decades of experience in designing, commissioning, and maintaining industrial electrical systems in large projects in the oil and gas industry. He has also participated in technical committees for writing and reviewing standards and recommended practices, such as the IEC 60079 series, and has published dozens of articles on various aspects of explosive atmospheres and electrical equipment.

I am certain that this book will contribute to the training of many professionals who also aim to design and maintain safer industry installations.

Caroline Maurieli de Morais,
Regulatory Specialist,
ANP—Brasilian National Agency of Petroleum, Natural Gas and Biofuels

Acknowledgements

Thanking my family for their understanding and support is a priority for the writer, as they were deprived of my company on several occasions, whether because I needed to describe a case study, do research, was planning to include a new chapter, or even looking for old inspection reports where rare non-conformities would be important to share with readers.

Thanking my colleagues is essential because our knowledge develops through interaction and the exchange of experiences at technical conferences.

I cannot forget to thank the instructors of the courses I attended because thanks to them I learnt about the subjects, had my doubts cleared, became interested in the topics, and acquired complementary literature to delve deeper into the topics.

The idea that a book writes itself is far from being true because even if it took me months to write, I could only do so because of the inspiration that my experiences have given me and which were intensified by the people who contributed to them.

At this point, for fear of forgetting someone, I would like to express my gratitude to all the people who have influenced my career, including the editors of technical magazines; conference organizers; professors; students who, with their curiosity, demanded that I always stay up to date to answer their questions; and also to my family, who gave me the necessary support and shared my joy when I finished the text.

Thanks a ton to everyone!

About the Author

Estellito Rangel Junior is a legally registered electrical engineer in Brasil, with extensive experience in electrical systems for the oil and gas segment, having worked for 33 years at Petrobras, the Brasilian state-owned oil company ranked among the largest ones in the world.

At Petrobras, he contributed by providing technical support to the following departments: Exploration and Production (offshore platforms and the floating, production, storage, and offloading—FPSO units), Refining (electrical systems up to 69 kV), Engineering (construction of gas compression stations, expansions of oil refineries, and performing compliance audits), and Safety (risk analysis and procedures for services in hazardous (classified) locations).

As an internal instructor at Petrobras, he provided training on explosive atmospheres and electrical safety to approximately 1,500 company's employees. Due to his participation in the Petrobras Technical Standardization Working Groups, he was nominated to participate in Study Commissions of the Brasilian Association of Technical Standards (ABNT), coordinating the development of Brasilian standards for explosive atmospheres.

He writes the first Brasilian technical column entirely dedicated to explosive atmospheres, published monthly since 2003 by the *Eletricidade Moderna* magazine, to address standardization, certification, legislation, and inspections. He was nominated as the first Brasilian expert to join the Technical Committee (TC-31—Equipment for Explosive Atmospheres) of the International Electrotechnical Commission (IEC), where he participated in the development of international standards for explosive atmospheres due to flammable gases and combustible dusts.

He has presented several technical papers at international conferences, including the Petroleum and Chemical Industry Conference (PCIC), in the United States, Canada, Mexico, Italy, and Brasil editions. With over 150 published technical papers, he has won the Abracopel Award (Brasilian Association for Awareness of the Dangers of Electricity) five times, for writing the best technical paper of the year about electrical safety.

He provides consultancy services on industrial electrical systems, electrical safety, and explosion prevention to large Brasilian and foreign companies in the food, pharmaceutical, oil, petrochemical, and gas segments; leads audits on technical and legal requirements for industries with hazardous areas; and conducts area classification studies for industries subject to explosive atmospheres of flammable gases and combustible dusts.

1 Gas and Vapour Cloud Explosions

1.1 INTRODUCTION

A major safety concern in industrial plants that process flammable materials is the occurrence of fires and explosions. The locations where flammable liquids, gases, or vapours (and combustible dusts) exist or are prone to exist, in sufficient quantities to produce an explosive atmosphere, are called "hazardous areas". In hazardous areas, specially designed electrical and electronic equipment must be used to ensure that these do not act as sources of ignition and specific installation requirements apply.

There are many safety requirements for the design of process equipment involving flammable substances at high pressures and flows. Despite them, the high number of explosions and fires has drawn the attention of researchers, not only due to the material losses and damage to the environment but also due to the number of victims.

Explosions cannot be totally prevented wherever an explosive environment, necessitated by the production processes, is present. Although human endeavours are fallible, we need to adopt preventive measures to an extent where explosions become possible only after many safety barriers are broken. The preventive measures should, in effect, reduce the risk of an explosion to a minimum level.

An explosion occurs when there is simultaneity of three elements:

- Fuel (as flammable liquids, gases, or vapours)
- Oxygen (available as 21% in air)
- Ignition source

Therefore, the implementation of an explosion safety management system in the industrial units where flammable materials' processes are present is essential for the health of the business.

1.2 DEFINITION OF KEY TERMS

Key concepts for understanding the explosion phenomenon are highlighted in the following subsections.

1.2.1 EXPLOSION

The term "explosion" is defined as a rapid expansion in volume of a given amount of matter, associated with an extreme outward release of energy, usually with the generation of high temperatures and release of high-pressure gases, causing the

DOI: 10.1201/9781003500001-1

development of pressure or shock waves. Explosions can be either detonations or deflagrations depending on their flame speed.

1.2.2 Deflagration

Deflagration refers to a type of explosion where the flame front moves through the combustible material at a speed less than the speed of sound (which is approximately equal to 340 m/s). It is characterized by the rapid but subsonic burning of a material, creating a significant amount of heat and gas pressure but without the shock waves typical of a detonation.

One of the most common types is the deflagration of a flammable mixture which is characterized by subsonic propagation rates relative to the unburned gas.

1.2.3 Detonation

Detonation is a type of explosion in which the combustion front moves at supersonic speeds, often exceeding the speed of sound. It generates shock waves with speeds of about 1 km/s which are much more intense and destructive than deflagration due to the rapid release of energy.

Detonation is an explosion of fuel–air mixture. Compared to deflagration, detonation does not need to have an external oxidizer, and it is more destructive than deflagration. Oxidizers and fuel mix when deflagration occurs.

1.2.4 Flammability Limits

A flammable atmosphere (gaseous mixture of combustible and oxidant, also known as explosive atmosphere) is one that, when ignited, will propagate flames beyond the influence of the ignition source. All combustible gases and combustible vapours (of liquids or solids) form flammable mixtures over a limited range of combustible concentrations, depending upon temperature, pressure, and nature of the oxidant.

The critical combustible concentrations are known as flammable limits (or explosive limits), where the fuel-lean concentration is called Lower Flammable Limit (also Lower Explosive Limit—LEL), and the fuel-rich concentration is called Upper Flammable Limit (also Upper Explosive Limit—UEL). Lower Explosive Limit is the limit below which the mixture will not burn due to lean mixture, and UEL is the limit above which the mixture will not catch fire, as it will be too rich.

In the petroleum industry, typical flammable gases are natural gas, propane, butane, methane, acetylene, and carbon monoxide.

These limits are usually in volume percent and refer to homogeneous combustible vapour-oxidant mixtures. As the flammable limits vary with temperature, and as the vapour formation of a combustible in its liquid or solid state is strongly dependent on temperature, it is useful to construct a flammable concentration-temperature diagram as shown in Figure 1.1.

In Figure 1.1, the flammable mixtures of a combustible liquid-air system may consist of mists (droplets + saturated vapour + air), saturated vapour–air mixtures (vapour pressure curve), or neat vapour–air mixtures (region beyond saturated

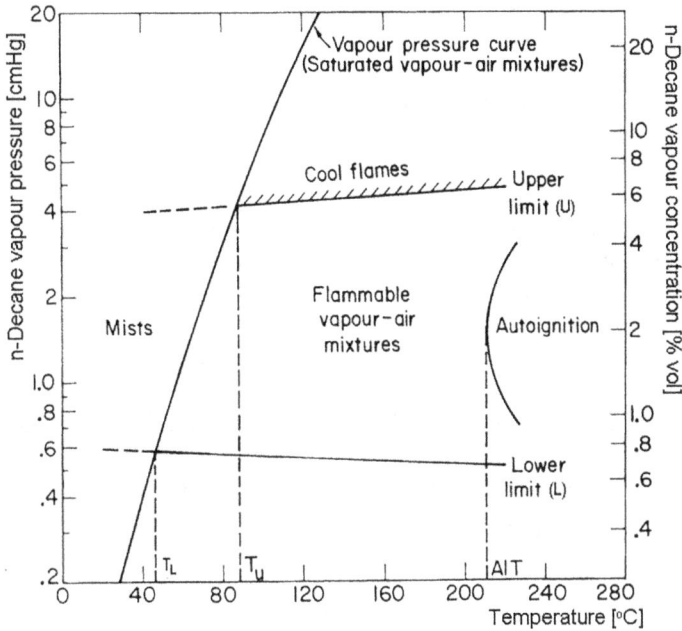

FIGURE 1.1 Flammability and vapour pressure diagram for the n-decane vapour–air system at atmospheric pressure.

vapour temperatures). The temperature range over which the liquid can form flammable vapour concentrations is defined by its temperature limits of flammability (T_L—lower and T_u—upper). Intersections of these temperature limits with the vapour pressure curve define the concentration limits of flammability (L and U) at liquid–vapour equilibrium conditions.

At higher temperatures, flammability domains for the neat vapour–air mixtures become widened, particularly on the fuel-rich side. Eventually, the temperature gets sufficient to produce the auto ignition of the mixtures, as illustrated in Figure 1.1 for n-decane.

Generally, limits of flammability are widened by increased temperature, pressure, oxygen concentration, and energy of the ignition source. The limits are also widened by turbulence and buoyancy effects, which increase the flame propagation rate.

Essentially, flammability limits may be classified according to their propagation mode: Upward or downward. Upward limits have the greatest practical value because they include the buoyancy effect and reflect the maximum flammability hazard; downward limits exclude buoyancy and are of greater fundamental significance.

In addition to buoyancy, the differences between molecular diffusivities of the combustible and oxidant can contribute to the wide variations often observed between these limits. As with ignition limits, flammability limits are narrowed by heat losses to vessel walls, and propagations can become impossible at some quenching diameter.

It is important to say that the tests to define explosive limits for methane (with air) were carried out in different laboratories and gave different results. There were cases in which the limits found varied more in the UEL than in the LEL.

In any case, the results were very dependent on the size and shape of the vessels used. The vessels used were not large enough to allow the mixture to be passed through the flame to be ignited solely by the heat of combustion itself after the heat of the initial ignition source had been removed.

The United States Bureau of Mines published test results for methane and many other gases in 1952 and 1965. Most of the results were obtained in tests conducted in the vertical glass tube apparatus, first described in detail in 1939.

At that time, it was already known that upward flame propagation in a tube of about 5 cm diameter by 1 m length gave, in most cases, the best performance in the explosive range, although there were some exceptions, particularly in the UEL, where larger diameter tubes gave higher UEL, as was typical for acetylene.

Limits of flammability may be used to determine guidelines for the safe handling of volatile chemicals. They are used particularly in assessing ventilation requirements for the handling of gases and vapours. NFPA 69 provides guidance for the practical use of flammability limit data, including the appropriate safety margins to use.

Although wider flammable explosive limits may be considered "more conservative", it is not the responsibility of the testing laboratory to include safety factors in the experiments, as they should not be inserted during the measurement.

Nowadays, there is a discrepancy between the explosive limits of methane obtained according to the ISO/IEC 80079-20-1 standard and the American ASTM E681 standard. While the ISO/IEC indicates the explosive range of methane as 4.4 to 17% vol., the American standard establishes 5 to 15% vol., which can impact the area classification studies.

The ASTM test methods aim to produce the best representation of flammability parameters and rely upon the safety margins imposed by the application standards, such as NFPA 69. On the other hand, European (ISO/IEC) test methods aim to result in a conservative representation of flammability parameters. For example, in ASTM E681, LEL is the calculated average of the lowest go and highest no-go concentrations, while the ISO/IEC test methods report the LEL as the minimum of the five highest no-go concentrations.

Note that for hydrocarbons, the break point between no flammability and flammability occurs over a narrow concentration range at the lower flammability limit, but the break point is less distinct at the upper limit. For materials found to be "non-reproducible" that are likely to have large quenching distances and may be difficult to ignite, such as ammonia and certain halogenated hydrocarbons, the lower and upper limits of these materials may both be less distinct. That is, a wider range exists between flammable and non-flammable concentrations.

The flammability limits depend on the test temperature and pressure. The ASTM test method is limited to an initial pressure of the local ambient or less, with a practical lower pressure limit of approximately 13 kPa (100 mmHg). The maximum practical operating temperature of this equipment is approximately 150°C.

It is important to say that the test methods are used to measure and describe the properties of materials, products, or assemblies in response to heat and flame under controlled laboratory conditions and should not be used to describe or appraise the fire hazard or fire risk of materials, products, or assemblies under actual fire conditions. However, results of the test method may be used as elements of a fire risk assessment that takes into account all of the factors pertinent to an assessment of the fire hazard of a particular end use.

1.2.5 MINIMUM IGNITION ENERGY

Figure 1.2 shows the relationship between the ignition energy and the concentration of air/gas mixtures of three gases: Propane–air, ethylene–air, and hydrogen–air. It can be noticed that the lowest point of each curve means the "Minimum Ignition Energy" (MIE) for each type of mixture, for the given environmental conditions, where ignition takes place.

1.2.6 AUTO-IGNITION TEMPERATURE

The ignition temperature of a substance, whether solid, liquid, or gaseous, is the minimum temperature required to initiate or cause self-sustained combustion, independently of the heating or heated element. Ignition temperatures observed under one set of conditions may be changed substantially by a change of conditions. For this reason, ignition temperatures should be looked upon only as approximations.

Some of the variables known to impact ignition temperatures are percentage composition of the vapour or gas–air mixture, shape and size of the space where the ignition occurs, rate and duration of heating, kind and temperature of the ignition source, catalytic or other effect of materials that may be present, and oxygen concentration.

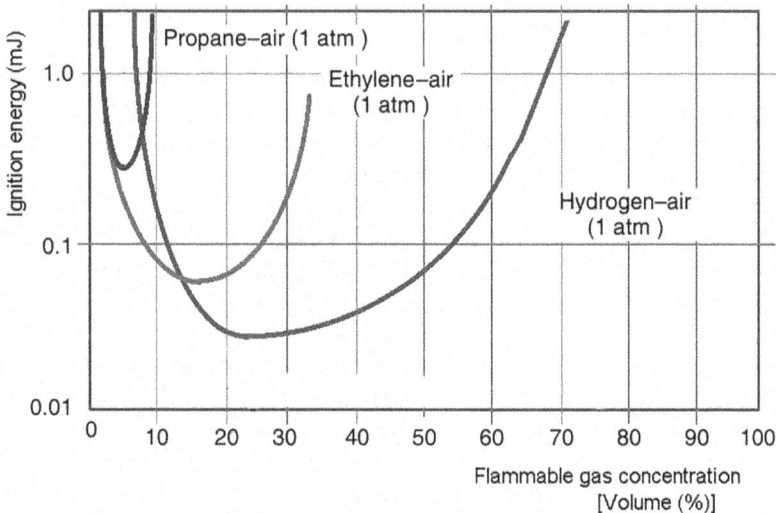

FIGURE 1.2 Curve showing flammability limits.

1.2.7 Flash Point

It is the lowest temperature at which liquid gives up enough vapour to maintain a continuous flame. An "ignitable mixture" is a mixture within the flammable range (between upper and lower limits) that is capable of the propagation of flame away from the source of ignition when ignited.

The test apparatus used for the measurement of flash point is normally one of two types, of which there are several variants. These are called generally open cup and closed cup flash point testers. For most liquids, the flash point determined by the closed cup method is slightly lower (in the region of 5%–10% when measured in degrees Celsius) than that determined by the open cup method.

1.2.8 Limiting Oxygen Concentration

It is the Minimum Oxygen Concentration (MOC) below which combustion is not possible with any fuel mixture. It is expressed as the volume percentage of oxygen. It is also called Maximum Safe Oxygen Concentration (MSOC).

1.2.9 Relative Vapour Density

The relative vapour density of a material is the mass of a given volume of the material in its gaseous or vapour form compared with the mass of an equal volume of dry air at the same temperature and pressure. It is often calculated as the ratio of the relative molecular mass of the material to the average relative molecular mass of air. A relative vapour density greater than 1 indicates that the gas or vapour is heavier than air and will tend to sink and fall towards the earth; a relative vapour density less than 1 indicates that the vapour is lighter than air, and the gas (or vapour) will rise in a still atmosphere.

1.2.10 Vapour Pressure

Vapour pressure is a quantified description of a liquid's ability to release vapours. Atmospheric pressure is the downward force exerted by the atmosphere on the surface of a liquid; vapour pressure is a measure of the opposing force exerted by vapour pushing upward from the surface of the liquid.

These pressures can be measured using a variety of units; atmospheres (atm), millimetres of mercury (mmHg), and pounds per square inch (psi) are some examples. Atmospheric pressure is relatively constant at any point above sea level; the sea-level value is 760 mmHg. Variation will occur at different elevations.

Atmospheric pressure is lower in the mountains than at the seashore. The pressure released by a liquid is a function of the characteristics of that liquid. For example, kerosene has a vapour pressure of 5 mmHg at 37.7°C, which indicates that it will release very little vapour at normal temperatures. Toluene, a common solvent, has a vapour pressure of 21 mmHg at 20°C, which indicates that it will release considerably more vapours than kerosene. Ethyl acetate, another common solvent, has a vapour pressure of 73 mmHg at 20°C, indicating that it will release more vapours than toluene.

1.3 MYTHS ABOUT FLAMMABLE GAS/CLOUD EXPLOSIONS

Lack of information produces myths like the following:

a) *Flammable gases and vapours only explode in confined spaces.*

   Myth! While confinement can increase the severity of an explosion, flammable gases and vapours can also explode in open spaces if the conditions are right, such as the presence of an ignition source and the correct fuel-to-air ratio.

b) *All flammable gases have the same explosive limits.*

   Myth! Different gases have different LELs and UELs. For example, propane has an LEL of 2.1% and a UEL of 9.5%, while methane has an LEL of 5% and a UEL of 17%.

c) *You can always smell a gas leak before it becomes dangerous.*

   Myth! Some gases are odourless, and even those with added odorants can sometimes go undetected. Relying solely on smell is not a safe method for detecting gas leaks.

d) *Flammable gas explosions are always caused by a spark or flame.*

   Myth! While sparks and flames are common ignition sources, static electricity, hot surfaces, and even certain chemical reactions can also ignite flammable gas mixtures.

e) *Once a gas explosion occurs, the danger is over.*

   Myth! After an explosion, there can still be lingering flammable gases, structural damage, and secondary fires, making the area dangerous even after the initial blast.

1.4 GAS/VAPOUR CLOUD EXPLOSIONS

There are many references to explosions in the oil and gas sector, and among the causes, these factors can be found: Poor maintenance, incorrect installation, and/or inadequate design of electrical equipment, as pointed out in the official incident reports.

In such a complex activity—which involves a large number of highly energized systems, flammable liquids, and combustible gases running through pipelines at high pressures and temperatures—the risks increase in proportion to the complex dynamics of devices interacting with each other. It is intriguing, however, that despite all the technological advances, they continue to occur. One hypothesis would be the lack of knowledge of risk scenarios.

The Brasilian Government Agency for Oil and Gas (ANP) points to the absence of safety procedures or inadequate identification of the actions necessary for risk mitigation and prevention as the main causes of accidents in the oil industry. The scenario is not very different from what occurs in other countries. According to the World Offshore Accident Databank (WOAD), 81% of reported accidents with human causes have their root cause in unsafe, or non-existent, procedures. Eighty-six per cent of accidents had no human causes identified, but this does not mean that human causes were not present but that there was a lack of systematic identification analysis, or that there were human organizational failures.

Understanding risk scenarios necessarily involves carrying out a comprehensive risk analysis of operational activities. However, these are not always treated with due seriousness. Sometimes, not even the recommendations generated in the risk analyses prepared in the design phase are effectively implemented; in other cases, the problem lies in the lack of periodic reviews. Changes made when the offshore platform, or refinery, is already operating can generate new and previously unknown risk scenarios—a common example is of equipment and systems dedicated to safety, which increase the complexity of the operation.

The *Out of Control*, published in 2003 by the Health and Safety Executive (HSE) in the UK, showed that 44% of accidents caused by problems in the control system of process installations occurred due to failures in specification. And it revealed, also, that in 20% of the incidents, the cause was related to changes after commissioning, and in 15% of cases investigated, the problem arose during the operation and maintenance phase.

Accident histories are full of examples caused by failures in change management. Engineers are prolific in designing improvements to optimize the efficiency of the installation, which begins to occur after some time of operation with the initially designed system.

In principle, this is good, but it brings with itself the need for great care, as many highly well-intentioned changes have been responsible for numerous major accidents. Hence the need to adopt strict procedures for change management, recommended by HSE management policies but often forgotten in practice. To investigate the causes, a basic item to ensure continuous improvement still faces the lack of willingness of companies to disclose internal investigations, which they classify as "confidential matters".

Most vapour cloud explosions offshore would fall into the category of deflagrations. A typical vapour cloud explosion on an offshore installation would start as a slow laminar flame ignited by a weak ignition source such as a spark. As the gas mixture burns, hot combustion products are created that expand to approximately the surrounding pressure.

As the surrounding mixture flows past the obstacles within the gas cloud, turbulence is created. This turbulence increases the flame surface area and the combustion rate and, further, increases the velocity and turbulence in the flow field ahead of the flame, leading to a strong positive feedback mechanism for flame acceleration and high explosion overpressures.

Table 1.1 shows four remarkable gas explosion events in the oil and gas sector with additional information.

The losses shown in Table 1.1 were estimated on the basis of initial court rulings and may, in fact, have reached much higher figures.

1.5   STATISTICS OF FIRES AND EXPLOSIONS

The oil and gas segment is the most representative of process facilities with flammable gases and vapours.

Table 1.2 presents only a portion of statistics related to fires and explosions in offshore facilities, limited at the U.S. Outer Continental Shelf (OCS), collected by the Bureau of Safety and Environmental Enforcement (BSEE), which aims to establish

TABLE 1.1

Some Explosion Events in the Oil and Gas Sector

Date	Event	Losses
July 6, 1988, North Sea, the UK	An explosion on the Piper Alpha platform, operated by Occidental Petroleum, after a gas leak, is considered the largest offshore disaster. Occidental Petroleum was found guilty, among other things, of having inadequate maintenance procedures.	167 deaths. At the time of the disaster, the platform accounted for approximately 1/10 of the North Sea's oil and gas production. Losses around US$3.4 billion.
March 15, 2001, Campos Basin, Brasil.	Explosion on the P-36 platform, operated by Petrobras. The analysis of the accident, which occurred less than a year after the platform went into operation, identified non-conformities related to procedures and, also, deficiencies in the area classification plan.	11 deaths. Losses around US$ 500 million at the time of the accident. In 2018, the penalties applied by government organisms reached the value of US$ 350 million.
February 11, 2015, Espírito Santo Basin, Brasil.	Explosion at FPSO Cidade de São Mateus, operated by BW Offshore. According to the accident report, the lack of planning and risk analysis during the maintenance of process pipelines in the pump room caused a gas leak that was probably ignited by an electrostatic discharge. The fact that employees were ordered to enter the pump room, even when the gas alarm was activated, caused the majority of deaths.	9 deaths and 26 injuries. Losses around US$ 100 million at the time of the accident, due to FPSO repairs.
March 23, 2005, Texas, US	BP Texas City refinery explosion. An electrical fault caused a vapour cloud explosion.	15 deaths and over 170 injuries. The initial damages were estimated at US$ 1.5 billion.

TABLE 1.2

Explosion Incident Statistics in U.S. OCS, by BSEE

	2018	2019	2020	2021	2022
Gas releases	19	20	73	44	57
Fires	77	84	87	117	126
Explosions	3	4	1	4	1
Injuries	171	222	160	164	199
Fatalities	1	6	6	2	1

appropriate actions to reduce the likelihood of recurrence of these incidents and to increase safety and environmental protection. Since its establishment in 2011, BSEE has been the lead federal agency charged with environmental protection related to the offshore energy industry, primarily oil and natural gas, on the U.S. Outer Continental Shelf (OCS).

What causes more serious accidents in the process industry: Failure to follow procedures or inadequate procedures? The IOGP Safety Performance Indicators provide some very interesting insights, including an answer to this question. They are based on the numbers of Tier 1 and Tier 2 Process Safety Events reported by participating IOGP member and companies, broken down for onshore and offshore, drilling and production, activities, and consequences and material released. The Tier 1 and Tier 2 data presented have been normalized against reported work hours associated with drilling and production activities to provide Process Safety Events rates.

"Tiers" are process safety performance indicators, with an emphasis on events and near-misses. There are four Tiers, and their meanings are:

- Tier 1—Process releases with significant consequences (reactive): Process safety events (PSE) that resulted in significant consequences, such as releases of hazardous materials, fires, or explosions. Tier 1 events are a key performance indicator (KPI) used to track and assess process safety performance, particularly in industries like oil and gas. They represent events with the highest potential consequences within a process safety framework.
- Tier 2—Potential severity but with less critical consequences (reactive): Incidents that did not cause serious consequences but involved failures in important safety barriers.
- Tier 3—Barrier performance indicators (proactive): Observations of failures or deviations indicating that control systems or barriers are not functioning as designed.
- Tier 4—Management indicators (proactive): Monitoring, auditing, inspection, and training activities aimed at preventing failures before they manifest.

Consistent monitoring and analysis of these indicators allow for the anticipation of systemic failures and help build a culture where prevention is at the core of decision-making.

According to the publication, as shown in Table 1.3, in the last ten years, 328 of the most serious process safety events, known as "Tier 1", were caused by "organizational conditions" due to "inadequate procedures", while 184 were caused by the "human act" of not following procedures.

This alerts that the probability of the failure being in the prescribed procedure is much greater than in its non-compliance. So, it is paramount that those responsible for issuing safety procedures receive greater training and knowledge of the processes, so that they can develop reliable procedures in order to promote effective safety.

The procedure must not only address the description of the tasks required to perform the service but must also correctly specify the required Personal Protective Equipment (PPE) and the tools that will be used, so that they do not act as sources of ignition in hazardous areas.

TABLE 1.3

IOGP Statistics of Safety Procedure Compliance in the Last 10 Years

	2014	2015	2016	2017	2018	2019	2020	2021	2022	2023	10 Years (Total)
Inadequate work/ standards procedures	18	37	44	30	32	40	22	25	28	52	**328**
Not following procedures (deviation)	8	29	13	23	8	33	11	15	26	18	**184**

Note: Bold terms are the sum of all year events, given the total number of events in 10 years period.

1.6 THE CAUSES OF EXPLOSIONS

Gas explosions are perhaps the most common type of explosion in the oil and gas sector. These occur when flammable gases, like methane, come into contact with an ignition source. Such incidents can be traced back to leaks in pipelines, malfunctioning equipment, or inadequate ventilation systems. It is not uncommon for gas explosions to result in escalation which is the release of more hydrocarbon inventory resulting in gas and liquid fires or another explosion.

Complex facilities and high-risk activities, such as start-up, testing, sampling, draining, flushing, and venting, can have the associated risks reduced by good design, as well as good working practices. Other high-risk activities like construction and maintenance, relief valve testing, and simultaneous operations can also have reduced associated leak risks through good design and planning.

Design has a big influence on the prevention of ignition. Hot surfaces, electrical equipment, and static electricity are usually fixed ignition sources that need to be avoided with the recognition of the potential explosion hazard.

Selection of a higher specification, different location, or gas-tight enclosure can reduce the ignition risk. Temporary ignition sources, such as hot work or poorly earthed scaffolding or equipment, need to be eliminated, or controlled, by design and operating procedures.

Explosions in the oil and gas industry can be caused by a variety of factors including human error and unforeseen events. Some of the most common causes of explosions are:

Hot work
Welding, cutting, and other hot work activities can produce sparks and heat, posing a risk of igniting nearby flammable materials.

Human error
Mistakes or negligence by employers or workers, such as improper maintenance, inadequate training, or failure to follow safety procedures, lead to explosions.

Gas leaks
Leaks in pipelines, valves, or storage tanks can release flammable gases or
liquids, which may ignite upon contact with an ignition source.
Natural disasters
Extreme weather events, such as hurricanes or earthquakes, can damage the
industry infrastructure and increase the risk of explosions.

1.6.1   HIGH-LOSS EVENTS

Incidents in the oil and gas industry often cause injuries or loss of life. But even
incidents that do not claim human lives have consequences, in the form of harm to
the environment and local wildlife, potential contamination of water supplies, and
negative outcomes for nearby communities. In order to illustrate the financial impact
that flammable gas/vapour explosions can generate, some of them are selected next.

a) **BP Texas City refinery**

This explosion took place on March 23, 2005, and occurred when a
hydrocarbon vapour cloud ignited, leading to a massive explosion. The blast
was so powerful that it caused nearby buildings and structures to collapse,
including trailers used as temporary offices for workers.

The explosion claimed the lives of 15 workers and injured more than 170
others, being one of the deadliest industrial accidents in the United States
in recent history. The primary cause of the explosion was determined to
be the overfilling of a raffinate splitter tower, which resulted in a release of
flammable liquid and vapour.

The vapour cloud then found an ignition source, causing the explosion.
Investigations revealed that there were deficiencies in process safety management systems, equipment maintenance, and employee training at the refinery.

Property damage from the explosion amounted to US$ 200 million; costs
of repairs and deferred production amounted to over US$ 1 billion; BP pled
guilty to federal environmental crimes, for which it paid US$ 50 million; BP
paid at least around US$ 2.1 billion in civil settlements; and, additionally,
BP paid US$ 100 million in fines to the Occupational Safety and Health
Administration (OSHA) and to the government of Texas for environmental
violations. This disaster is the world's costliest refinery accident.

b) **Deepwater Horizon platform**

Due to this explosion, which occurred on April 10, 2010, BP has been the
target of multiple lawsuits, several of them brought by the government, for
both criminal violations and civil regulatory violations, such as the Clean
Water Act. In a settlement considered the largest of its kind in American
history, BP agreed to pay about US$ 20 billion to the federal government
and the five states (Louisiana, Texas, Mississippi, Alabama, and Florida)
impacted by the environmental disaster.

After the accident, the company had its rating downgraded by rating
agencies, saw its shares plummet, and had to sell billions of dollars in assets.

All the instalments and interest are estimated to exceed US$ 70 billion.
This is because many lawsuits filed by individuals and small businesses

impacted by the disaster are still pending, and these must be added to the expected payments by 2032.

c) **Buncefield Oil Storage Depot**

The final report of the Major Incident Investigation Board (MIIB) of this explosion occurred on December 11, 2005, was written in 2008 and released in February 2011. The investigation found that Tank 912 at the Buncefield oil storage depot was being filled with petrol. The tank had a level gauge that employees used to monitor the level manually and an independent high-level switch which would shut off inflow if the level got above a certain setpoint.

On Tank 912, the manual gauge was stuck, and the independent shut-off switch was inoperative, meaning that the tank was being "filled blind" with petrol (i.e., being filled without a clear indication of the level). Eventually, Tank 912 filled up completely, the petrol overflowed through vents at the top, and formed a vapour cloud near earth level, which ignited and exploded. The fires from the explosion then lasted for five days. Total estimated losses were around US$ 1.2 billion.

1.7 PREVENTIVE MEASURES

Once the statistics of fire and explosion events in units that process flammable gases/vapours are known, it is necessary to verify whether the conditions that gave rise to those events could be present in our industrial unit and thus implement the appropriate preventive measures.

There are three categories of gas-related risks:

- Risk of fire and/or explosion due to flammable gases
- Risk of poisoning due to toxic gases
- Risk of suffocation due to the lack of oxygen.

In this chapter, we will discuss the first category: Risk of fire and explosion preventive measures for facilities with flammable liquids and gases. The preventive measures are composed of administrative controls and technical measures.

1.7.1 Administrative Controls

Administrative controls focus on safety policies and procedures, providing the necessary training to workers as well. This involves designating a dedicated safety team for the worksite, with the duty of conducting the safety management system and organizing periodic safety audits. Among the administrative controls, we highlight the points given as follows.

1.7.1.1 Training of Workers

It is important that only properly trained workers are authorized to perform services in hazardous areas. The employer must provide the principles of explosive atmospheres, hazards of vapour clouds and flammable gases, as well as of appropriate safety procedures, included as part of the technical training on the equipment and processes used in the industry, considering they are complex and require special training to be operated safely.

The objective must be a promotion of a safety culture, where all employees are actively involved in identifying and mitigating risks. As a monitoring tool, a periodic audit programme, which is recommended to be carried out by an external company with professionals specialized in explosion prevention, must be in place.

1.7.1.2   Written Instructions and Work Permits

Work in hazardous areas must be carried out in accordance with written instructions provided by the employer, and a work permit system must be applied, authorizing the execution of services that may cause direct, and also, indirect risks when interacting with other operations. Work permits must be issued by a competent person before the start of service, taking into consideration the risks that may be present during the development of the tasks.

1.7.1.3   The Area Classification Study

It is up to the area classification study, based on the estimation of the release rate, the local ventilation, and the chemical characteristics of the flammable products, to identify how far from each probable source of release the concentration of the flammable mixture will remain above its LEL and mark it in the plant design documents. This is important to allow the correct specification of the electrical and electronic equipment to be installed in those regions, ensuring the plant safety against explosions.

As the area classification drawings are also consulted for the elaboration of safety procedures in the operation and maintenance of the industrial unit, it is essential that they be elaborated on the basis of the real characteristic data of the industrial plant. More details on area classification will be shown in Chapter 3.

1.7.1.4   Signage of the Hazardous Areas

A considerable number of workers in industrial plants are outsourced, with considerable turnover. This highlights the need for a signage system that alerts to the risks of explosive atmospheres formation. In compliance with the directives that are established to alert employees on the risks existing in the workplace, industries must promote a signage campaign for areas subject to explosive atmospheres. An example of signage for hazardous areas, complying with the ISO recommendations, is shown in Figure 1.3.

The sign has the same symbol for hazardous areas adopted in Europe (a yellow triangle with the letters Ex in black, inside). The red background highlights the risk of fire, and it also contains important information, such as the type of Zone, the Equipment Group, and the Temperature Class of the electrical equipment that can be safely used on that location.

In this way, this sign complies with international principles of avoiding texts on safety warning signs (like: "no open flames", "no smoking", etc.), which require time to read and are inefficient, especially when dealing with foreign workers and visitors, and immediately provides an alert that only suitable electrical equipment may be used in the region to avoid explosions. On the bottom line, it indicates the hazardous area drawing number of that region, where the workers can consult in order to get more information, if necessary.

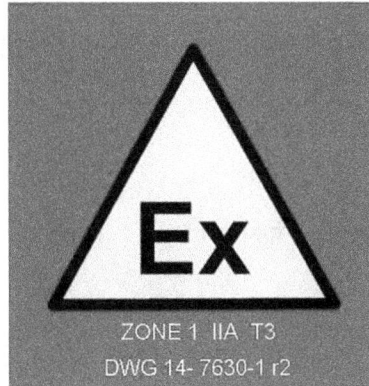

FIGURE 1.3 Alert sign for a hazardous area Zone 1, Equipment Group IIA, and Temperature Class T3.

Unfortunately, it is a common mistake to find signs with only phrases about what can and cannot be done in hazardous areas, because the worker will need to read all the text and only then, he can know what he has to do. Not to say that foreign workers usually would not be able to read a different language. It should be noted that the signage does not replace the necessary training for all workers assigned to perform services in hazardous areas.

If necessary, workers must be alerted by means of optical and/or acoustic alarm signals also and evacuated before explosion conditions are reached.

1.7.2 TECHNICAL MEASURES

The technical measures consist of the preparation of a risk assessment, the control of ignition sources, and the implementation of an inspection programme.

1.7.2.1 Risk Assessment

It is performing detailed risk analysis to identify potential sources of flammable vapour or gas cloud formation. To begin the assessment of the risk of fire or explosion at a facility and to determine the necessary measures for ensuring safety, it will be necessary to gather information on the flammability characteristics of the liquids (and their vapours) and gases on site. This information relates to conditions for forming a flammable vapour atmosphere, the vapour's ease of ignition, oxygen levels to prevent explosion, and the pressure of explosion.

Note that processing conditions such as temperature, pressure, and the presence of second phases/mixtures influence the flammability characteristics of liquids. Unfortunately, very limited data are available on the flammability properties of liquids or liquid mixtures at low or elevated temperature and pressure conditions. Therefore, if the facility is processing flammable liquids at conditions other than ambient temperature and atmospheric pressure, it will be needed to experimentally determine their flammability properties at representative process conditions.

Risk analysis should not usually be entrusted to workers performing the tasks. Only experienced professionals with expertise in the applicable techniques can perform reliable risk analysis.

1.7.2.2   Risk Management

Properly dealing with fire and explosion hazards requires the control of three key elements:

a) **Vapour concentration**

Methods of controlling the vapour concentration at a safe level include maintaining the liquid below its flash point temperature or providing adequate ventilation to keep the concentration of vapour under its Lower Flammable Limit (LFL). Providing adequate ventilation would maintain the concentration of vapour in the work area, and perhaps within the process equipment, under the LFL. Ventilation should include all floor areas and pits where flammable vapours can collect.

Another preventive measure is to provide local exhaust ventilation for capturing vapours at the sources of release and transport it to a safe exhaust point, scrubber, thermal oxidizer, or filter.

Installation of gas detection systems to continuously monitor the presence of flammable vapours or gases and the implementation of alarm systems that immediately alert operators to the gas detection in a situation close to dangerous concentrations will be required.

b) **Oxidant concentration**

Limiting oxidant concentration (LOC) is the minimum quantity of an oxidant that is required for flame propagation through a homogenous gas–vapour mixture. Replacing oxidant with an inert gas can keep oxidant under the LOC. Inerting does not replace the flammable vapours but, instead, reduces the concentration of oxygen to a level insufficient to support combustion. Note that LOC depends on the nature of the fuel as well as on the type of inert gas employed. In the context of combustion, oxygen is the most common oxidant, but other substances like chlorine or ozone can also act as oxidants.

c) **Ignition sources**

Control of ignition sources involves identifying and eliminating all credible ignition sources that would have sufficient energy to ignite the flammable atmosphere under both normal and foreseeable abnormal operations. Typical ignition sources include open flames, friction/impact sparks, electrical arcs and sparks, and electrostatic discharges. Open flames in a hydrocarbon facility usually result from welding or hot work operations, smoking, and the facility flare. These sources are normally controlled through operational measures, supplemented by physical separation.

Facility electrical systems and components can provide a convenient source of ignition within hydrocarbon occupancies whenever the design, installation, or maintenance is substandard. Failures in electrical systems can present themselves as available ignition sources for hydrocarbon vapour releases. Due to this, all electrical and electronic equipment must be certified under the relevant standards, such as IEC 60079 series.

In processes where heat is applied (e.g. drying of solvent wet materials), it is necessary to keep the temperature below the auto-ignition temperature (AIT). Additionally, it is necessary to verify the necessity of insulating/shielding hot surfaces.

Regarding the prevention of accumulation of static charges on conductive items, it is important perform an effective bonding and earthing, in order to put them very close to the zero electrical potential. A resistance to earth less than 10^6 ohms generally suffices for this purpose, but it is necessary to verify if a local regulation defines a different maximum value, and a periodic inspection and testing of the bonding and earthing system will be required.

1.7.2.3 Maintenance and Inspections

Industrial installations with hazardous areas, before starting operations, must be verified against explosion safety. Among the necessary conditions to ensure protection against explosions, an inspection under the orientations given at IEC 60079–14 and IEC 60079–17 is included. These verifications must be carried out by persons who are competent in the field of electrical and electronic equipment for hazardous locations.

Some attention points, where releases of a combustible vapour into the atmosphere during normal operations can occur, are:

- **Safety or pressure relief valves, which release to an atmospheric vent**
  They produce a vapour cloud, which, even though the release may be remote from the facility, may drift. Atmospheric storage tanks are normally fitted with pressure-vacuum relief valves to reduce vapour emission and evaporation losses to atmosphere.
- **Pumps and compressors seals**
  The more the pump seal wears, the more difficult it is to prevent leakage through the seal. Historically, the industry has much evidence of the problems with pump seals, and fire hazards areas are required on almost all pump-handling hydrocarbon products.
- **Process system drains**
  Process equipment liquid drains should be provided with a sealed drainage system where it is practical, and backpressure from the system is not a concern. Open drain ports should be avoided, and separate sewage and oil water drains provided. Surface drainage should be provided to remove liquid spills immediate from the process area.
- **Sample points**
  Open vessel collection means should be avoided, as spillage may occur due to an inappropriate operation of the sample valve. Automatic sampling methods are commonly available which can eliminate the need of manual hand sampling processes. The sampling points should be located where an adequate dispersion of released vapours will occur and is easily accessible.

In Chapter 11, the requirements for an inspection and maintenance plan on electrical installations in hazardous locations are presented.

1.8   LEGAL REQUIREMENTS

The employer should implement a complete, duly documented safety management system, customized to the specific conditions of the company and communicated to all employees. Explosion prevention will then constitute a subsystem of the company's corporate safety management process, which encompasses all situations that may offer a fire and/or explosion risk, including activities like electrical services, confined spaces, and hot work.

In the EU, it was declared that explosion protection is essential to ensure the safety of workers, as described in the Directives ATEX 2014/34/EU (which established requirements to equipment and protective systems intended for use in potentially explosive atmospheres) and ATEX 1999/92/EC (which established the minimum requirements for improving the safety and health protection of workers potentially at risk from explosive atmospheres). It is recognized that in the event of an explosion, the uncontrolled effects of flames and pressures, the presence of harmful reaction products, and the consumption of ambient oxygen required for breathing endanger the life and health of workers.

Pursuant to Directive ATEX 1999/92/EC, the employer must have an assessment of the risks to the health and safety of workers. This requirement is made more precise in this Directive by imposing on the employer the obligation to draw up and keep up to date an explosion protection document or a series of documents which satisfy the minimum requirements laid down in this Directive. The explosion protection document includes the identification of the assumptions, the assessment of the risks, and the definition of the specific measures to be taken to ensure the health and safety of workers exposed to the risks arising from explosive atmospheres.

In the United States, the OSHA has regulations that require owners of industries handling flammable products to implement safety measures to prevent explosions. Specifically, OSHA's 29 CFR 1910.106 regulation addresses the handling, storage, and use of flammable liquids and outlines requirements for design and construction, ventilation, ignition sources, and storage to mitigate explosion and fire hazards. The United States Code of Federal Regulations (CFR) contains the major laws (statutes and regulations) issued by federal agencies. These laws govern private sector employers in order to protect workers from health and safety hazards.

A flammable vapour or gas cloud explosion occurs when a vapour or gas mixture reaches a concentration within the explosive limits (LEL and UEL) and is exposed to an ignition source.

The rapid exothermic oxidation reaction that occurs in the presence of oxygen results in the instantaneous release of energy in the form of heat and pressure.

1.9   CLOSING REMARKS

Vapour cloud and flammable gas explosions are catastrophic events that can be prevented with the appropriate implementation of preventive measures. Understanding the processes' characteristics, learning from significant historical events, and applying robust preventive measures are essential to ensuring safety in the plant.

To prevent explosions in industrial plants prone to the occurrence of vapour cloud and flammable gases, it is essential to integrate technical measures and operational procedures into a corporate safety system, where the various disciplines have their interdependencies duly characterized, and employees are duly aware of the risks inherent to the industrial unit.

This includes:

- Knowing the characteristics of the flammable products processed
- Identifying the hazardous areas, where there is a risk of explosive atmospheres forming
- Selecting electrical equipment appropriately
- Implementing a maintenance and inspection plan
- Developing procedures where safety takes priority over production
- Conducting periodic safety audits
- Training conscientious and motivated professionals
- Managing emergencies and process changes.

Hazard identification, control of ignition sources, continuous monitoring and effective training are essential steps to mitigate risk and protect human lives, the environment, and industrial infrastructure from the dangers of vapour cloud and flammable gas explosions. It is important to realize that no guide or framework, by itself, is strong enough to guarantee operational safety. Appropriate management practices can reduce risks. To manage them properly, the operational safety management model must be adapted to your environment.

The problem lies in the effective implementation of existing procedures. The responsibility, first and foremost, lies with the company's senior management and, then, with the operational management, including the health and safety system management. If the company is certified according to the ISO 45001 standard, the certifying entity also has some responsibility. Third, there are the regulatory bodies (in Brasil, the ANP) which are legally responsible for verifying compliance with the regulations.

Asset management is a tool that can contribute to greater safety in facilities, but, today, as its use has been limited to contain costs, it is necessary to establish stricter requirements to avoid major accidents, typical of the oil and gas segment, as the high number of accidents and deaths shows that worker safety has been compromised.

BIBLIOGRAPHY

API RP 752—*Management of hazards associated with location of process plant permanent buildings*. API, 2024.

ASTM E681-09—*Standard test method for concentration limits of flammability of chemicals (vapours and gases)*. ASTM International, 2015.

ATEX 1999/92/EC—Directive 1999/92/EC of the European Parliament and of the Council of 16 December 1999 on minimum requirements for improving the safety and health protection of workers potentially at risk from explosive atmospheres.

ATEX 2014/34/EU—Directive 2014/34/EU of the European Parliament and of the Council of 26 February 2014 on the harmonization of the laws of the Member States relating to equipment and protective systems intended for use in potentially explosive atmospheres.

Coward, H. F. and Jones, G. W.—*Bulletin 503—Limits of flammability of gases and vapours.* U. S. Bureau of Mines, 1952.

Fatal Accident Investigation Report—*Isomerization unit explosion, Texas City, 23rd March, 2005.* BP America, Inc., Dec. 9, 2005.

Health and Safety Executive (HSE)—*Buncefield: Why did it happen?* Competent Authority Strategic Management Group, UK, 2011.

IEC 60079-14—*Explosive atmospheres—Part 14: Electrical installation design, selection and installation of equipment, including initial inspection.* Ed. 6.0. International Electrotechnical Commission, 2024.

IEC 60079-17—*Explosive atmospheres—Part 17: Electrical installations inspection and maintenance.* Ed. 6.0. International Electrotechnical Commission, 2023.

ISO 45001—*Occupational health and safety management systems.* International Organization for Standardization, 2018.

ISO 7010—*Graphical symbols—safety colours and safety signs—registered safety signs.* International Organization for Standardization, 2019.

ISO/IEC 80079-20-1—*Explosive atmospheres—Part 20-1: Material characteristics for gas and vapour classification—Test methods and data.* International Organization for Standardization/International Electrotechnical Commission, 2017.

Kletz, T. A.—*What went wrong—case histories of process plant disasters.* Gulf Publishing Company, 1985.

Major Incident Investigation Board (MIIB) —*The final report of the Major Incident Investigation Board.* Health and Safety Executive, UK, Dec., Vols. 1 & 2, 2008.

Morais, Caroline M., Almeida, D. P. S. and Alves, V. F. B.—Regulating human factors in Brasilian Oil & Gas offshore installations. In: *Offshore Technology Conference Brasil,* Proceedings, IBP, Oct., 2023.

NFPA 69—*Standard on explosion prevention systems.* National Fire Protection Association, 2024.

Offshore Incident Statistics, 2018–2022—*Bureau of Safety and Environmental Enforcement.* Available at: www.bsee.gov/stats-facts/offshore-incident-statistics

OSHA 29 CFR 1910.106—*Code of federal regulations—flammable and combustible liquids.* Occupational Safety and Health Administration, 2023.

Out of control: Why control systems go wrong and how to prevent failure. Health and Safety Guidance 328, Health and Safety Executive, UK, 1999.

Process safety—*Recommended practice on Key Performance Indicators.* International Association of Oil and Gas Producers (IOGP), Report 456, Nov., 2018. Available at: www.iogp;org

Rangel Jr., Estellito—Evaluation of explosive atmospheres formation in industries with flammable materials—A critical risk analysis. *In: 5º Congresso Geral da ABRISCO, (Brasilian association of risk analysis, process safety and reliability),* on line. Recife, Brasil, 2021.

Rangel Jr., Estellito—Explosion risks management for oil & gas installations. In: *Rio Oil & Gas Conference,* Paper 2021. Rio de Janeiro, Brasil, 2018. Available at: https://bit.ly/3C7VckE

Rangel Jr., Estellito—As causas das explosões. *Section EMEx, Eletricidade Moderna Magazine,* São Paulo, ed. 573, Sep/Oct. 2023, p. 68.

Rangel Jr., Estellito—Limites de explosividade. *Section EMEx, Eletricidade Moderna Magazine,* São Paulo, ed. 471, Jun. 2013, p. 124.

Rangel Jr., Estellito, Naegeli, Guilherme S. T., Esposte, Jorge L. D.—Conscientização para atmosferas explosivas. In: *I Encontro Petrobrás sobre Instalações Elétricas em*

Atmosferas Explosivas—EPIAEx, Anais. Petrobras, CEN-SUD, Rio de Janeiro, 1994, pp. 12–17.

Safety Performance Indicators—*Process safety events—2023 data.* International Association of Oil and Gas Producers (IOGP). Available at: www.iogp.org

Sesseng, C., Storesund, K. and Steen-Hansen, A.—*Analysis of 985 fire incidents related to oil and gas production on the Norwegian continental shelf.* RISE Fire Research AS, 2015.

Silva, Bruno D. da, Morais, Caroline Maurieli, Almeida, Alex de, Ferreira, Nayara and Pires, Thiago—Explaining the explosion onboard FPSO Cidade de São Mateus from regulatory point of view. In: *Risk, reliability and safety: Innovating theory and practice—Proceedings of ESREL 2016,* 1st edition, Glasgow, Scotland, 25–29 Sep. Routledge, 2016.

The 100 Largest Losses 1974–2015—Large property damage losses in the hydrocarbon industry, 24th edition. Marsh & McLennan Companies, 2016.

Worldwide Offshore Accident Databank (WOAD)—*Det Norske Veritas Germaniscer Lloyd.* DNV GL, 2017.

2 Combustible Dust Explosions

2.1 INTRODUCTION

It is a priority for the safety of companies that process combustible dusts (such as those in the chemical, pharmaceutical, food, and timber sectors) to know the properties of their products and the characteristics of their processes in order to control the risks of fires and explosions. A presentation of combustible dusts and typical industries where they can be found is shown in Table 2.1.

To produce a dust explosion, the following conditions must occur simultaneously:

- Combustible dust must be in suspension.
- The concentration of dust in suspension must be above its Lower Explosive Limit (LEL).
- The dust must contain particles of adequate size.
- Air (oxygen) must be present.
- Ignition source should be with adequate energy.

In general, it can be said that it is more difficult to initiate a dust explosion than to initiate an explosion of a flammable gas or vapour atmosphere, because the energy required to ignite the dust is greater (with a magnitude of mJ) than that required to ignite gases (with a magnitude of μJ).

2.2 THE DUST EXPLOSION PARAMETERS

Industrial units that handle combustible dust need to be fully aware of the risks of explosion. Certain points in the process may be subject to the formation of clouds of combustible dust in adequate concentrations, such as grinding and unloading areas. Among the products with such characteristics are wheat flour, corn flour, soybean flour, sugar flour, and even metal powders. Table 2.2 shows statistics on these occurrences, collected by the Dust Science Academy, from accidents recorded in the United States, Canada, and other countries.

The phenomenon starts when the combustible dust deposited in various locations of the industrial plant is agitated and it causes clouds of dust to appear, which, in the presence of an ignition source with sufficient energy, can explode, causing what is known as a primary explosion. This first explosion causes other areas with deposited dust to enter into suspension and secondary explosions to occur, each one releasing more energy than the previous one, causing irreversible damage to property and taking lives. To properly control this risk, the first step is to correctly classify the area,

DOI: 10.1201/9781003500001-2

TABLE 2.1
Industrial Segments with Combustible Dusts

Industry	Products
Food	Flours, starches, sugars, powdered milk, corn meal, cocoa powder, whey, cereal, spices, and gluten.
Agriculture and grains	Wheat, corn, barley, oats, and soybeans.
Metals	Aluminium, iron, magnesium, nickel, niobium, tantalum, titanium, zinc, and zirconium dusts.
Wood	Sawdust and cellulosic dusts, including dusts from paper, cardboard, paper towels, facial tissue, and other paper products.
Plastics	Resins (melamine, epoxy, phenolic, etc.), polymers (polyethylene, polyvinylchloride, polyacrylamide, polyacrylonitrile, etc.), and copolymers.
Chemicals	Adipic acid, anthraquinone, dextrin, lactose, paraformaldehyde, sodium stearate, and sulphur.
Coal and other carbonaceous materials	Bituminous, subbituminous, lignite, charcoal, petroleum coke, and carbon black.
Other	Textiles, biosolids, soap, and pet food.

TABLE 2.2
Statistics of Events in Combustible Dust Environments

	2018	2019	2020	2021	2022
Fires	213	250	165	163	163
Explosions	68	75	60	52	50
Injuries	114	118	88	215	89
Fatalities	23	8	10	14	49

i.e., to conduct an engineering study, where professionals with the appropriate technical training, combined with experience in the process, knowledge of the physical and chemical properties of combustible powders, and expertise in the ventilation characteristics of the unit, identify the regions subject to the occurrence of explosive concentrations of the processed powders.

Once these regions have been properly identified, they are recorded in specific documents, called "area classification documents", which, in turn, will allow the electrical installation designer to select electrical equipment with the types of protection suitable for safe use in these situations. In audits carried out in companies with such processes, we found several area classification studies that were considered unreliable, which gave to the user a "pseudo-safety", as they assume that the project was undertaken considering a correct area classification.

The significance of hazards for any combustible dust depends on its characteristics. The important parameters that impact the behaviour of combustible dust explosions are presented in the next sections.

2.2.1 Dust Chemical Composition

It is important to identify the chemical composition of the dust to determine if it is combustible. Different compositions present in dust materials can lead to varying combustion behaviours and ignition potentials. For instance organic materials like wood, flour, sugar, or certain metals in powder form can be highly combustible; catch fire; and cause a devastating dust explosion. Conversely, inorganic substances such as minerals, or certain non-metallic compounds, may have lower combustibility tendencies for dust explosions. Analysing the composition aids in identifying the inherent risks associated with different types of combustible dust, enabling industries to assess and manage potential hazards effectively, by adopting appropriate safety protocols and preventive measures, based on the specific properties of the materials present.

Many dusts are known to be non-combustible because they include chemicals and/or compounds that do not react with oxygen (i.e. do not combust). For example, sand (SiO_2) will not combust and, therefore, will not be subject to combustible dust controls and abatements, regardless of other conditions or characteristics.

2.2.2 Oxidation Potential

Combustible dust oxidation, which includes fires, flash fires, deflagrations, and explosions, can and will eventually occur when sufficient energy is applied. This is referred to as the Minimum Ignition Energy (MIE) or, in chemical terms, the Activation Energy. The combustion will continue until the materials have reached their stable state, which is typically the lowest energy state. Sand, as noted earlier, is in its stable and lowest energy state, so it will not oxidize any further. However, many dusts are not in their lowest energy state and are combustible even in partially oxidized form, as an example, many iron oxides (such as rust particles) can oxidize further when a sufficient ignition source and oxygen are available.

2.2.3 Dust Size

Combustible dust includes finely divided combustible particulate solids that present a flash-fire or explosion hazard when suspended in air. Combustible particulate solids include dusts, fibres, fines, chips, chunks, or flakes. The term combustible dust generally refers to powders, fines, and fibres. Particulate smaller than 500 microns (μm), approximately the size of white granulated sugar, is considered to have the potential to present a flash-fire hazard or explosion hazard when suspended in air. Smaller particles are more easily dispersed, have longer airborne times, and are more susceptible to combustion. Smaller dust particles require lower energy and will reduce the minimum concentration and the minimum temperature of ignition. This is because smaller particles have a larger surface area per mass.

Combustible particulate solids having a minimum dimension greater than 500 μm generally have a surface-to-volume ratio too low to pose a deflagration hazard. However, exceptions occur, and combustibility should be determined by testing. This is especially true where dust combustibility can vary significantly depending on a number of conditions such as the type of equipment involved, ambient weather conditions, equipment maintenance status, and changes in product specification.

Dust particles' size influences a potential dust hazard. The key three elements in the fire triangle necessary for a dust explosion are: Fuel (the combustible dust), oxygen (present in the air), and ignition sources. The fire triangle represents the primary explosion. If one key element is missing from the fire triangle, such as open flames, then an explosion would not occur.

The addition of small particle, dust dispersion in a confined space, such as dust collectors, generates the "dust explosion pentagon" (oxygen, heat, fuel, dispersion, and confinement). If one element of the dust explosion pentagon is missing, a combustible dust explosion cannot occur.

Combustion occurs when oxygen is available in sufficient concentration to cause the combustible material to oxidize (burn) when ignited. In general, the finer the dust the greater the total dust cloud surface area. Therefore, a dust cloud with finer particulates will be more combustible and explosive.

2.2.4 MOISTURE CONTENT

Moisture in dust particles raises the required Minimum Ignition Temperature (MIT) of the dust because of the heat needed to vaporize the moisture. However, once ignition has occurred, the moisture in the air surrounding a dust particle has little effect on the course of a deflagration.

Moisture content cannot be considered an effective explosion preventive safeguard as most ignition sources provide more than enough energy to vaporize ambient moisture in the air and to ignite the dust. In order for moisture to prevent the ignition of dust by common sources, the dust typically has to be so damp that a dust cloud could not be formed.

To determine the moisture content, a representative sample of the material is weighed accurately by a balance. The weight of the initial mass of the sample (before drying) is taken together with the weight of the weighing boat using a balance.

The sample is placed in an oven set at a specified temperature and maintained at this temperature until the sample reaches a constant weight. The temperature and drying time may vary based on the material being tested and the requirements outlined in the specific test method. The drying is typically conducted under vacuum. After drying, the sample is allowed to cool in a desiccator before reweighing it. The weight of the dried sample is then taken to determine the final moisture content.

There is, however, a direct relationship between moisture content and the MIE, Minimum Explosive Concentration (MEC), Maximum Pressure (P_{max}), and maximum rate of pressure rise (K_{st}). For example, the ignition temperature of corn starch may increase as much as 50°C with an increase of moisture content from 1.6% to 12.5%.

2.2.5 Percent Combustible Dust

Many dusts are a combination of dust materials with only some of the dust particles being combustible. The percentage of total combustible dust present will increase or decrease the hazards. Thus, the percent combustible dust in the total dust cloud or pile impacts the overall combustibility. This is an important reason why dust sample testing should be conducted, before any area classification study.

2.2.6 Bulk Density

Bulk density is a measure of the mass of the dust per unit of volume and is used when there is a need to adjust the layer depth criterion for housekeeping in the National Fire Protection Association (NFPA) standards. A dust's bulk density is dependent upon, among other things, the particle size, shape, and chemical content of the dust. When combined with the measured volume of dust in a space, this provides an estimate of the total amount of combustible dust in that space.

For example, light-fluffy dusts have a low bulk density. Therefore, the amount of light-fluffy combustible dust in a space is less than the amount of a heavy-thick dust with a higher bulk density. For instance, paper tissue dust has a bulk density of approximately 320 kg/m^3, while the bulk density of carbon black is approximately 1,121 kg/m^3. As a result, the layer depth of paper tissue dust will have to be 3.5 times the layer depth of carbon black to be equal in mass.

2.2.7 Minimum Explosive Concentration

Minimum Explosive Concentration (MEC) is the minimum concentration of combustible dust suspended in air (measured in mass per unit volume) that will support a deflagration. MEC is analogous to the Lower Flammable Limit (LFL) for flammable gases. MEC is dependent on many factors, including particulate size distribution, chemistry, moisture content, ignition energy, and shape. Necessary MEC data is often determined by testing. Generally, the lower the MEC, the greater the hazard.

2.2.8 Minimum Ignition Energy

MIE is the lowest capacitive spark energy (electrical or electrostatic discharge) capable of igniting the most ignition-sensitive concentration of a combustible dust–air mixture. Some solids in dust form can ignite with very little energy—as low as 1 to 3 millijoules (mJ) in some cases. Necessary MIE data is often obtained by testing. Generally, the lower the MIE the greater is the potential for a dust cloud to ignite.

2.2.9 Minimum Ignition Temperature

This is the lowest temperature at which a dust cloud can ignite when exposed to an external ignition source such as a hot surface or flame. The MIT is determined by the ISO/IEC 80079–20–2, which specifies two test methods for determining the MITs of dust, for the purpose of selecting apparatus for safe use in such environments.

Method A

It is applicable to the determination of the minimum temperature of a pre-scribed hot surface which will result in the decomposition and/or ignition of a layer of dust of a specified thickness deposited on it (Minimum Ignition Temperature of a Dust Layer—MITDL). The method is particularly rele-vant to industrial equipment with which dusts are present on hot surfaces in thin layers exposed to the atmosphere.

Method B

It is applicable to the determination of the minimum temperature of a pre-scribed hot surface which will result in the ignition of a cloud of given sample of dust or other particulate solid (Minimum Ignition Temperature of a Dust Cloud [MITDC]). The method is intended to be carried out as a complementary test after determining the MITDL by method A of ISO/IEC 80079–20–2.

Note 1: Concerning method B—because the method of operation of the fur-nace gives short residence times for dust particles within it, this method of test is applicable to industrial equipment where dust is present as a cloud for a short time. This method of test is of small scale, and the results are not necessarily representative of all industrial conditions.

Note 2: Concerning method B—the method is not applicable to dusts which may, over a longer period of time than provided for in the test method, pro-duce gases from deposits generated during pyrolysis or smouldering.

ISO/IEC 80079–20–2 requires that approximately 0.1 g of combustible dust is placed in a dust holder at the top of a temperature-controlled furnace with an open bottom. The dust is dispersed by compressed air downwards past the hot surface of the fur-nace to see if ignition occurs and flames are produced. If the dust does not ignite, the furnace temperature is increased and the test repeated until ignition of the dust occurs. Once ignition has been established, the mass of the dust sample and injection pressure are varied to find the most vigorous explosive flame discharge.

The temperature of the furnace is then reduced incrementally until flame propa-gation is no longer observed. At this temperature, the dust mass and injection pres-sure are varied to confirm that no ignition is found over ten consecutive tests.

The MIT is the lowest temperature of the furnace at which flame is observed minus 20°C for furnace temperatures over 300°C or minus 10°C for furnace tempera-tures under 300°C. For items of plant such as driers, testing the MIT is important to prevent a dust explosion occurring through contact with a hot surface.

There is a difference between the MIT and the Minimum Auto-Ignition Tem-perature (MAIT) for combustible dusts, even though they are related concepts. In essence, MIT involves an external ignition source, while MAIT involves sponta-neous ignition.

2.2.10   Layer Ignition Temperature

Dust Layer Ignition Temperature (LIT) is the minimum temperature required to cause self-sustained combustion (catching fire), independent of any other source of heat (such as a spark or flame), for settled dusts on hot objects such as heaters, motors, and dryers. Dust LIT is a function of time, temperature, and the thickness of the layer and can be several hundred degrees below the dust cloud ignition temperature.

The LIT is determined with about 12.7 mm layer of the material placed on a pre-heated hot surface. A K-type thermocouple is stretched across the material on the hot surface. The temperature of the hot surface is kept constant, while observing the reaction of the material for about 2 hours.

Ignition of the material is defined as an observation of a flame, incandescence, or a rise in temperature of the material to at least 50°C above the temperature of the hot surface. At the ignition of the material, the temperature of the hot surface is lowered with a fresh sample until there is no ignition. ASTM E2021–15 recommends the particles' size of the material to be tested to be at least 95% less than 75 microns and having less than 5% moisture. In certain unique cases where there is no chance of segregation of the material during normal operation, the material can be tested as received.

2.2.11   Minimum Auto-Ignition Temperature

MAIT is the minimum temperature at which a dust cloud will auto-ignite when exposed to heated air. This temperature represents the lowest threshold at which a dust–air mixture, suspended in the atmosphere, can ignite spontaneously without the need for an external ignition source.

The test is used to assess the maximum operating temperature for electrical and non-electrical equipment used in areas where test material is present. The test uses a small quantity of the material, dispersed under defined air pressure in a Godbert-Greenwald furnace, which is preheated to a defined temperature. At the ignition of the material, the temperature of the furnace is reduced until there is no ignition at which point the concentration of the material and the pressure of dispersion are varied until the non-ignition is independent of the sample concentration and air pressure. ASTM E1491–06 recommends the particles' size of the material to be tested to be at least 95% less than 75 microns and less than 5% moisture.

2.2.12   Deflagration Index

The deflagration index, K_{st}, is a universal value used to characterize and compare the relative explosion severity of a dust cloud. It is a maximum, normalized rate of deflagration pressure rise and is obtained by multiplying the measured change in pressure over time (dP/dt) during the testing event by the testing vessel's volume to the 1/3 power, as in Equation 2.1.

$$K_{st} = \sqrt[3]{V}\left(\frac{dP}{dt}\right)max \qquad\qquad (2.1)$$

It is expressed in units of bar.m/s (pressure × velocity). The value is typically used in the design of process equipment and the associated deflagration/explosion, protective/preventive systems.

The $(dP/dt)_{max}$ is the maximum rate of pressure rise, before an explosion runs out of oxygen and fuel. It seeks to understand the result of pressure combustion acceleration inside a vessel and the rapid increase after an explosion has occurred. As the rate of pressure rise reduces after an explosion runs out of oxygen or fuel, the maximum rate of pressure rise analyses the point just before the explosion reduces. When looking at the test data via a graph, the maximum rate of pressure is the point at which an increasing gradient is at its highest. The measured $(dP/dt)_{max}$ is a function of the vessel size used in dust testing.

K_{st} values change with varying factors such as particle size (the smaller the particle, the higher the K_{st}), dust cloud concentration, and particle agglomeration. Dust agglomeration may result in a decreased K_{st}. K_{st} is not a value inherent to the substance because it varies based on the shape of the test instrument, ignition energy, time delays, and dispersion system that impact the turbulence degree. The conditional constant K_{st} is mainly determined by the chemical nature of the dust and its reaction surface area by keeping these boundaries.

Table 2.3 shows the classification of combustible dust hazard, based on K_{st} values.

Note 1: K_{st} is a measurement of explosion pressure, not of combustibility. A low K_{st} does not mean that your dust cannot burn and cause catastrophic damage. K_{st} only tells us how strong the potential explosive force is, and not how flammable the dust is. A K_{st} value of 0 means that dust is not combustible; its P_{min} and P_{max} are 0, and in a testing chamber, it cannot produce any explosion. Any dust with any K_{st} value above zero is potentially combustible and can cause an explosion; testing P_{max} can create an explosion in the testing chamber. From 0 to 200 (which includes many metal dusts), the explosion Class (under EN 14034–2) is 1—a "weak explosion".

Note 2: A "weak explosion" does not mean "no damage"! The catastrophic Imperial Sugar explosion that destroyed a building and killed 14 people in 2008 was caused by sugar with a K_{st} of 1! While such explosion has not a high pressure, it can create multiple large low-pressure explosions that blow apart the building and cause numerous deaths. A low K_{st} does not mean the facility is safe from combustible dust explosions! A K_{st} from 201 to 300 is a strong explosion (Class 2) and could include things like cellulose dust, other organic fine dust, and some metals and plastics. A K_{st} over 300 is a very strong explosion (Class 3). Aluminium and magnesium dust fall in this category.

TABLE 2.3

Combustible Dust Hazard Classes

Hazard Class	K_{st} [bar m/s]	Characteristic
ST0	0	No explosion
ST1	1–200	Weak explosion
ST2	201–300	Strong explosion
ST3	> 300	Very strong explosion

2.2.13 MAXIMUM PRESSURE

The maximum measured pressure of the deflagration, P_{max}, is also an indication of the explosion severity, along with the K_{st} value. P_{max} is normally expressed in units of barg. The higher the pressure (P_{max}) developed by a dust deflagration, the greater the hazard. Pressure resulting from a deflagration in a vessel will first result in deformation, followed by an explosion, if the P_{max} exceeds the strength of the vessel.

The significance of P_{max} lies in its use for categorizing dust as either deflagration-dust or detonation-dust, helping industries and regulatory bodies classify and manage the associated risks. Different combustible dusts have varying P_{max} values. P_{max} is measured by dispersing a material in a 20-litre spherical testing chamber while initiating ignition using strong ignition sources such as a chemical ignitor. Data created via the 20-litre spherical testing chamber demonstrates what the maximum explosion pressure would be under optimum concentration. The test is conducted by increasing the concentration of dust inside the closed chamber and measuring the pressure of the explosion until a maximum is reached.

P_{max} measurement is usually independent of vessel size. So, when testing materials and their hazardous properties in a 20-litre testing vessel, the same pressure will usually occur in a similar real-world environment.

Industries handling combustible dust must conduct thorough risk assessments to identify the types of dust present, their respective P_{max} values, and the conditions under which deflagration could occur. This information is instrumental in designing and implementing effective explosion protection systems such as venting, suppression, and containment measures.

2.2.14 PRESSURE RATIO

The Pressure Ratio (PR) is a parameter used to assess the explosiveness of a combustible dust cloud during a deflagration test. This ratio is determined by comparing the maximum pressure generated during the test deflagration to the initial atmospheric pressure measured before ignition.

It is essential to account for the rise in pressure caused by the igniter in the air at atmospheric pressure. The PR provides insights into the force exerted by the combustible dust when ignited, helping to gauge the potential risk and explosiveness of the material. The maximum pressure is adjusted for the rise due to the igniter itself in air at atmospheric pressure.

In the context of assessing whether a material is explosive, the ignition criterion relies on traditional values of PR. A commonly accepted threshold is set at a minimal value of PR equal to or exceeding 2.0. This criterion acts as a safety measure, indicating that materials with a PR equal to or exceeding 2.0 are considered potentially explosive. The use of this criterion serves as a standard for evaluating the hazardous nature of combustible dust, assisting in the development of safety protocols and preventive measures to mitigate the combustible dust explosions and risks associated with dust deflagrations in industrial settings.

Table 2.4 summarizes the key indexes for dusts classification, associated with their relevant test standards.

It is important to know some facts about the relevant standards to avoid misconceptions on what is been tested.

2.2.14.1 ASTM E1226

Its purpose is to provide standard test methods for characterizing the "explosibility" of dust clouds in two ways: First by determining if a dust is "explosible"—meaning a cloud of dust dispersed in air is capable of propagating a deflagration, which could cause a flash-fire or explosion. If a dust is found to be explosible, the test provides a way to quantify the potential explosion hazard by measuring parameters like maximum explosion pressure (P_{max}), maximum rate of pressure rise $(dP/dt)max$, and the explosibility index (K_{st}).

Results obtained by the application of the methods of ASTM E1226 pertain only to certain combustion characteristics of dispersed dust clouds. No inference should be drawn from such results relating to the combustion characteristics of dusts in other forms or conditions (e.g. ignition temperature or spark ignition energy of dust clouds,

---

TABLE 2.4
Some Test Methods Defining Properties of Combustible Dust

Property	Definition	Relevant Standards	Application
K_{st}	Dust deflagration index	ASTM E1226 / ISO/IEC 80079–20–2	Measures the relative explosion severity compared to other dusts
P_{max}	Overpressure generated in the test chamber	ASTM E1226	Used to design enclosures and predict the severity of the consequences
$(dP/dt)_{max}$	Maximum rate of pressure rise	ASTM E1226	Predicts the violence of the explosion; used to calculate K_{st}
MIE	Minimum Ignition Energy	ASTM E2019 / ISO/IEC 80079–20–2	Predicts the ease and the likelihood of ignition of a dispersed dust cloud
MEC	Minimum Explosive Concentration	ASTM E1515 / ISO/IEC 80079–20–2	Measures the minimum amount of dust, dispersed in air, required to spread an explosion. It is similar to the LEL of flammable gas–air mixtures.
MITDC	Minimum Ignition Temperature of a dust cloud	ISO/IEC 80079–20–2	The lowest temperature at which a dust cloud can ignite when exposed to an ignition source
MITDL	Minimum Ignition Temperature of a dust layer	ISO/IEC 80079–20–2	The lowest temperature at which a dust layer can ignite when exposed to an ignition source
LIT	Layer Ignition Temperature	ASTM E2021 / ISO/IEC 80079–20–2	The lowest temperature required to cause self-sustained combustion

ignition properties of dust layers on hot surfaces, ignition of bulk dust in heated environments, etc.) It is intended that results obtained by application of this test be used as elements of a Dust Hazard Analysis (DHA) that takes into account other pertinent risk factors and in the specification of explosion prevention systems.

2.2.14.2 ASTM E2019

This test method determines the MIE of a dust cloud in air by a high-voltage spark. The MIE of a dust cloud is primarily used to assess the likelihood of ignition during processing and handling, which is used to evaluate the need for precautions such as explosion prevention systems. The MIE is determined as the electrical energy stored in a capacitor which, when released as a high-voltage spark, is just sufficient to ignite the dust cloud at its most easily ignitable concentration in air. This test method described does not optimize all test variables that impact MIE. Smaller MIE values might be determined by increasing the number of repetitions or optimizing the spark discharge circuit for each dust tested.

In this test method, the test equipment is calibrated using a series of reference dusts whose MIE values lie within established limits. Once the test equipment is calibrated, the relative ignition sensitivity of other dusts can be found by comparing their MIE values with those of the reference dusts or with dusts whose ignition sensitivities are known from experience.

2.2.14.3 ASTM E1515

This test method covers the determination of the MEC of a dust–air mixture that will propagate a deflagration in a near-spherical closed vessel of 20 L or greater volume. The MEC is also referred to as the LEL or LFL. Data obtained from this test method provide a relative measure of the deflagration characteristics of dust clouds.

This test method should be used to measure and describe the properties of materials in response to heat and flame under controlled laboratory conditions and should not be used to describe or appraise the fire hazard or fire risk of materials, products, or assemblies under actual fire conditions. However, results of this test may be used as elements of a fire risk assessment that takes into account all of the factors that are pertinent to an assessment of the fire hazard of a particular end use.

2.2.15 Dust Explosibility Properties

Dust explosion risks are among the most easily overlooked in industry; they often go unnoticed until tragedy strikes. Despite their destructive potential, the science behind dust explosions remains misunderstood, leading to a lack of preparedness and unnecessary risk.

Dust combustibility testing should be a periodic process, not a one-off task. As materials, suppliers, processes, and operations change, so do the risks.

Facilities often undergo changes in materials, suppliers, processes, or equipment, all of which can impact the characteristics of the dust being generated. A dust that may have been safe previously could become unsafe under new conditions. "Ignition sensitivity" and "Explosion severity" testing, when there are changes to the process,

TABLE 2.5
Properties of Some Combustible Dusts Taken from NFPA Standards

Dust Material	LEL [g/m³]	MIT [°C]		Pmax [psi]	dP/dt [psi/s]
		Cloud	Layer		
Corn	30	400	250	136	6,000
Soy	35	520	190	133	6,500
Wheat	55	480	220	103	3,600
Flour	60	380	360	120	3,700
Coal	60	610	180	133	2,300

operation, and raw material sources, ensures that your safety measures remain effective as your operations evolve.

Therefore, it is essential to consider the physical and chemical characteristics of the combustible dust present in your industrial plant. Do not rely only on tables available in the literature, as shown in Table 2.5, because they serve to illustrate the differences. The same product can vary greatly, depending on the conditions of the sample to be tested.

2.3 THE RISKS OF COMBUSTIBLE DUSTS

When the risks of combustible dusts are underestimated, serious accidents and workers' health can be compromised. Based on a report by OSHA—Occupational Safety and Health Administration (US government agency), the questionnaire given next was prepared to allow readers to assess their knowledge of the risks of dusts. The answers are published at the end of this chapter.

1) How many American industries are at risk of forming explosive atmospheres from combustible dusts, according to OSHA?
   a) 1,000
   b) 5,000
   c) 10,000
   d) 30,000
   e) 50,000
2) Which of the risks listed below is not caused by the release of combustible dust into the workplace in pharmaceutical industries?
   a) Cross-contamination of active ingredients
   b) Fire or explosion due to combustion
   c) Allergic reactions in workers
   d) Multiple sclerosis
   e) Respiratory diseases

3) If you "discovered" that your industry process contained silica dust, what would be your first action?
   a) Buy respirators for all employees
   b) Send a sample of the dust for analysis
   c) Sweep or vacuum the dust that has settled on the floor
   d) Order a dust collection system
   e) Buy waterproof uniforms
4) What risk is present in the manufacture of metal articles involving cutting, grinding, pressing, welding, etc.?
   a) Workers breathing toxic fumes such as hexavalent chromium and beryllium
   b) Workers with irritated eyes and skin
   c) Machine coolants being sprayed and falling onto the floor
   d) Combustible dust explosion
   e) All of the above

2.4 MYTHS ABOUT COMBUSTIBLE DUSTS

Combustible dusts present considerable risks, which, strangely, are unknown to industrial plant managers. There is also considerable misinformation about the risks involved, which can result in events with serious consequences. Among the situations found in compliance audits, some of the "myths" that compromise the safety of companies, including their facilities and the people who work in them, are described next:

"Data on combustible dusts, necessary for the safe design of installations, can be easily obtained from tables published in standards, or even from websites on the Internet."

Myth! Not having the actual data is as dangerous as not knowing it. Small variations in particle size and composition can result in considerable differences in the explosiveness of the dust. And, in addition to being necessary to establish the basis for explosion protection design, correct data will allow the feasibility of inerting or other protective measures to be assessed.

"A single sample of combustible dust is sufficient. If industrial plants are 'similar', a sample from one of them will be valid for all others."

Myth! This is a very common myth among managers who are unaware of the risks of adopting incorrect data. Due to the savings obtained by carrying out only one test, the safety of the plant is put at risk since the protection of the equipment may be undersized in relation to the energy released in a possible explosion.

"Studies on the classification of areas for combustible dusts are simple to carry out, as everything is considered as Zone 21."

Myth! The classification of areas must indicate not only the regions where explosive atmospheres may form under normal operating conditions but also those resulting from abnormal operating conditions. This requires that the professional responsible for the study demonstrate in the Descriptive Report how the classified areas were defined. If "everything is Zone 21", there is a strong indication that the person responsible for the study does not have the necessary experience and that his or her training was probably a self-taught training, which has been proven as insufficient to provide reliable studies for the safety of industrial plants.

"Fans are prohibited in plants with combustible dust, as they disperse it, increasing the estents of classified areas."

Myth! Depending on the geometry of the environment and the specifications of the fans, mechanical ventilation may be sufficient to prevent the accumulation of combustible dust, especially in the upper parts of the plant. When the plant is first put into operation, yes, there may be an increase in suspended dust, requiring the development of a safety procedure; however, an adequate engineering project will allow the air flow to be controlled in such a way as to promote faster deposition of dust, facilitating the cleaning of the environment.

"Our dust is non-combustible because it has a low K_{st}."

Myth! The K_{st} value defines that the powder is combustible. A "low" value means that the explosion will be less severe (under the test conditions) compared to another powder with a higher K_{st}, but both will explode. A K_{st} value lower than 45 bar.m/s requires special precautions, also.

"A large amount of combustible dust is required for an explosion to occur."

Myth! Technical literature states that even a layer of 1 mm thickness can cause a room to explode. Several laboratory tests have been carried out with corn starch powder, which have confirmed that with a layer of powder, with a thickness of 0.254 mm on the floor, the explosion propagated.

The mechanics of secondary explosions that follow the original event also occur in the interconnections of vessels through piping. Flame propagation was recorded through a 27 m long pipe from a 0.64 m^3 vessel containing corn starch. In tests, flame propagation distances ranged from 6 m to 27 m, even with a clean air duct.

And a recent research has shown that flame propagation can be even more easily achieved when the dust concentration in the pipes is as low as 50 g/m^3, such as wood dust.

"A dust explosion cannot propagate against the pneumatic process flow."

Myth! Propagation tests on a pneumatic conveying system consisting of a powder feeder, a 40 m transfer pipe, a cyclone, and a suction fan have shown that this claim is false.

Using corn starch, lycopodium, and wheat flour, in concentrations appropriate for each product (ranging from 75 to 450 g/m^3), it was demonstrated that the explosion is capable of travelling both with and against the flow of the process, even over long distances. This finding is particularly important with regard to industrial conveying systems.

It is clear that misinformation is still widespread, which requires the implementation of training programmes for the entire workforce. Accidents reported in the media and on specialized websites should draw the attention of those responsible for industrial plants that process combustible powders to the potential for destruction of an explosion.

2.5 STATISTICS OF FIRES AND EXPLOSIONS

In recent years, we have heard of fires and explosions at grain storage facilities that have resulted in millions of dollars in losses in several countries. A study on combustible dust hazards conducted by the U.S. Chemical Safety and Hazard Investigation Board (CSB) found that at least 281 dust fires and explosions have occurred at U.S. industrial facilities between 1980 and 2005, resulting in at least 119 deaths and 718 injuries.

Statistics show a considerable number of explosions in facilities with combustible dusts. Table 2.6 shows the numbers of explosions, fires, injuries, and fatalities recorded in 2019 by the Dust Safety Science (DSS) Incident Report. In this table, the "other" class includes pulp and paper industry, coal handling, and educational facilities.

The breakdown between fires, explosions, injuries, and fatalities by the type of involved equipment, occurred in 2019 and recorded in the DSS Incident Report, is summarized in Table 2.7.

Although equipment labelled under "Other" only had 7.6% of the total fires, this category has 47.4% of the injuries and 37.5% of the fatalities. Some of these include: A wood dust fire on top of an exhaust manifold, a wood dust explosion in ductwork, a dust explosion in a titanium shredding machine, a mixing drum explosion at a cosmetics manufacturer, an explosion in a grinding machine at a packaging plant, an explosion in a wood pressing machine, an explosion in a spice grinder, and an explosion in a paint mixing machine.

Table 2.7 also shows that dust collectors demonstrate the highest percentage of combustible dust incidents with 59 fires and 12 explosions reported in the 2019 DSS

TABLE 2.6
Events Involving Combustible Dusts

	Fires	Explosions	Injuries	Fatalities
Food	107	24	30	2
Wood	62	24	40	3
Metal	29	6	12	0
Coal	11	5	17	2
Paper	13	2	1	0
Plastic	0	2	2	0
Other	3	7	6	1
Unknown	25	5	10	0

TABLE 2.7
Dust Events by Equipment Type

	Fires	Explosions	Injuries	Fatalities
Dust collector	59	12	9	0
Storage silo	29	13	18	4
Other storage	27	7	7	1
Elevator/conveyor	30	5	0	0
Dryer	52	2	4	0
Other	19	16	56	3
No details	34	20	24	0
Total	250	75	118	8

Incident Report. This is consistent with the range given in previous incident reports between 2016 and 2018 but lower than the historic data from the CSB, which suggests up to 40%. Although more incidents occur in dust collectors, they appear to be less severe than fires and explosions occurring in storage silos, bins, buckets, and hoppers.

2.6 THE CAUSES OF EXPLOSIONS

As combustible dusts are present in various types of industries, explosions occur due to a wide variety of causes. It is recommended that explosion events be known and their causes discussed, so that lessons learnt can prevent their recurrence. Some explosions will be discussed later in the chapter, along with the causes indicated in the investigation reports, to raise awareness about the risks of combustible dusts to the plant managers.

2.6.1 HIGH-LOSS EVENTS

Some examples that show the severity that a combustible dust explosion can achieve are given here.

1) **Imperial Sugar Refinery, US, 2008**
   In this event, which occurred on February 7, 2008, 14 people died, and 38 were injured. The CSB investigation report indicated as follows:
   - Sugar and corn starch conveying equipment was not designed or maintained to minimize the release of sugar and sugar dust into the work area.
   - Inadequate housekeeping practices resulted in significant accumulations of combustible sugar and sugar dust on the floors and elevated surfaces throughout the packing buildings.
   - Airborne combustible sugar dust accumulated above the minimum explosible concentration level inside the newly enclosed steel belt assembly under silos 1 and 2.

- An overheated bearing in the steel belt conveyor most likely ignited a primary dust explosion.
- The primary dust explosion inside the enclosed steel conveyor belt under silos 1 and 2 led to massive secondary dust explosions and fires throughout the packing buildings.
- The 14 fatalities were most likely the result of the secondary explosions and fires.
- Imperial Sugar emergency evacuation plans were inadequate. Emergency evacuation drills were not conducted, and prompt worker notification to evacuate in the event of an emergency was inadequate.

2) **West Fertilizer Company, US, 2013**

This event that occurred on April 17, 2013, killed 15 people, injured more than 260, and is one of the most destructive incidents ever investigated by the U.S. CSB. More than half of the structures damaged during the explosion were demolished to make way for reconstruction. The demolished buildings include an intermediate school (168 m southwest of the facility), a high school (385 m southeast), a two-storey apartment complex with 22 units (137 m west) where two members of the public were fatally injured, and a 145-bed nursing home (152 m west where many of the seriously injured civilians resided).

The investigation into the accident pointed to possible failures in the electrical systems as contributing to the explosion.

Among the key findings in the CSB final report are the following:

- The presence of combustible materials used for construction of the facility and the fertilizer grade ammonium nitrate (FGAN) storage bins, in addition to the West Fertilizer Company (WFC) practice of storing combustibles near the FGAN pile, contributed to the progression and intensity of the fire and likely resulted in the detonation.
- The WFC facility did not have a fire detection system to alert emergency responders nor an automatic sprinkler system to extinguish the fire at an earlier stage of the incident.

3) **Copersucar Sugar Terminal, Santos, Brasil, 2013**

This event, which occurred on October 18, 2013, resulted in 4 injuries, and it is noteworthy that the Copersucar Sugar Terminal had been expanded just 4 months earlier. The fire started in the supply lines of Warehouse 20/21 and impacted four Copersucar warehouses. At the time, the conveyor belts were operating normally. The fire burned 180 thousand tons of sugar. Copersucar was accounted for 25% of all sugar exported through the Port of Santos.

The improvements on the renovated warehouses were:

- The conveyor belts received an automatic sprinkler system, activated by heat and flame detectors, acting above and below them.
- Sprinklers were installed inside the bucket elevators and in bag filters.
- The interior of the warehouses received firefighting monitors for water.
- New lighting system for warehouses was set up, using Ex luminaires and projectors.

- Conveyor belt misalignment monitoring system was installed.
- Greater control over the humidity and granulometry of the sugar was achieved.

4) The Didion Milling Facility, US, 2017

On May 31, 2017, combustible dust explosions at the Didion Milling facility in Cambria, Wisconsin, killed 5 of the 19 employees working on the night of the incident. The other 14 were injured.

The CSB determined that the cause of the dust explosions and collapsed buildings was the ignition of combustible corn dust inside process equipment which transitioned to multiple explosions. Contributing to the severity of the explosions was Didion's lack of engineering controls, which allowed the fire and explosions to propagate through the facility uncontrolled. The CSB recommendations given at the Investigation Report were:

- Contract a competent third party to develop a comprehensive combustible dust process safety management system, such as OSHA Process Safety Management standard, or the requirements in NFPA 652, which includes, at a minimum, the following elements:
    - Management of change for combustible dust plants
    - Process safety information management
    - Management of audits and inspections
    - Fugitive dust management
    - Incident investigation
    - Dust hazard analysis
    - Management of engineering controls for combustible dust
    - Personal protective equipment
    - Emergency preparedness.

Note 1: The characteristics of sugar types produced in Brasil are shown in Table 2.8.

Note 2: NFPA 652 was replaced by NFPA 660 in 2024.

TABLE 2.8
Most Common Sugar Types Produced in Brasil

Class	Types	Parameters Humidity (% max.)
White	Sanding sugar	0.1
	Granulated	0.3
	Caster sugar	0.05
	Confectioners' sugar	0.3
Raw	Demerara	1.2
	VHP (Very High Polarization)	0.15
	VVHP (Very Very High Polarization)	0.15

It is interesting to know that the VHP/VVHP sugar is simple to produce and easy to handle.

However, the number of fires in warehouses and port terminals has increased due to the high concentration of fine powders resulting from various factors, such as low-quality products and handling equipment that becomes hotter when in operation. Consequently, as of 2016, new requirements were established for the quality of sugar to be delivered to Brasilian terminals, with lower levels of fine powders, which contributed to the reduction in the number of fires in sugar terminals.

Granulometry is another important parameter for assessing dust explosion risks. Sugar particles with a diameter of 420 microns can serve as fuel and contribute to the spread of a fire. Sugar is sold in particle sizes ranging from 850 to 150 microns, with the majority being between 350 and 450 microns.

2.6.2 INVESTIGATIONS

An investigation aimed at determining the real cause of an explosion cannot be limited to a visual inspection of the debris, as it is unlikely that any evidence will still be intact at the site. It is necessary to search maintenance records, operational procedures, training plans, employee testimonies, and other records prior to the tragedy.

In the case of fires in recently renovated or expanded industrial units, the following questions are usually asked by investigators, in order to clarify what happened:

- Which company executed the expansion project? Had it trained professionals in the requirements of facilities subject to explosion risks?
- Was an area classification plan drawn up?
- Had the company responsible for the area classification plan, a trained team?
- Had all electrical equipment specified for hazardous areas in the expansion, their certificates of conformity?
- Why explosion safety procedures, if existent, were not effective?
- Had the professionals who worked on the electrical equipment, the required technical training?
- Was the installation put into operation after the completed commissioning of their systems or was it operating with "minor" issues?

These are just a few points that experts need to know to determine the causes of explosions. Generally, an explosion does not result from a single cause but from a series of violations of technical principles and safety procedures.

The propagation of explosions through industrial process ducts is a reality, and the implementation of preventive measures is a must! In addition to explosion protection techniques, such as suppression, it is essential to have an adequate area classification study (unfortunately, we have detected in audits the erroneous application of considerations exclusive to flammable gases in environments with combustible dusts), as well as attesting that the installations were carried out in accordance with current standards. The propagation of explosions is impacted by many parameters, but a correct design of the installations, a programme to control the ignition sources, and a plan of periodic audits coordinated by an experienced professional will ensure that they do not occur.

2.6.3   THE DUST BEHAVIOUR

With most grains, rapid combustion is possible when the particle size is small enough. Under confinement, this combustion will reach explosive conditions that produce hot gases, which in turn will increase the pressure inside the compartment.

As in the case of gas explosions, a dust explosion is caused by the simultaneous presence of an ignition source and an explosive atmosphere. Sparks produced by electrical equipment (motors, control stations, switches, etc.) are one of the most common ignition sources found in industrial installations subject to explosive atmospheres.

It is important to note that the behaviour of combustible dust atmospheres is quite different from gaseous atmospheres. While flammable gases dispersed in the air seek to reach a homogeneous concentration, dust particles tend to settle and accumulate in layers. Dust particles can also remain in suspension for a certain period of time, depending on their density and particle diameter. Dust particles can also travel from the point where they were released to other more distant points. Leakage from one piece of process equipment to another piece of equipment or component, such as an electrical terminal box, is therefore possible.

Dust can also accumulate on floors, in pipes, on equipment surfaces, in cable trays, and on electric motors, for example. Accumulations of combustible dust on horizontal surfaces in the work area are a serious fire hazard because they can ignite and burn. However, dust on horizontal surfaces will not explosively ignite, even when it has accumulated to a depth of 1 inch.

Combustible dusts must become airborne and achieve a concentration above the MEC in the air to ignite explosively. In fact, industrial hygiene-based dust concentration limits (concentrations in air that can pose a health hazard) are four to five orders of magnitude lower than the minimum dust concentrations necessary to propagate a dust explosion.

A fireball will likely result when airborne combustible dust concentrates above the MEC and if comes in contact with an ignition source. The likelihood of explosive ignition increases if the airborne combustible dust is concentrated in a semi-confined or confined workspace.

As example, when confined, concentrated, and ignited, airborne sugar dust can generate overpressure sufficient to cause explosive destruction, as indicated by its maximum pressure, P_{max} A demonstration of a hazardous accumulation of combustible dust is illustrated in Figure 2.1.

Dust particles can come into contact with ignition sources when they accumulate in layers or form a cloud during a normal activity in the location (e.g. sweeping). If an explosive cloud of dust comes into contact with an ignition source (just few millijoules are sufficient), an initial ignition will occur. This would be the primary explosion, usually at subsonic speed (also known as "deflagration"), which generates

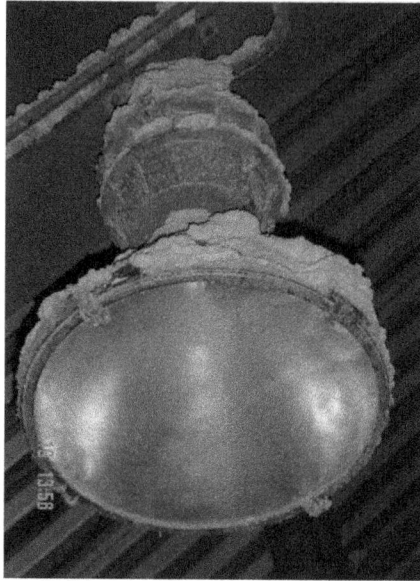

FIGURE 2.1   A luminaire covered with an excessive dust layer.

a considerable volume of hot gases, which in turn generate a pressure wave. After this, the dust that has settled nearby is easily suspended, forming a new dust cloud before the first flash. The initial flame now becomes a new ignition source for the newly formed cloud (flammable mixture). The process is then repeated continuously, rapidly producing a sequence of new secondary explosions, with increased energies that can potentially cause the destruction of the entire plant.

An assessment carried out by an experienced professional should indicate the most appropriate safety measures for each type of plant, as well as guidance on risk reduction, aiming to ensure the health and safety of workers. The applicable solutions can be both technical and organizational and should be combined or complemented, where necessary, with measures to contain the spread of possible explosions.

2.7   PREVENTIVE MEASURES

The measures to prevent dust explosions can be divided into two classes: administrative and technical.

2.7.1   Administrative Controls

Measures that are not based on devices to be installed in the plant are considered organizational. Examples are given in the next subsections.

2.7.1.1   Safety Procedures

The main prevention technique is organizational. Specific risk work must not be allowed to be carried out without the corresponding work permit, and even less so if it is to be carried out in a hazardous area. A "permit-to-work" procedure is a means of achieving effective control of a system of work through formal written documentation, known as a permit-to-work form or certificate. So, it is necessary to comply with safety requirements and to implement a robust work permits system for:

- Hot work (cutting and welding)
    Hot work always presents a significant fire/explosion risk factor if carried out on premises where the locations and the equipment have not been adequately cleaned before the hot work is initiated.
- Hot surfaces
    The measures taken to prevent ignition by hot surfaces must cover both layer and cloud dust ignition. The measures include removal of all combustible dust before performing hot work; removal of dust accumulations on hot surfaces; isolation, or shielding, of hot surfaces; and inspection and maintenance procedures that minimize the risk of overheating.
- Use of compressed air
    Use of compressed air for blowing spilled dust away should be prohibited. By this method, dust is not removed but only transferred to another location in the same place. Besides, dust explosions can result if the dust concentration in the cloud that is generated is in the explosive range, and an ignition source exists in the same location.
- Work permit for confined spaces
    Some confined spaces are not easily identified but may be equally dangerous in the presence of hazardous conditions, for example: Trenches, parts of buildings during construction, storage silos, and unventilated or inadequately ventilated compartments. The general state of the confined space should be assessed to identify what substances or conditions might be present and cause an explosion risk. The physical dimensions and layout of the confined space, as well as the type of combustible dust present, can impact air quality and/or combustible dust concentrations, which may not be impacted by ventilation. Depending on the structure of the confined space and the type and duration of the work being carried out, mechanical ventilation may be required to provide sufficient fresh air to replace the oxygen that is being used up by people working in the space and to dilute and remove gas, fume, or vapour produced by the work.

2.7.1.2   The Area Classification Study

The purpose of the area classification study is to indicate the regions subject to possible formation of an explosive atmosphere, that is those where combustible dust suspended in the air may occur in a concentration greater than the product's LEL. By marking these regions on the layout drawings of the industrial unit, it is possible to correctly specify the electrical and electronic equipment considered safe for use in such environments.

The first step in preparing the study is to obtain the characteristics of the combustible dusts expected for the industrial unit, because if incorrect data are considered, the study will not be reliable. In audits, it has been very common to find data on combustible dusts obtained from publications such as electrical equipment manufacturers' catalogues. However, it is important to note that these values are not "typical", as each sample of combustible dust has specific moisture and particle size levels, which modify its properties and behaviour, in the event of an explosion. Therefore, it is completely wrong to think that the classification of areas can be easily carried out by copying "typical" figures, because as stated earlier, only a correct engineering assessment of the processes present in the industrial unit can provide a valid study.

The history of explosions has highlighted the relevance of electrical installations in the safety of industrial units and reinforces the importance of correctly specifying electrical equipment designed to prevent the ignition of explosive atmospheres in normal and abnormal operating situations. In conclusion, the integrity of electrical systems in industries with classified areas is a crucial part of safety management, which must include periodic independent inspections and audits conducted by specialists, as we are dealing with extremely dangerous facilities!

In Chapter 3, the area classification study will be discussed.

2.7.2 Technical Measures

Technical measures are related with devices and sensors that can detect abnormal situations.

Some examples are explained in the next subsections.

2.7.2.1 Control of Static Electricity

It is important to consider the generation of static electricity during the transfer processes of gases with particles and solids, which may cause the ignition of the explosive mixture. For gases or vapours, the minimum ignition energies are in microjoules (µJ), while for solids they are in millijoules (from 1 to 1,000 mJ).

Monitors of electric potential can be used, installed on key points of the processes. The use of antistatic materials and to have an effective earthing with an adequate periodic inspection programme, carried out periodically by a specialist in this field, are highly recommended.

2.7.2.2 Spark Extinction

Sparks are normally generated by mechanical action in pneumatic transport or in mechanical, horizontal, or elevating conveyors. They can also be generated in heat exchangers (dryers, etc.)

Spark extinction systems will always be recommended to be used in pipelines. This technique consists of detecting incandescent particles using infrared sensors, placing extinguishing nozzles downstream of the detectors at a distance that will depend on the flow speed. Depending on the material, different extinguishing agents can be used, although water is the most commonly used.

2.7.2.3   CO Detection

Carbon monoxide (CO) detection is the fastest way to identify the start of a fire, since when a slow combustion phenomenon occurs, such as self-combustion by oxidation, an initial emission of CO occurs. The reading must be noted by the difference in concentration of CO between the input to the process and the output. For example, this technique is widely used for detection in large electrical panels, or in spray drying processes, such as in the production of powdered milk. This system allows for early detection of a fire, which in turn could cause an explosion.

A recommended design is to use three alarm levels. The first one warns that CO is being generated, which allows personnel to act. If the concentration of CO continues to increase, the second level is activated, which determines the process to stop, and if the concentration of CO continues to increase, it activates the automatic extinguishing system.

2.7.2.4   Vibration Measurement

When a machine contains elements that rotate at high speed, it is crucial to ensure that this movement occurs in a balanced manner and to avoid sparks that could cause explosions. An imbalance in the rotating parts is preceded by an increase in vibration, so that if it is monitored, breakage due to wear of components, such as bearings, shafts, or transmission belts, can be prevented.

2.7.2.5   Alignment Gauges

Alignment gauges are widely used in bucket elevators, especially those with great heights. They operate on the basis of having a constant reading of the metal—of the bucket itself. When the bucket is made of a plastic material, the detection is carried out on the screws that fasten it to the belt.

2.7.2.6   Jam Sensors

This is a highly recommended technique for those processes that move a large amount of product, as a jam can cause frictional heating and cause a fire and, subsequently, an explosion. Using it in the silo-filling process makes it possible to monitor whether the product is still flowing through the pipeline. If it stops, the anomaly can be detected and identified.

2.7.2.7   Temperature Sensors

Temperature monitoring is a very effective preventive technique, especially in processes involving high-speed movements, such as the case of bearings.

2.7.2.8   Speed Meters

Industrial processes are carried out at a controlled speed; therefore, if the machine does not rotate or accelerates, anomalies in the system are indicated. Such control is simple, as it involves measuring the frequency using a magnetic component.

2.7.2.9   Inerting

For any type of combustible dust, and a given type of inert gas added to the air, there is a limiting oxygen content below which the dust cloud is unable to propagate a self-sustained flame. So, keeping the oxygen content below this limit throughout the process system, dust explosions are effectively excluded.

Among the types of inert gases that are used for this purpose are the following:

- Nitrogen
- Carbon dioxide
- Water vapour

In some cases, the protective method called "intrinsic inerting" can be used, as when the required quantity of inert gas is produced in the plant itself, consisting of nitrogen, carbon dioxide, and water vapour. However, normally such combustion gases are not clean enough for being used in connection with food materials, pharmaceuticals, etc.

The choice of inert gas depends on considerations as availability, cost, and possible contaminating effects on products. The design of gas inerting systems depends on whether the process is continuous, or is of the batch type, the strength of the process equipment, and type and source of the inert gas. Two main principles are used for establishing the desired atmosphere in the process:

1) **Pressure variation method**

    This operates either above or below the atmospheric pressure. In the former case, the process equipment, initially filled with air at atmospheric pressure, is pressurized to a given overpressure by inert gas. When a good mixing of air and inert gas has been obtained, the process equipment is vented to the atmosphere and the cycle repeated until sufficiently low oxygen content has been reached.

2) **Flushing method**

    This is used when the process equipment has not been designed to withstand the significant pressure increase or vacuum required by the pressure variation method.

There are two extreme cases of the flushing method:

- The replacement method (plug flow), where in order to maintain the plug flow, the flow velocity of inert gas into the system must be low (< 1 m/s) and the geometry must be favourable for avoiding mixing. In practice, this is very difficult to achieve.
- The stirred tank method, using high gas velocities and turbulent mixing, is normally employed. It is essential that the instantaneous through-mixing is complete over the entire volume, otherwise pockets of unacceptably high, oxygen concentrations may form.

2.7.2.10 Humidification

The conductivity of many materials is impacted by humidity, such as fabric, wood, paper, and concrete. Relative humidity greater than 65% may reduce the surface resistivity of these types of materials. The surface resistance of insulating materials should be less than 10^{11} Ω/square.

However, increasing the relative humidity may not be adequate for some materials such as certain types of plastics, because they do not absorb enough moisture to provide a conductive layer. Also, in some cases, humidification may adversely impact processing and therefore may not even be an option.

2.7.2.11   Ionization

Equipment is available which makes air conductive enough to dissipate static charges on insulating materials. This type of application is typically limited to small areas. The majority of ionization devices are not suitable for use in hazardous locations, but there are few radioactive devices certified for such application.

2.7.3   MANAGEMENT'S ROLE ON PREVENTING EXPLOSIONS

The stance for preventive actions is highlighted on the investigation reports of combustible dust explosions, as cited next:

- Companies should ensure that the pneumatic transport and the dust collection ductwork are designed to maintain a minimum transport speed and define it based on the particular characteristic of their dust.
- To ensure effective prevention and mitigation of combustible dust deflagrations, engineering controls (detection, suppression, isolation, venting, and pressure containment) must be utilized when designing a dust safety system.
- Companies should ensure that the standards applied are appropriate to the hazards inside the facility. Food safety standards, for example, are appropriate for preventing food hazards such as pathogens and contaminants from reaching consumers, but they are not intended to address workplace or process hazards such as combustible dust. Appropriate tools such as a DHA should be used to address process hazards like combustible dust.
- Employers should utilize the findings of external audits to identify and correct hazards that could result in significant incidents, resulting in injuries and property damage.
- A fire inside an enclosed combustible dust handling process should not be considered an "isolated case", because it cannot be characterized without the risk of increasing the severity of the incident.
- Safety regulations, only, do not guarantee the safety of the process!
- Safety signage.
  A worrying factor is the absence of particular safety signalization in industrial plants. The direction is:

  Safety signs should be clear to all professionals involved in services on hazardous areas, providing information on the Zone classification, the equipment group related to the combustible dust atmosphere, and the allowed temperature class for equipment to be used there.

Unfortunately, it can be found that many signs do not comply with ISO requirements. It is not a matter of ingenuity; as the signage has to be clear and initiate a quick action, texts should be avoided, in order to be understandable also by foreign workers. Figure 2.2 shows a sign with all information needed for safe services in hazardous locations due to combustible dusts.

In color, this safety sign contains the "Ex" (meaning explosive atmosphere) in a yellow triangle (warning) against a red background (fire risk), that are the ISO recommendations for safety alert signs and that is already in use by many companies. It warns employees about a probable existence of a dust explosive atmosphere in abnormal operating conditions (Zone 22); that such atmosphere is composed by combustible flyings (IIIA); and that the maximum surface temperature for all electrical Ex equipment approved for safe use in that area is 125°C (T 125°C).

The number of the area classification drawing is also indicated, allowing workers to get more information on the size and shape of the classified area, if necessary. It is important to highlight that the signage is only an alert device; it does not replace the technical training required for all workers designated for tasks in hazardous locations.

As a conclusion, regulations are the minimum threshold for maintaining safe operations. The combustible dust hazards must be controlled beyond the existing regulatory requirements through the adoption of current industry standards and through the development of new regulations to prevent safety incidents.

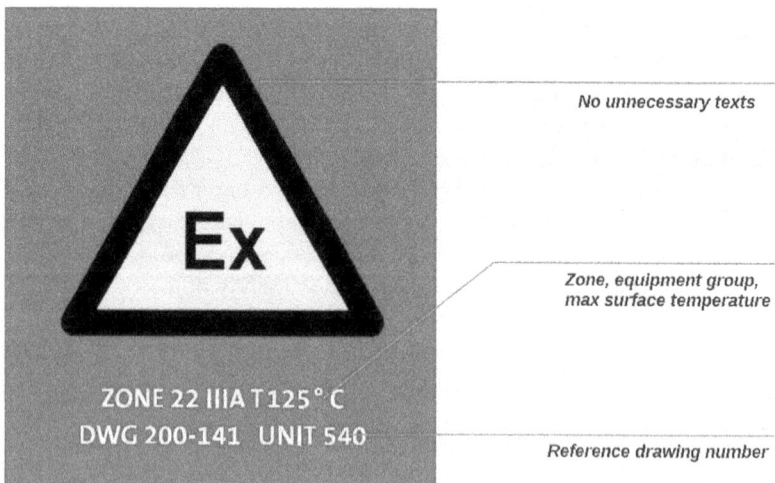

FIGURE 2.2   Effective safety sign for hazardous locations due to combustible dusts.

2.8 ANSWERS TO QUESTIONNAIRE IN SECTION 2.3

Have you answered all of them? Let us know the answers!

In the 2010s, combustible dust explosions in the United States alone were responsible for more than 100 deaths and hundreds of injuries. Such events could have been avoided if the companies involved had adopted best practices for fire and explosion protection, including an adequate training programme for all their employees.

1) How many American industries are at risk of forming explosive atmospheres from combustible dusts according to OSHA?

   Answer: d) 30,000.

   Several establishments are subject to this risk, such as food industries (sweets, sugar, starch, flour, etc.); synthetic materials manufacturing (plastics, rubber, etc.); timber (furniture, metalworking, etc.); metallurgical (aluminium, iron, magnesium, etc.); and recycling (various materials involved).

2) Which of the listed risks is not caused by the release of dust into the workplace in pharmaceutical industries?

   Answer: d) Multiple sclerosis. All other risks are due to the release of dust in these environments.

3) If you "discovered" that your industry process contains silica dust, what would be your first action?

   Answer: b) Send the dust for analysis.

   To solve a problem, you have to understand what it is. The test will tell you about the particle size, moisture content, etc., allowing you to select the most effective control system. Silica kills about 1,000 workers a year in the UK from silicosis, lung cancer, and other diseases. It can be found in dental laboratories, foundries, shipyards, and many other industries. In the United States, it is present in about 676,000 establishments, impacting about 2.3 million workers.

4) What risk is present in the manufacture of metal articles involving cutting, grinding, pressing, welding, etc.?

   Answer: e) All of the above.

   Sprayed machine coolants fall onto the floor and make it slippery. Combustible dust can accumulate in hard-to-reach places and ignite, causing an explosion. In the United States, between 1980 and 2005, the Chemical Safety Board reported 281 explosions caused by combustible dusts, resulting in 199 deaths and 718 injuries.

2.9 DAMAGE CONTROL FOR DUST EXPLOSIONS

The risk of a combustible dust explosion is considerably high, and preventive measures depend on the strict control over the operation and maintenance of equipment. In the event of a failure in the control and/or implementation of preventive measures, there are some resources that make it possible to avoid an explosion or, if one does occur, to limit the damage to the plant. The limitation of damage due to a combustible dust explosion is achieved by suppression, isolation, containment, and venting actions.

2.9.1 SUPPRESSION

In the field of protection of industrial process equipment against dust explosions, chemical suppression systems are based on the detection of the deflagration in its incipient state and extinguishing the deflagration by means of automatically activated extinguishing agents, applied directly to the source of the explosion. The most common extinguishing agent is $NaHCO_3$—the baking soda.

Explosion suppression systems act to preserve workers and the processing equipment where a dust cloud of suitable concentration and particle size can form an explosive atmosphere—as for example a filter, a mill, a bucket elevator, a hopper, or a silo. The suppression system has three components: Detector, controller, and extinguisher, as shown in Figure 2.3.

The detection can be done by pressure or optical sensors, which are sensitive to radiation.

In figure 2.3:

    1 is the ignition source.
    2 is the wave front.
    3 is the extinguishing agent.
    4 is the UV radiation detector.
    5 is the controller.
    6 is the gate valve.
    7 is the high-pressure extinguisher.

The suppression must consider K_{max}, as it is a technique that requires specialized technical service, and it is not cheap. In general, it requires 2 kg of agent per m^3 of equipment. The most common referenced standard is EN 14373.

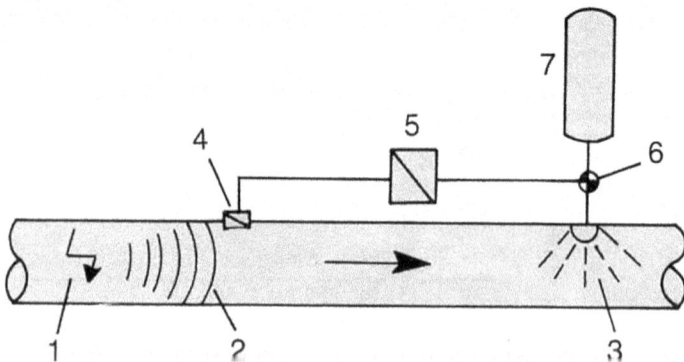

FIGURE 2.3 Explosion suppression system installed in duct.

2.9.2 ISOLATION

Isolation is a concept that is completely complementary to suppression.

It consists of stopping the combustion front, preventing propagation and secondary explosions. It can be done in three ways:

- Chemical: Using an extinguishing agent
- Mechanical: Using a fast-acting valve
- Flame diverter: Using a device that slows down the flame front.

Through suppression, we can act directly on the source of the explosion to extinguish it. However, part of the flame, or the blast wave, could spread through the installation to other equipment connected to the protected machine. An example is bucket elevators, shown in Figure 2.4, which are connected at their bases to the feed, which can be of the redler, rotary valve, or worm screw type.

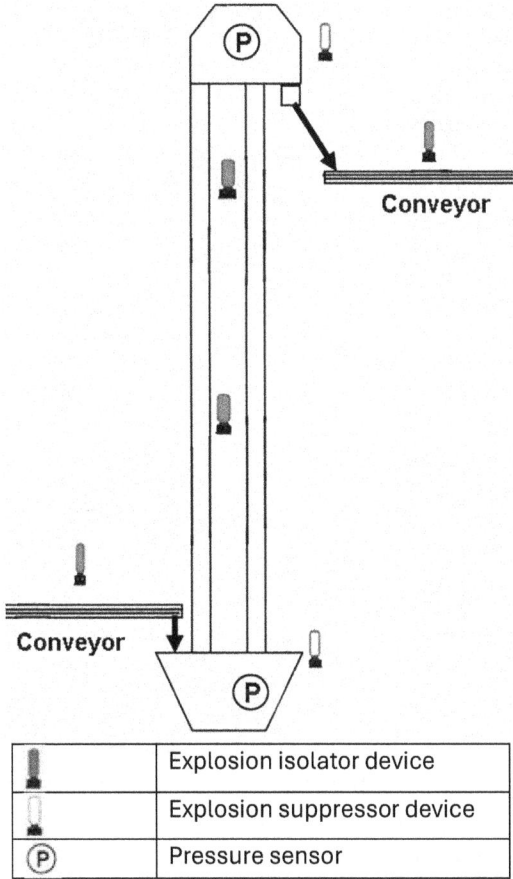

	Explosion isolator device
	Explosion suppressor device
(P)	Pressure sensor

FIGURE 2.4 Explosion isolation system in a bucket elevator.

TABLE 2.9

Summary of the damage controls for dust explosions

Method	Main Objectives	Components	Considerations	Standards
Suppression	To stop and extinguish the deflagration.	– Use of extinguishing agents: • $NaHCO_3$: Absorbs free radicals from the reaction and shields combustion • H_2O: Cooling and inerting.	– The minimum and maximum cubic capacity of a device to be protected – The working temperature of the protected system – Pmax – Kst	EN 14373 NFPA 69
Isolation	– To stop the spread of a deflagration – To prevent secondary explosions, protecting upstream and downstream equipment.	– Chemical barrier agent – Quick-acting barrier valves – Flame front diverter – Dust flame arrester	– The barrier valve can only be used on pneumatic transport lines or circular ducts of dust extraction systems – The valves require frequent inspections, to avoid operation failures.	EN 15089 NFPA 69
Containment	Certain pieces of equipment, such as hammer mills, can be built sufficiently robust to contain an explosion.	The protected equipment is designed to resist the Pmax	– Verify the protection of adjoining equipment as containment will increase the explosion severity – The containment design should be validated by a dust explosion testing.	EN 14034-1 NFPA 69
Venting	– To relieve the pressure – To divert the flames to a safe area.	Venting panels	– Not suitable for use in confined spaces – Kst – Pmax – There must be no other equipment or people in the venting path during normal operation of the system, as the resulting fireball can be up to 75 times the volume of the enclosure itself – Only used with non-toxic substances.	EN 14491 EN 14494 EN 14797 EN 15233 NFPA 68

In turn, the head of the elevator is also connected to a silo, a hopper, a mill, etc. In the event of an explosion, both the blast wave and the flame front seek the shortest and most immediate path to spread. In the case of bucket elevators, this path is found at the inlets and outlets (base and head).

In case of an explosion in a bucket elevator that feeds a series of silos, even if the suppression system works correctly and immediately suppresses the explosion, it may happen that a hot ember is transported through the ducts that connect the elevator to the silos. This possible ignition source could reach one of the silos and start a new explosion in unprotected equipment. To prevent this, chemical or mechanical isolation is used.

Chemical isolation consists of bottles identical to the suppression bottles, which discharge the extinguishing agent directly into the ducts, preventing the spread. Mechanical isolation is achieved by means of ultra-fast acting valves, either passive or active.

Parameters to be taken into account when designing an explosion insulation system are as folows:

- K_{max}
- The duct diameter (maximum 1,400 mm)
- Minimum and maximum distance to the isolated volume
- Minimum time to close
- P_{red} (pressure that the system can withstand at any given time, without suffering any deformation) maximum and minimum, for valves that close
- Valve resistance
- The volume of the isolated container.

Where:

In an installation with protection and isolation, the fact that an explosion occurs in one of the equipment does not directly impact the surrounding equipment, allowing production to continue. Isolation can be considered a fundamental complement in explosion suppression systems.

2.9.3 CONTAINMENT

When contained in a defined space, the possible destruction caused by a dust explosion to the neighbouring surroundings can be averted. Containment is an appealing alternative due to its largely passive approach. However, the complete design of dust treatment equipment to withstand the pressure generated by the explosion of dust, especially on large factories, is unfeasible.

For specific equipment, such as a grinder, it is possible to design it appropriately strong to endure an explosion of dust, considering that the maximum pressure of explosion, for much combustible dust and gaseous substances, is in the range of 7–10 bars. However, the solitary principle is not static pressure, as the upward pressure rate is great during the explosion of dust. Hence, the designed apparatus must be able to endure the dynamic load during some explosion stage. Furthermore, the equipment must be designed according to the revolving symmetry to avoid large flats and corners.

Specific consideration is needed at the point where the dust is injected, or recovered, from the production facility and the point of connection in the devices. For highly toxic powder or dust, there is a need for reliable restrictions.

2.9.4 VENTING

Further related to poisonous dust, ventilation can lessen the destructive possibility of a dust explosion. In the beginning glance, venting seems to be a reasonably easy operation. If any section of the construction is predisposed to the explosion of dust, the pressure resistance of the remainder of the structure is considerably reduced through the use of a slimmer panel or can be allowed to fail in the initial phases of the blast. By properly determining the size of this vent, the operator can ensure that the vents operate as quickly as possible.

Hence, the overpressure can exceed a certain safety threshold and rapidly release a sufficient amount of gas and particulates to prevent the pressure inside the protected area from reaching a destructive level. For size, the venting location needs to understand all the elements that control the rigorousness of the exploding dust. The consideration includes the device's geometry that supplies the explosive venting, preliminary turbulence, primary pressure, dust concentration, original temperature, and detonation source for the combustible dust or inert gases.

All the defined elements impact on the pressure of the explosion, particularly in terms of pressure rise rates. Furthermore, the abridged pressure of explosion along with the area, pressure, distribution, and panel of the vents must be considered. In addition, the inclusion of protective ventilation is not the most critical consideration in ventilation for dust explosion.

The applicable European standards on designing of venting devices are EN 14491, EN 14494 (both similar to NFPA 68), EN 14797, and EN 15233.

The explosive waves and flames that vent into the surrounding environment can be dangerous. Not to mention the risk of unburnt dust, fumes, and soot, which can suddenly spray out of a vent in many different ways.

The maximum dimensions of the flame cleared out of the vent could be ten times the cube root of the ventilation container volume. If the vents emit too much unburned flammable material, it may even be ignited by a ventilated flame causing a secondary explosion.

The task of designing the vent spreads to removing the dangerous effects of venting substances, particularly the flames. Therefore, it is critical to reiterate the importance of providing safe and effective ventilation. For example, if the duct is installed at the vent, it may safely discharge the ventilation quality and energy. However, the process can prevent the complete suppression of the explosion pressure.

The summary of the methods for damage control of combustion dust explosions is shown in Table 2.9.

2.10 MAINTENANCE

All methods for limiting damages in case of eventual dust explosion shown earlier require a maintenance plan, when at least once a year, the following items should be verified:

- The initial design is still being met (same product, process conditions, safety distances, etc.).

- The maintenance manual is been followed, which must be in the end user's language.
- The components that had been replaced comply with the initial specification.
- Annual inspection report, from a specialized company, confirms the compliance with the original specifications.

The summary of the methods for damage control of combustion dust explosions is shown in Table 2.9.

2.11 CLOSING REMARKS

The propagation of dust explosions through industrial process ducts should be taken into account by professionals, as preventive measures are required.

A proper area classification study is essential, as equipment that could act as ignition sources cannot be placed in Zones 20, 21, and 22.

Organizations across multiple industries must have their combustible dust and powders tested to ensure regulatory compliance and to guarantee that the equipment in their processes can be installed and/or modified to ensure safety and business continuity.

Effective K_{st} and P_{max} dust testing data will support the creation of explosion hazard mitigation strategies. As risk levels can vary between processes, ongoing attentiveness and effective management of change strategies are required to identify conditions in the plant that might cause a potential safety problem. Therefore, it is essential to ensure that the dust testing strategy and maintenance plans are consistently reviewed and kept up to date.

The propagation of explosions is impacted by many parameters, but a correctly executed design, together with a robust ignition source control programme, will ensure that they do not occur.

BIBLIOGRAPHY

Abbasi, T. and Abbasi, S. A.—Dust explosions—cases, causes, consequences, and control. *Journal of Hazardous Materials*, vol. 140, 2007, pp. 7–44.
Amyotte, P. R.—Facing the pentagon. *Industrial Fire Journal*, First Quarter, 2010, pp. 34–35.
Amyotte, P. R. and Eckhoff, Rolf K.—Dust explosion causation, prevention and mitigation: An overview. *Journal of Chemical Health & Safety*, vol. 17, 2010, pp. 15–28.
ASTM E1226-19—*Standard test method for explosibility of dust clouds*. ASTM International, 2019.
ASTM E1491-06—*Standard test method for minimum auto ignition temperature of dust clouds*. ASTM International, 2019.
ASTM E1515-14—*Standard test method for minimum explosible concentration of combustible dusts*. ASTM International, 2022.
ASTM E2019-03—*Standard test method for minimum ignition energy of a dust cloud in air*. ASTM International, 2019.
ASTM E2021-15—*Standard test method for hot-surface ignition temperature of dust layers*. ASTM International, 2023.

Barton, J., editor.—*Dust explosion prevention and protection. A practical guide.* Institution of Chemical Engineers, 2002.

Cen, Kang, Song, Bin, Huang, Yan and Wang, Qingsheng. CFD simulations to study parameters affecting gas explosion venting in compressor compartments. *Mathematical Problems in Engineering*, vol. 2017, Article ID 1090561, 2017. https://doi.org/10.1155/2017/1090561

CSB Investigation Report No. 2008-05-I-GA—*Sugar dust explosion and fire: Imperial Sugar Company.* Chemical Safety and Hazard Investigation Board, Sep. 2009. Available at: www.csb.gov/userfiles/file/imperial%20sugar%20report%20final%20updated.pdf

CSB Investigation Report No. 2006-H-1—*Combustible dust hazard study.* Chemical Safety and Hazard Investigation Board, Nov. 2006. Available at: www.csb.gov/combustible-dust-hazard-investigation/

CSB Investigation Report No. 2013-02-I-TX—*West Fertilizer Company fire and explosion: Final.* Chemical Safety and Hazard Investigation Board, Apr. 2013. Available at: www.csb.gov/file.aspx?DocumentId=5983

CSB Investigation Report No. 2017-07-I-WI—*Fatal combustible dust explosions at Didion Milling Inc.* Chemical Safety and Hazard Investigation Board, Dec. 2023. Available at: www.csb.gov/file.aspx?DocumentId=6240

Di Benedetto, A. and Russo, P.—Thermo-kinetic modelling of dust explosions. *Journal of Loss Prevention in the Process Industries*, vol. 20, 2007, pp. 303–309.

DSS Combustible Dust Incident Report 2019 – *Dust safety science.* Available at: http://dustsafetyscience.com/

Dust incidents 2006–2017. U.S. Chemical Safety and Hazard Investigation Board. Available at: www.csb.gov/assets/1/6/csb_dust_incidents.pdf

Eckhoff, Rolf K.—Differences and similarities of gas and dust explosions: A critical evaluation of the European 'ATEX' directives in relation to dusts. *Journal of Loss Prevention in the Process Industries*, vol. 19, 2006, pp. 553–560.

Eckhoff, Rolf K.—*Dust explosions in the process industries*, 3rd edition, Gulf Professional Publishing, 2003.

Eckhoff, Rolf K.—Understanding dust explosions. The role of powder science and technology. *Journal of Loss Prevention in the Process Industries*, vol. 22, 2009, pp. 105–116.

EN 14034-2—*Determination of explosion characteristics of dust clouds—Part 2: Determination of the maximum rate of explosion pressure rise (dP/dt)max of dust clouds.* European Committee for Standardization, 2006.

EN 14373—*Explosion suppression systems.* European Committee for Standardization, 2021.

EN 14491—*Dust explosion venting protective systems.* European Committee for Standardization, 2012.

EN 14494—*Gas explosion venting protective systems.* European Committee for Standardization, 2007.

EN 14797—*Explosion venting devices.* European Committee for Standardization, 2006.

EN 15233—*Methodology for functional safety assessment of protective systems for potentially explosive atmospheres.* European Committee for Standardization, 2007.

EN 50281-2-1—*Electrical apparatus for use in the presence of combustible dust. Test methods. Methods of determining minimum ignition temperatures.* European Committee for Standardization, 1999.

Fernandes, Marcelo Eloy, Namba, Camila Eiko Yazawa, and Gozzi, Marcelo Pupim—Estudo de prevenção de acidentes por explosões verticais para abastecimento de cereais. In: *XXXI Encontro Nacional de Engenharia de Produção.* Associação Brasileira De Engenharia de Produção, ABEPRO, Belo Horizonte, Brasil, Oct. 2011.

Hattwig, M. and Steen, H.—*Handbook of explosion prevention and protection.* Wiley-VCH, 2004.

ISO 7010—*Graphical symbols—Safety colours and safety signs—Registered safety signs.* International Organization for Standardization, 2019.

ISO TS 20559—*Graphical symbols—Safety colours and safety signs—Guidance for the development and use of a safety signing system.* International Organization for Standardization, 2020.

ISO/IEC 80079-20-2—*Explosive atmospheres. Part 20–2: Material characteristics—Combustible dusts test methods.* International Organization for Standardization, 2016.

Kletz, T.—Equipment and procedures that cannot do what we want them to do. Workshop notes and slides. In: *Hazards XXI.* Institution of Chemical Engineers, Manchester, UK, Nov. 2009.

Kletz, T. and Amyotte, P.—*Process plants. A handbook for inherently safer design*, 2nd edition. CRC Press, Taylor & Francis Group, 2010.

Mendes, Celso Pereira—Métodos de proteção contra explosões industriais. *Eletricidade Moderna Magazine*, São Paulo, ed. 204, pp. 46–49, Mar. 1991.

NFPA 660—*Standard for combustible dusts and particulate solids.* National Fire Protection Association, 2024.

NFPA 921—*Guide for fire and explosion investigations.* National Fire Protection Association, 2021.

NIOSH, Information Circular 9529—*Best practices for dust control in coal mining.* National Institute for Occupational Safety and Health, 2021.

OSHA 1910 Subpart H—*Hazardous materials, 1910.119, Process safety management of highly hazardous chemicals.* Occupational Safety and Health Administration, 2015.

Rangel Jr., Estellito—As causas das explosões. *EMEx Section, Eletricidade Moderna Magazine*, São Paulo, ed. 573, Sep/Oct. 2023, p. 68.

Rangel Jr., Estellito—Avoiding dust explosions: Preventive measures. In: *2nd CCPS Latin American process safety conference.* São Paulo, 2010. Available at: http://bit.ly/2yo7Uey.

Rangel Jr., Estellito—Considerations for the prevention of combustible dust explosions. *Fire-and-explosion Magazine*, no. 1, Bulkmedia, Germany, pp. 48–54, May 2019. Available at: https://bit.ly/3Ki3rio.

Rangel Jr., Estellito—Explosividade dos pós. *EMEx Section, Eletricidade Moderna Magazine*, São Paulo, ed. 508, Jul 2016, p. 90.

Rangel Jr., Estellito—Explosões de pós. *EMEx Section, Eletricidade Moderna Magazine*, São Paulo, ed. 538, Jan. 2019, p. 50.

Rangel Jr., Estellito—Mitos dos pós combustíveis. *EMEx Section, Eletricidade Moderna Magazine*, São Paulo, ed. 568, Nov/Dec. 2022, p. 61.

Rangel Jr., Estellito—Pó combustível. *EMEx Section, Eletricidade Moderna Magazine*, São Paulo, ed. 477, Dec. 2013, p. 122.

Rangel Jr., Estellito—Pós: como evitar explosões. *Química e Derivados Magazine*, no. 464, Editora QD, Brasil, pp. 64–69, Aug. 2007.

Rangel Jr., Estellito—Pós e eletricidade: mistura explosiva. *Lumière electric Magazine*, ISSN: 0100-834X, ed. 137, Editora Lumière, Brasil, p. 30, Sep. 2009.

Rangel Jr., Estellito—Riscos dos pós. *EMEx Section, Eletricidade Moderna Magazine*, São Paulo, ed. 520, Jul. 2017, p. 61.

Rangel Jr., Estellito—Riscos dos pós (II). *EMEx Section, Eletricidade Moderna Magazine*, São Paulo, ed. 521, Aug. 2017, p. 73.

Rangel Jr., Estellito and Nägle, Rainer—Dust explosion protection in Brasil. *Ex Magazine*, no. 32, R. Stahl Schaltgeräte GmbH, Germany, pp. 54–58, Aug. 2006. Available at: https://bit.ly/3jBKzP1

Sa, Ary. Efeito devastador. *Revista Protecao, Sao Paulo*, no. 181, p. 63, Jan. 2007.

Thirty experts share largest problems facing the combustible dust community. Dust Safety Academy, US, 2017. Available at: https://bit.ly/2UpBXeN. Accessed on July 20, 2024.

3 Area Classification Studies

3.1 OBJECTIVES

The basic objective of an area classification (also known as "hazardous area classification") study is to identify the possibility of an explosive atmosphere existing in a given location (due to flammable gas, vapour, mist, or combustible dust), and estimate its extents, in a three-dimensional scenario. The meaning of "area" in this context is "volume".

Area classification study is a tool that allows to correctly specify the electrical and electronic equipment to be installed; gives orientation to improve the safety of any industrial plant, process, or facility; and, also, serves as basis for elaborating operational safe procedures. To this end, area classification study can begin as soon as the plant's process flowgrams are available, and be constantly refined throughout the plant design development, until the as-built drawings are issued.

When evaluating a given location, the probability of the presence of a flammable gas or vapour is a significant factor in determining the classification of the area, and a distinction must be made: The presence of the flammable mixture is said to be under "normal conditions" (when the process is operating within its design parameters) and under "abnormal conditions" (a situation that can generate a release of the flammable product to the ambience, but excluding catastrophic events such as the violent rupture of a process piping or vessel).

As for an explosion or fire to occur, the gas–air mixture must have a concentration within its explosive range, it is very important to estimate the quantities of flammable mixture that may arise and evaluate the local ventilation available, in order to determine the extents of the probable explosive atmosphere. The greater the possible quantity, the larger the impacted volume will be.

3.2 EXPLOSIVE ATMOSPHERES

Let us emphasize the three basic elements required for a fire or explosion to occur:

(1) A fuel
(2) An oxidizer (e.g. air)
(3) An ignition source with sufficient energy to ignite the flammable mixture (e.g. electrical equipment, open flames, or hot surfaces)

DOI: 10.1201/9781003500001-3

In addition to these elements, there are two additional conditions that must be considered:

(4) The fuel concentration must be between the upper and lower explosive limits (its explosive range)
(5) All three first elements must be present at the same time.

So, the only presence of a flammable gas or vapour in the air is not sufficient to cause an explosion. An explosion will occur only when the gas or vapour has mixed with air or oxygen in a ratio in which the gas or vapour concentration is above its Lower Explosive Limit (LEL) and below its Upper Explosive Limit (UEL), expressed in terms of percentage by volume of gas or vapour in air.

Between the explosive limits, the range is known as the explosive range. This range may vary from a few percent to 100%.

Below the LEL, the mixture is too lean for combustion, because there are insufficient gas or vapour molecules. Above the UEL, the mixture is too rich for combustion, because there are too many gas or vapour molecules. However, within the LEL and UEL range, combustion is possible, and the flame will spread throughout the mixture when it is ignited. This is known as flame propagation, and if it is very rapid, it is popularly called "explosion".

The ignitable limits are based on normal atmospheric temperature and pressure. There may be variations in the explosive limits at temperatures and pressures above or below normal.

An increase in temperature of the mixture will cause the flammable range to shift downwards, and a decrease in temperature will shift it upwards.

Under favourable conditions, an electrical spark will ignite a flammable mixture. The minimum amount of energy required to ignite a flammable mixture varies with the type of fuel. The minimum arcing energy to ignite hydrocarbon–air mixtures varies from 0.017 to 0.3 millijoules. Hydrogen gas, for example, can be ignited by 0.017 millijoules.

3.3   AREA CLASSIFICATION FOR GAS/VAPOUR

The area classification study does not take into account the effects of consequential damage after an explosion; its objective is limited to define the extents of a probable explosive atmosphere, as defined in NBR IEC 60079–10–1.

3.3.1   GENERAL DEFINITIONS

The IEC standards refer to the hazardous locations as "Zones", differing if the explosive atmosphere is due to flammable gases and vapours, from those due to combustible dusts. In explosive atmospheres due to flammable gases, the following terms are defined in IEC standards.

Source of release

Point or location from which a flammable gas, vapour, or liquid may be released into the atmosphere in such a way that an explosive gas atmosphere could be formed.

Flammable gas or vapour

Gas or vapour which, when mixed with air in certain proportions, will form an explosive gas atmosphere.

Explosive atmosphere

Mixture with air, under atmospheric conditions, of flammable substances in the form of gas, vapour, mist, or dust in which, after ignition, combustion spreads throughout the unconsumed mixture.

Extent of Zone

Distance in any direction from the source of release to the point where the gas/air mixture has been diluted by air to a value below its lower explosive limit.

Flash point

This is the liquid temperature at which a liquid releases sufficient vapour to form an ignitable mixture with air, either near the liquid surface or within a vessel. Note that flash point is determined by a flash point tester and is the basis for hazard classification.

There is a direct connection between volatility and flash point. A liquid with a flash point near normal temperatures, without being heated, will produce vapour that can be ignited by a small ignition source such as a spark or pilot flame. Flash point and boiling point temperatures are reduced as altitude increases, as liquid volatility increases with reduced atmospheric pressure. Liquids that are combustible at sea level may be more hazardous as atmospheric pressure is reduced.

Hazardous area

Volume in which an explosive gas atmosphere is present, or may be expected to be present, in quantities such as to require special precautions for the construction, installation, and the use of equipment.

Non-hazardous area

Area in which an explosive gas atmosphere is not expected to be present in quantities such as to require special precautions for the construction, installation, and use of apparatus.

Ignition temperature of an explosive gas atmosphere

The lowest temperature of a heated surface at which, under specified conditions, the ignition of a flammable substance in the form of a gas or vapour mixture with air will occur.

Lower Explosive Limit

The minimum amount of gas or vapour that must be mixed with air to produce an explosive mixture.

Relative density of a gas or a vapour

Density of a gas or a vapour relative to the density of air at the same pressure and at the same temperature (the density of air is equal to 1.0).

Source of release

Point or location from which a flammable gas, vapour, or liquid may be released into the atmosphere in such a way that an explosive gas atmosphere could be formed.

Vapour pressure

This is the pressure, measured in kilopascals (kPa), being 1 pascal equal to 1 newton per square metre ($N \cdot m^{-2}$ or $kg \cdot m^{-1} \cdot s^{-2}$), exerted by vapour against the atmosphere. Just as the atmosphere exerts pressure on the liquid surface, the liquid pushes back. Vapour pressure is normally less than atmospheric pressure and is a measure of the evaporation or change in state from liquid to the gaseous state. This characteristic is often termed "volatility", and liquids that easily evaporate are termed as "volatile" liquids. The safety concern is that the higher the vapour pressure, the more the liquid evaporates, and the lower the boiling point, resulting in more vapours with an increased risk.

Zone 0:

Area in which an explosive atmosphere consisting of a mixture with air of flammable substances in the form of gas, vapour, or mist is present permanently or for a prolonged period of time or frequently.

Examples of Zone 0 may include:

- The interior of closed storage containers containing flammable liquids. This refers to containers that are not under pressure, into which atmospheric air can enter through vent pipes, or vents, or through the opening of covers or registers during filling and emptying operations, etc. For the same reason, the environment close to the outlet of the aeration pipes of atmospheric tanks for flammable liquids could also be considered Zone 0.

Zone 1

Area in which, under normal operating conditions, an occasional explosive atmosphere consisting of a mixture with air of flammable substances in the form of gas, vapour, or mist is likely to form. Examples of Zone 1 may include:

- The exterior of containers that may be opened occasionally or the immediate proximity of feed openings, loading mouths, and sample outlets
- The open air outlet holes of hydraulic flame arresters (devices with a water column that acts as a flame-return valve in devices with flammable gases)
- Ends of articulated arms and flexible loading hoses of tank vehicles and other containers.

Zone 2

Area in which an explosive gas atmosphere is not likely to occur in normal operation, but, if it does occur, it will exist for a short period only. Examples of Zone 2 may include:

- Areas containing vessels and equipment which are so well maintained that leaks can be assumed only to occur rarely (in abnormal operation) and with relief valves which only operate rarely.

3.3.2 DIFFERENCES IEC × NEC

It is important to know that in North America, the classification system most widely utilized is defined by NFPA 70 (National Electrical Code – NEC), establishing area classifications in Groups, Classes and Divisions. The IEC (International Electrical Commission) uses the "Zone" system for classifying locations where fire or explosion hazards may exist due to flammable gases, vapours, or liquids or combustible dusts may exist. The NEC adopted this approach in 1996 and expanded it in 1999 as an alternate to the Class and Division method.

Table 3.1 shows the correspondence of Groups IEC and NEC, and Table 3.2 shows the correspondence between Divisions (NEC) with Zones (IEC). The Classes define the type of explosive or ignitable substances which are present in the atmosphere, and they are defined in the NEC as follows:

- Class I locations are those in which flammable vapours and gases may be present
- Class II locations are those in which combustible dust may be found
- Class III locations are those which are hazardous because of the presence of easily ignitable fibres or flyings.

TABLE 3.1
Correspondence NEC × IEC Groups

Gas	Group IEC	Group NEC	Minimum Ignition Energy [μJ]	MESG (Maximum Experimental Safe Gap)[mm]
Propane	IIA	D	250	0.92
Ethylene	IIB	C	80	0.65
Hydrogen	IIC	B	17	0.29
Acetylene	IIC	A	17	0.29

TABLE 3.2
Correspondence between NEC Divisions × IEC Zones

NEC	IEC	Definitions
Division 1	Zone 0	Location in which an explosive atmosphere consisting of a mixture with air of flammable substances in the form of gas, vapour, or mist is present continuously or for long periods or frequently.
	Zone 1	Location in which an explosive atmosphere consisting of a mixture with air of flammable substances in the form of gas, vapour, or mist is likely to occur in normal operation occasionally.
Division 2	Zone 2	Place in which an explosive atmosphere consisting of a mixture with air of flammable substances in the form of gas, vapour or mist is not likely to occur in normal operation but, if it does occur, will persist for a short period only.

In Class I areas (the Division concept), four distinct groups are based solely on the liquid or gas ease of ignitability, and its corresponding range of flammability, as follows:

Group A—atmospheres that contain acetylene
Group B—flammable gas or vapour atmospheres having either an MESG less than or equal to 0.45 mm or a Minimum Ignition Current (MIC) ratio less than or equal to 0.40 mm
Group C—flammable gas or vapour atmospheres having either an MESG greater than 0.45 mm, and less than or equal to 0.75 mm, or an MIC ratio greater than 0.40 mm, and less than or equal to 0.80 mm
Group D—flammable gas or vapour atmospheres having either an MESG greater than 0.75 mm or an MIC ratio greater than 0.80 mm

The explosive ranges of each flammable gas, that is to say, the concentration between their LEL and the UEL, are based on normal atmospheric pressure and temperature. As the mixture temperature increases, the flammable range shifts downward. As the mixture temperature decreases, the flammable range shifts upward. The mixture volatility is much greater for Group A mixtures compared to Group D mixtures.

Table 3.2 is well known as a practical approximation, but the definitions adopted by each side have particularities that do not have direct correspondences, such as:

NEC defines Division 1 as a location:

(1) In which ignitable concentrations of flammable gases or vapours can exist under normal operating conditions
(2) In which ignitable concentrations of such gases or vapours may exist frequently because of repair or maintenance operations or because of leakage
(3) In which a breakdown or faulty operation of equipment or processes might release ignitable concentrations of flammable gases or vapours and might also cause a simultaneous failure of electric equipment.

On the IEC side, Zone 0 is defined as an area in which an explosive gas atmosphere is present continuously or for long periods. And, Zone 1 is defined as an area in which an explosive atmosphere is likely to occur in normal operation. From these definitions, it can be seen that there is a subtle difference between these three scenarios, although they were united on the first row of Table 3.2.

NEC uses "can", "may", and "might", which are related with the concept of "possibility", and is about if something can happen or not; and IEC uses "likely", denoting a "probability", that is how much chance (by percent) it has to happen. Possibility can answered with yes or no; probability can be answered with percentage.

The terms "likely" and "unlikely" are imprecise. Various sources have tried to place time limits on to these Zones, but none have been officially adopted. The most common values suggested were:

- Continuous release, generating a Zone 0: Explosive atmosphere for more than 1,000 h/yr
- Primary release, generating a Zone 1: Explosive atmosphere for more than 10 but less than 1,000 h/yr
- Secondary release, generating a Zone 2: Explosive atmosphere for less than 10 h/yr but still sufficiently likely as to require controls over ignition sources.

3.3.2.1 Normal Operation

Another significant definition to compare is the definition of IEC for normal operation. IEC defines it as "The situation when the equipment is operating within its design parameters", with the three following observations:

1. Minor releases of flammable material may be part of normal operation. For example, releases from seals which rely on wetting by the fluid which is being pumped are considered to be minor releases
2. Failures (such as the breakdown of pump seals, flanges gaskets or spillages caused by accidents), which involve urgent repair or shutdown, are not considered to be part of normal operation.
3. Normal operation includes start-up and shutdown conditions.

There is no formal definition of "normal operation" in the NEC, but, there is a note regarding motors: "Unless otherwise specified, normal operating conditions for motors shall be assumed to be rated full-load steady conditions", which put starting a motor as not a normal operation in the NEC.

3.3.3 CLASSIFICATION FRAMEWORK FOR GAS ATMOSPHERES

To elaborate the area classification study, in order to define the type of the Zones and their extents, it is necessary to follow the steps described ahead:

3.3.3.1 Step 1: Get the Materials' Data

All the flammable materials that are, or may be present, and their properties need to be registered. Taking the situation in respect of gases, vapours, and mists, the following information in respect of the flammable materials is necessary as a minimum:

- **The flash point:** This is the minimum temperature at which a fuel begins to release vapours, which ignite when in contact with a heat source. However, the flames do not sustain themselves, as there is not enough vapour produced to do so. Below this temperature, the gas or vapour and air mixture is not normally ignitable although a mist may still be capable of forming an explosive atmosphere below this temperature.

- **The ignition temperature:** The minimum temperature at atmospheric pressure necessary to ignite an ideal mixture of the flammable gas or vapour in air. This determines the limit of maximum temperature of equipment permitted within the hazardous area.
- **The explosive limits:** The range of percentage mixtures of the flammable gas or vapour with air within which ignition is possible at atmospheric pressure. The Lower Flammable Limit (LFL) is normally the important one, as it determines the extent of the explosive atmosphere. Above the Upper Flammable Limit (UFL) and below the LFL, explosions do not occur.
- **The gas/vapour density:** Normally quoted as relative to air which is assumed as for the flash point. This parameter has an effect upon the dispersion of the flammable gas or vapour insofar as it may be lighter or heavier than air. Values less than 1 indicate that the vapour is lighter than air. Density greater than 1 characterizes a vapour that is heavier than air.
- **The boiling point:** The temperature at which a flammable liquid will boil at atmospheric pressure. If it is below the actual temperature at the point of release, then the released liquid will vaporize quickly.

In order to promote a computer simulation, the following properties will be also necessary to be known:

- The molecular weight
- The latent heat of vaporization
- The heat capacity.

The technical literature provides the values of these properties for various substances. However, it is necessary to confirm whether the chemical composition of the substance tested is exactly the same as that found in the industrial unit under study. One example is gasoline, which in the United States has a certain octane rating, but in Brasil, "gasoline" can have up to 27% ethanol in its composition, which means that its properties are very different from those in the U.S. product.

The evaluation of the characteristics of flammable substances by an experienced professional is essential for a reliable area classification study. Table 3.3 is a suggestion to keep all the materials' relevant properties registered.

3.3.3.2 Step 2: Identify the Sources of Release

With all material properties' data registered, the next step is to identify the sources of the potential releases. The various potential sources of release (including e.g. pumps and compressors in flammable liquid and gas plants) must be considered within the context of relevance, together with the grade of release, the degree of ventilation, and the topography of the plant.

Without the knowledge of the environmental conditions and the characteristics of the sources of release, the location will most certainly be given a reliability level too low (which impairs safety) or much too high (which is not economically justified). This is the approach that must be avoided.

TABLE 3.3
Example of a Table to Register Characteristics of Flammable Materials

Material	Composition	Boiling point [°C]	Auto Ignition Temperature [°C]	Relative Density (air = 1)	Ignition Temperature [°C]	Vapour Pressure @ 20°C [kPa]	LEL [kg/m³]	Molar Mass [kg/kmol]	Flash point [°C]	Notes

Properties of Flammable Materials—Gas/Vapour/Liquid

The grades of release can be continuous, primary, or secondary.

Continuous
A release that is continuous, or nearly so, or one of relatively short duration that
   occurs frequently.
Primary
A release that is likely to occur periodically or occasionally in normal operation.
Secondary
A release that is unlikely to occur in normal operation and, in any event, will
   do so only infrequently and for short periods.

Grade of release is dependent solely on the frequency and duration of the release. It
is completely independent of the rate and quantity of the release, the degree of ventilation, or the characteristics of the fluid. However, these factors determine the extent
of vapour travel and, hence, the dimensional limits of the hazardous area.

3.3.3.3 Step 3: Identify the Characteristics of the Location

The locations where flammable substances are present must be known in detail, identifying their ventilation degree and availability. In fact, the area classification study is
a process that must be monitored throughout the life of the plant from the first draft
of the design, verifying any changes that are made after the unit goes into operation, in order to refine the data accuracy, to obtain a reliable and updated classification. The degrees of ventilation define their influence on the explosive atmosphere
formation, as:

   High: Can reduce the concentration at the source of release virtually instan-
      taneously, resulting in a concentration below the lower explosive limit.
      A Zone of small (even negligible) extent results.
   Medium: Can control the concentration, resulting in a stable situation in which
      the concentration beyond the Zone boundary is below the LEL, while the
      release is in progress and where the explosive atmosphere does not persist
      unduly after release has stopped. The extent and type of Zone are limited to
      the design parameters.
   Low: Cannot control the concentration while the release is in progress and
      cannot prevent the undue persistence of a flammable atmosphere after the
      release has stopped.

And, also, ventilation needs to be assessed in terms of its availability:

   Good: Is present virtually continuously
   Fair: Is expected to be present during normal operation. Discontinuities are
      permitted, provided that they occur infrequently and for short periods
   Poor: Does not meet the definitions of fair or good, but discontinuities are not
      expected to occur for long periods.

Using the ventilation and the source of release characteristics, Table 3.4 was proposed by NBR IEC 60079–10–1 to define the type of the Zones.

TABLE 3.4

A Proposal to Describe the Influence of the Ventilation on the Zone Classification

Source Type	Degree of Ventilation						
	High			**Medium**			**Low**
Continuous	Unclassified (Note 1)	Zone 2 (Note 1)	Zone 1 (Note 1)	Zone 2 (Note 1)	Zone 1 (Note 1)	Zone 0	Zone 0
Primary	Unclassified (Note 1)	Unclassified (Note 1)	Zone 2 (Note 1)	Zone 1	Zone 1 (Surrounded by Zone 2)	Zone 1 (Surrounded by Zone 2)	Zone 0 or 1 (Note 2)
Secondary	Unclassified	Unclassified (Note 1)	Zone 2 (Note 3)	Zone 2	Zone 2	Zone 2 (Note 3)	Zone 0 or 1 (Note 2)
Availability of Ventilation	Good	Fair	Poor	Good	Fair	Poor	

Notes:

1. These classifications actually have a small area around the point source with a higher level of classification, but the extent of this transition area is negligible. Care should be applied in assessing risk whenever ventilation is used as criteria for reducing the classification around Zone 1 and Zone 2 locations.

2. The area will be Zone 0 if the low ventilation is so weak and the release is such that, in practice, an explosive atmosphere exists virtually continuously (i.e., approaching a "no ventilation" condition).

3. The Zone 2 area created by the secondary grade of release may exceed that attributable to a primary or continuous grade or release, in which case, the greater distance should be taken.

4. The approach of a "Zone 1 surrounded by a Zone 2" is commonly found in API RP-505 but without a reasonable justification. By principle, the extents of a Zone 2 will be greater than those of a Zone 1, only if the emission of flammable vapour in the properly identified abnormal situation is greater than the estimated emission in normal operation.

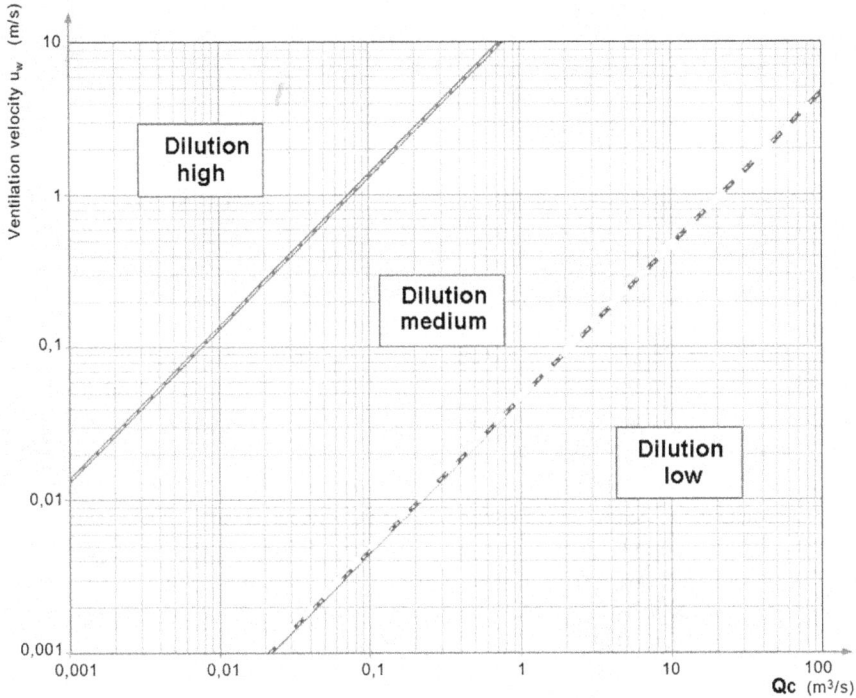

FIGURE 3.1   Chart for assessing the flammable gas emission dilution.

3.3.3.3.1   Negligible extents

The NBR IEC 60079–10–1: 2022 removed the safety factor "k" in the denominator of Equation 3.1 (in the previous edition it was a safety factor attributed to LEL), and the "NE" (Negligible Extents) area since then depends on the characteristic emission parameter Wg. Figure 3.1 shows the logarithmic diagram for assessing the mixture dilution, and Qc is defined in Equation 3.1:

$$Qc = \frac{Wg}{\rho g \times LEL} \tag{3.1}$$

Where:

Qc is the volumetric release characteristic of the source [m³/s];

$pg = \dfrac{pa \times M}{R \times Ta}$ is the density of the gas/vapour [kg/m³];

Wg is the mass release rate of the flammable gas [kg/s];

LEL is the Lower Explosive Limit [vol/vol];

pa is the atmospheric pressure (101 325 Pa);

M is the molar mass of gas or vapour (kg/kmol);
R is the universal gas constant (8314 J/kmol.K); and
T_a is the absolute ambient temperature (K);

Figure 3.1 was drawn considering the following:

- The curves are based on an initial zero concentration value and are not applicable to low dilution situations in indoor environments.
- The graph was developed on the basis of continuous equations and models selected from computational fluid dynamics (CFD) simulations, assuming a dispersion distance proportional to the square root on the x-axis, and the results were moderated for the purpose of NBR IEC 60079–10–1.
- It does not identify different Zones, and these should be analysed on the basis of ventilation around the source of the release and possible variations in the release conditions.
- "High dilution" refers to situations where the concentration near the source of release can be quickly reduced, and there will be no persistence after the release is stopped. If, under the appropriate conditions, an area is considered as "NE", this would put it in the same status as "non-hazardous".
- "Medium dilution" applies to situations where the concentration of the release is controlled, resulting in a stable boundary when the release is in progress, and the explosive gas atmosphere does not persist after the release has stopped. A medium dilution scenario is expected to lead the classification to a Zone 2.
- "Low dilution" applies to situations where there is a significant concentration while the release is in progress, and the flammable atmosphere persists after the release is stopped. A low dilution scenario is expected to lead the classification to a Zone 1.
- It does not include a specific safety factor. A suitable factor should be determined by the user based on the application and any safety factors applied to other parameters used in the assessment, e.g. assumed release rate.
- The degree of dilution is obtained by finding the intersection of respective values displayed on horizontal and vertical axes. The line dividing the chart area between "dilution high" and "dilution medium" represents a flammable volume of 0.1 m³, so any intersection point left to this curve implies an even smaller flammable volume. The line dividing the chart area between "dilution medium" and "dilution low" represents a flammable volume of approximately 100 m³, so any intersection point right to this curve implies an even larger flammable volume.

But, after the NBR IEC 60079–10–1: 2022, to designate an area as "NE", in addition to verifying whether there will be a "high" dilution of flammable gases, and "good" ventilation availability, it will be also necessary to ensure the following:

- The formation of the explosive atmosphere will not cause damage due to overpressure.

- The ignition of the explosive atmosphere will not generate enough heat to cause damage or fires in the surrounding material.
- A risk assessment is made for gas emission sources at pressures above 10 barg.
- Gas emission sources at pressures above 20 barg will not be assigned as NE areas, unless supported by a detailed risk assessment.

The influence of the plant's topography can be decisive in defining the extent of classified areas, as it can promote interesting wind flow directions, transporting the explosive atmosphere released through the industrial plant. Depending on the relative density, it can either continue along the floor or follow the contour of processes' equipment.

The example given in NBR IEC 60079–10–1: 2022, clause 4.3, for "NE" is "a natural gas cloud with an average concentration that is 50 % by volume of the LEL and that is less than 0,1 m³ or 1,0 % of the enclosed space concerned (whichever is smaller)". In the same clause, NBR IEC 60079–10–1: 2022 says that in some cases, an NE may arise, and it "may be treated as non-hazardous. Such a Zone implies that an explosion, if it takes place, will have negligible consequences". This is a consideration in NBR IEC 60079–10–1: 2022 which conflicts with its clause 1—Scope, where it is defined that "area classification did not take into account the consequences of ignition of an explosive atmosphere".

It is interesting to note that two situations rose from the above considerations:

1) The return to "Vz", the hypothetical volume of the explosive atmosphere (which never completely disappeared, as highlighted in diagram C.1 of NBR IEC 60079–10–1:2022).
2) The indication, from the diagram, of a maximum volume equal to 2.2 litres of natural gas as a reference measure for the consequences. It was noted, however, that diagram C.1 of NBR IEC 60079–10–1 calculates the release in LEL (and not LEL/2, which corresponds to 50%), and, therefore, using this parameter, the limit becomes 4.4 litres of "pure" natural gas.

So, a question rose: Do we need to consider 2.2 or 4.4 litres?

The answer is: It depends, because, regardless of the actual volume involved, the NE Zone is determined not only on the basis of the explosive atmosphere volume (also known as "degree of dilution") but also on the basis of the availability/reliability of ventilation, in the case of first degree or continuous emissions.

The NBR IEC 60079–10–1 states in 4.4.2 that the NE Zone "may be considered non-hazardous". But, is the vice versa true?

In this regard, NBR IEC 60079–10–1 clause 3.3.2 says that "non-hazardous area" is: "Area in which an explosive gas atmosphere is not expected to be present in such quantities that special precautions are necessary for the construction, installation and use of the equipment". So, NE Zones, as "non-hazardous", are areas in which common mechanical and electrical equipment, i.e. not certified, can be installed. But, are "non-hazardous areas" equivalent to "safe areas"?

No, because "non-hazardous areas" inherit this property because they were considered by NBR IEC 60079–10–1 as NE Zones, i.e. areas where the mixture, in the event of ignition, would cause "negligible consequences".

The influences of the emission flow, the effectiveness of ventilation, and its availability over time create an NE Zone. If only one of the above factors were to change, the "NE area" can become a hazardous area, as Zone 2 IIA T3 or Zone 1 IIC T1 or whatever.

To be considered as a "safe zone", it is necessary to be consistent with IGEM/SR/25: "An area in which an explosive atmosphere is never present". This is just one of the conflicts present in NBR IEC 60079–10–1, a document that, when it included a specific model for estimating the hazardous areas' extents, went beyond its mission as a standard, which was to establish the minimum requirements that should be met by those responsible for area classification studies.

The teachings of "how to do it" must be learned from books and training; the standards must define only the minimum requirements stipulated for materials and services, as there are several ways to obtain results, which must be chosen by the responsible professional, aiming to obtain the best option, case by case. So, as there are several models for estimating gas dispersions, each with its own advantages and limitations, NBR IEC 60079–10–1 should have been issued as a Technical Report (TR), leaving it up to those responsible for the area classification study to choose the most appropriate option for each case.

There are some special situations that can result in atypical volumes of Zones, which highlight that it is essential for the professional team responsible for the study to have a solid knowledge on these phenomena.

3.3.3.4   Step 4: Define the Classified Area Extents

There are two methods for performing this task: One is called "by example", and the other is known as "by the source of release". The first one is which has been widely used, although not in a reliable way. It is based on just applying illustrations from documents that usually refer the process equipment as "typical", without any warranty that the real process characteristics and the local ventilation were respected.

This is the quickest way to define the extensions of a classified area, and it is wrongly considered "conservative" because it indicates large extensions, ranging from 3 m to 30 m. However, as the extents considered "normal" are not known, nor are known the justifications for the extensions given in the typical figures, such designation is actually just a myth!

Extents greater than those estimated by mathematical simulations increase the installation and maintenance costs of a given plant and may even lead to conflicts with regulatory bodies and fire insurance companies. Another concern is that if the frequency of release of flammable products enough to create an explosive atmosphere is significant to classify an area as Zone 1 (under the informal estimative between 10 and 1,000 h/yr), considering that many flammables can be toxic or stuffy, and the exposure of workers may not be considered acceptable by local authorities. Additionally, when the adoption of "typical figures" for classifying locations is executed mechanically, it can also lead to certain hazardous areas not being adequately identified, compromising the correctness of the area classification study.

The source of release method for area classification was introduced progressively from around 1970 onwards. This method is very different from the "by example" one. Each source of release and the mode of release at that point have to be identified. Release quantities and dispersal criteria have to be identified, and, from this information, the extent of a hazardous area emanating from a particular release is determined.

The method of determination of this extent is mainly based on the use of mathematical approaches. For large plants, with many sources of release, it is recommended to use specialized software for analysing ventilation systems in classified areas. Such tool can help to determine ventilation requirements, taking into account parameters such as air renewal rate, air flow distribution, differential pressure, and dispersion of flammable gases.

The "source of release method" provides the necessary adherence to the characteristics of the plant under study, as their fluid temperatures, ventilation, and system pressures characteristics will be considered; that is it reveals the risks in the real plant conditions, and it avoids the use of blind conditions shown in drawings where the parameters that were considered for their preparation are unknown.

With the area extent defined, the area classification study for flammable gases and vapours is completed, and all considerations must be registered in the data sheet, where a suggested model is shown in Figure 3.2. Once an area has been classified, the IEC 60079–14 (in the United States, the NEC) provides very specific requirements about the electrical equipment and associated wiring that can be installed within that area. The requirements are intended to prevent electrical equipment from being the ignition source for a flammable mixture. Obviously, a facility's area classification study must be known before any electrical equipment can be specified, designed, or installed.

On many industrial projects, special-purpose mechanical equipment with long lead times must be specified and ordered early. Failure to determine the classified areas for the facility and such equipment in a timely fashion can result in unsafe installations, rework, confusion, delays, and cost overruns.

1	2	3	4	5	6		7	8				9	10	11	12	13
	Source of release				Product			Ventilation					Classification			
N	Name	Tag	Type (a)	Name (b)	Temperature and Pressure		State (c)	Type (d)	Grade	Availability		Zone (e)	Zone extents [m]		Ref. (f)	Notes
					°C	kPa							Vertical	Horizontal		
1																
2																
3																
4																
5																
7																
8																

FIGURE 3.2 A model for the data sheet of the area classification study's considerations.

To specify the electrical and electronic equipment that can be safely installed in each Zone, the most suitable types of protection for the design of the industrial unit will be chosen then or, if a consequence analysis is carried out, they can be defined by Equipment Protection Level (EPL), as provided for in IEC 60079–14, taking into account the requirements given for the Group and Temperature Class.

In Figure 3.2, the observations are:
 (a) C is continuous, P is primary, and S is secondary
 (b) Its formula can be also used
 (c) G is gas, H is hybrid, and L is liquid
 (d) N- is natural, and A is artificial
 (e) U is unclassified
 (f) S is simulation, and E is by example

3.3.3.5 Step 5—Report

As stated, the Hazardous Area Classification study is the key to identify areas where dangerous substances may lead to explosive atmospheres. Disclosing this information to all employees will help avoid ignition sources within hazardous areas. In hazardous areas, only Ex certified equipment may be installed, with the exception of European countries governed by ATEX, where a manufacturer's self-declaration of conformity with the standards is accepted for equipment intended to be installed in Zone 2 areas.

A Hazardous Area Classification Report (HACR) needs to be issued, providing clear information and guidance to ensure the safety of personnel, equipment, and facilities. It should typically contain varying information, including a site overview, dangerous substance identification, suitable control measures, and mitigation response. Depending on the country's legislation, it should also state how the personnel are protected from harm from fires and explosions and what can be included under explosion protection such as explosion relief, containment, or suppression. A well-structured HACR ensures that all stakeholders understand the potential risks associated with explosive atmospheres and provides a basis for implementing effective control measures to enhance safety.

It should include the following:

 • The physical and chemical information on the plant's flammable gases and liquids, obtained from the safety data sheets
 • A detailed list of all emission sources indicated on the drawings
 • The pressure and flow data of the processes analysed
 • Any decisions made on exceptional situations
 • The results, or findings, obtained from the site assessment audit
 • The justifications for the distances defined in the area classification study
 • The safety signage plan, indicating on the drawings the location of the signs containing the Zone, Group, and Temperature Class.

All area classification documentation should be placed under the protection of the facilities management of change process control. As modifications are made to the facility, these documents should be periodically reviewed to assess the impact of the modifications.

3.3.3.6   Step 6—Signage of the classified areas

Zone marking will enable us to classify Zones where appropriate electrical installation can be made and a marking should be shown on the plant layout to easy access of zoning of hazardous areas, and a sign board shall also be displayed at prominent place for the awareness of workers and visitors. It is important that the sign shows in a clear way:

- The Ex triangle (classified area)
- The red background (fire and explosion risks)
- The Zone of the location.
- The adequate equipment group
- The adequate Temperature Class for equipment intended to be used
- The area classification drawing number.

The Ex triangle sign is suggested in IEC standards, and it is already used in many countries. It means that the region is a hazardous area. Such information is important but not enough for services being executed with safety, as the Zone, the Temperature Class, and the equipment group are required for selecting adequate instruments and devices to perform the tasks. So, these information need to be included in alert signs.

ISO standards highlight that a warning sign must have a minimum amount of text, as the aim is to extract an immediate reaction from readers, therefore phrases such as "no smoking" and "caution: dangerous area" are not effective and would be duplicated, as the Ex inside the triangle with a yellow background already means this. And as the plate cannot provide an idea of the three-dimensional extent of the Zone, the reference to the area drawing number allows workers to quickly refer to the details of the extent of the hazardous area on the corresponding drawing.

An example of a sign with all the important information for services in hazardous locations, which follows the ISO recommendations, is shown in Figure 3.3. It is important to say that a warning sign does not replace the necessity of a required safety training, and it is designed to be recognized by trained and authorized workers who perform services in hazardous areas.

3.3.4   STANDARDS × RECOMMENDED PRACTICES

Several documents can serve as guidance for developing area classification studies. However, contrary to what many people think, there is no "one-size-fits-all" recipe.

One common misunderstanding is to consider "standard" and "recommended practice" as synonyms. A standard establishes requirements, but a recommended practice only offers suggestions, and it is up to the user to assess whether they are applicable in his case.

A programme of audits on electrical projects and installations based on IEC standards related to explosive atmospheres, carried out in several petrochemical

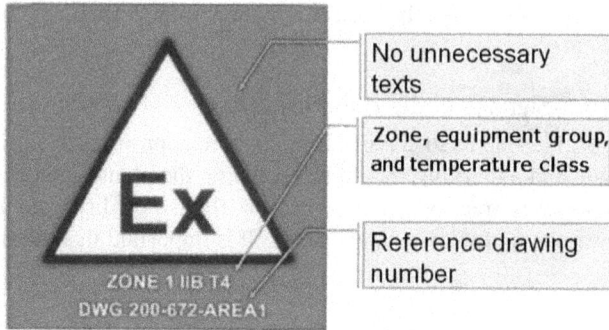

FIGURE 3.3   The effective warning sign for hazardous locations due to flammable gases.

companies, found, with worrying frequency, a lack of coherence in area classification documents, as an example, with conflicting citations, like referencing simultaneous compliance with NEC and IEC area classification standards, which, in fact, follow two different philosophies.

One of the reasons given to explain the causes of such conflicts is the style of American "Recommended Practices" on this subject. The profusion of "typical figures" on documents as API RP-505 and NFPA RP-497, where the figures with extents of classified areas are mistakenly interpreted as "ready to use", apparently has contributed to the misconception that area classification is done just copying them, without paying attention to the many warning notes highlighted in the text of those documents. Sections 3.3.4.1 and 3.3.4.2 will discuss the characteristics of the two methodologies used to define area classification extents: The "by example" and the "by source of release".

3.3.4.1   The "By Example" Method

API RP-500/505 are the most used documents to define the extents and shape of classified areas in the oil and gas industry, under the "by example" method but, unfortunately, without the required care.

Area classification drawings are only one part of the area classification study; the basis and the justifications for the defined extents must be recorded in the area classification report, with all the considerations adopted by the responsible team, as the parameters of the processes, the products, and the environmental conditions.

And this is duly written in all sections of API RP-500/505 (RP-500 deals with the classified areas under the Class/Division scheme in the United States; RP-505 deals with the IEC Zone scheme).

For example, the following note is found below most of the illustrations represented there:

Distances given for a typical refinery: they should be used with discretion, taking into account all the factors addressed in the text. In some situations, greater or lesser distances may be necessary.

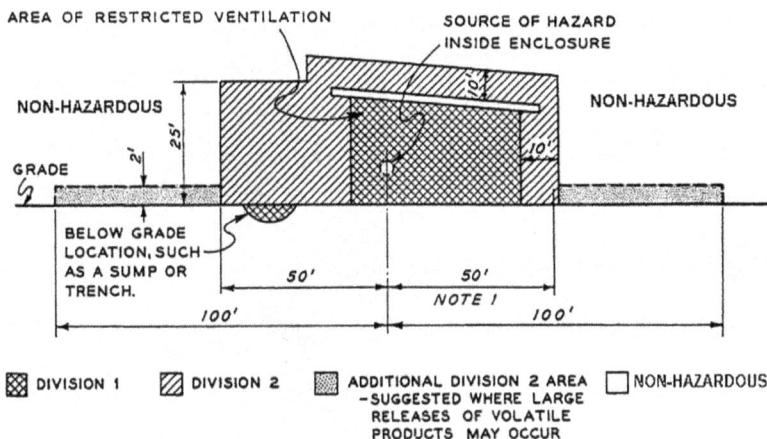

FIGURE 3.4 "Recommended" area classification for process area with restricted ventilation, API RP-500, 1957.

In order to detail the scope of these notes, we can quote the introduction section of these documents:

> API publications may be used by anyone who wishes to do so. Every effort is made by the Institute to ensure the accuracy and reliability of the data contained therein; however, the Institute expressly assumes no responsibility for loss or damage resulting from their use, or violation of any law that this publication may conflict with.

Another issue of API RP-500/505 is about the so-called "typical refinery". Considering that the figures in those documents have remained virtually unchanged over decades, would the so-called "typical refinery" today be one that currently uses the same process technology as it did seven decades ago? Or, would "typical" be related to the refinery's production capacity, such as 20.000 barrels/day on those early years? This is illustrated in Figures 3.4 and 3.5.

Figure 3.5 is very well known, as it can be found in manufacturers' catalogues, blogs, books, and, unfortunately, in area classification drawings of many companies also. But, to apply it in area classification studies, first, it is necessary to confirm that the process under study is the same as that depicted in the illustration. Considering that no information is given about the plant reproduced in figure, nor the dimensions of the shelter, some questions need to be answered beforehand, as example:

- Is the flammable material processed in the plant the same as shown in the figure?
- Are the characteristics of the process, the same as that in the figure?
- Are the walls of the shelter constructed exactly as specified in the figure?
- If the vapour emission source is 4 m further to the right, will Division 2 on the right side, external to the shelter, be extended beyond the 3 m shown?

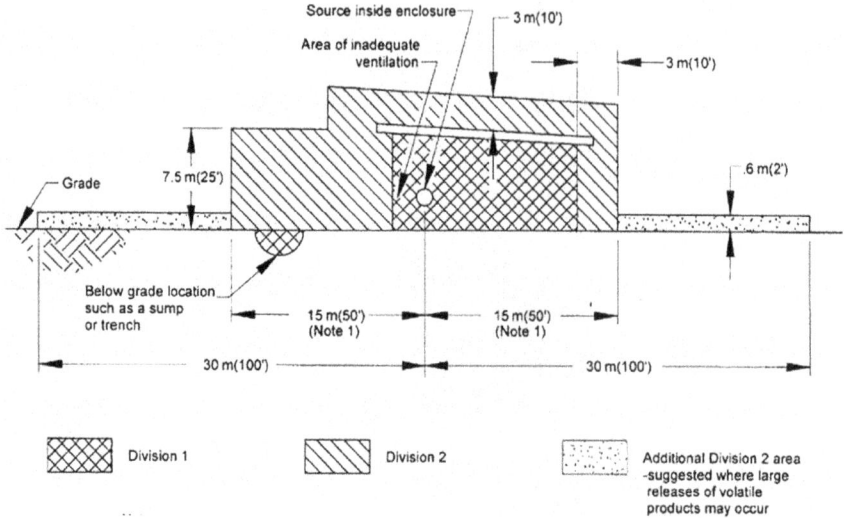

FIGURE 3.5 "Recommended" area classification for inadequately ventilated process location with heavier than air vapour source, API RP-500, 2012.

- If the shelter is larger than 12 m, will Division 2 above the roof keep the 3 m height throughout its entire length?
- Will the 3 m extent above the roof also be maintained if the vapour emission source in the plant under study is only 1 m below the roof?

The oil industry is subject to several technological changes, which requires periodic review of standards. The API RP-500: 2012 brought important additional considerations, but it is important to highlight that the scope of the American document is only to provide generic guidance, as it is impossible to cover all possible variations in the construction and operation of industrial plants.

The comparison of Figures 3.4 and 3.5 shows that exactly the same dimensions are considered, despite the 55-year difference between them. This similarity would only be justifiable if the two facilities were, operationally speaking, identical.

However, in practice, a modern plant is very different from another one built six decades ago. In Brasil, for example, in addition to new processes, the processing in refineries is of heavy oil, which requires different parameters not mentioned in recommended practices.

Considering the various changes in the oil industry processes over those 55 years between the 1957 and 2012 editions, it is expected that one figure could not be identical to the other. And, as the recommended practice does not contain any identification of which flammable substances were considered, nor the characteristics of the respective processes, nor the ventilation considered in the preparation of the distances shown in the figures, they cannot be simply reproduced for every unit of the oil industry, as highlighted in there.

Unfortunately, many companies have their area classification documents composed by copies of the figures taken from API RP-500/505, and as consequence, their real risks are not correctly identified, what could impact the specification of electrical equipment, and as consequence, it can lead to fires and explosions. Understanding the processes and relevant parameters to determine to what extent a release would form an atmosphere with a concentration above the LEL is an engineering activity that can only be carried out by experienced professionals.

The purposes of the API RP-500/505 are written in their texts:

> [T]o provide guidance for the classification of Class I Division 1 and Division 2 (in RP-500; Class I, Zone 0, Zone 1 and Zone 2 in RP-505) locations in petroleum facilities for the selection and installation of electrical equipment. This publication is only a guide and requires the application of sound engineering judgment.

So, it is clear that their figures are not ready to be just "copied-and-pasted" in any area classification study of industrial plants in the oil and gas sector, especially because no information is given about on the safety factors, the characteristics of the processed products, their flows, pressures, and temperatures that were considered in the preparation of the figures.

The method "by example" based on just copying figures from the recommended practices are accepted in very few specific situations, as an example, vapours emanated due to a liquid pool in open air, ensuring that the atmospheric pressure of the location is considered. As mechanically copying the figures, without a sound engineering judgement, as requested by API RP-500/505, does not guarantee that the real risks of the industrial plant are represented, and as consequence, such area classification studies are not reliable.

In Brasil, for example, in addition to new technologies, the processes in oil refineries are of heavy oil, which requires higher pressures and temperatures, not represented by the typical figures from API RP-500/505. This assertion can be proven by observing Figure 3.6, where different technical documents used as guides for area classification related to the storage of flammable products in fixed-roof tanks define different extents, because the processes where they were based had different parameters and, also, different safety factors. Such figures are intended only to help the responsible professional team on visualizing the probable conditions and not to assign a final, definitive, and unchangeable situation to all possible cases of combination of products, process parameters, and ventilation conditions.

Other examples are from naval classification societies that issue requirements for petrol tankers and FPSO (floating, producing, storage, and offloading). An FPSO is essentially a combination of a merchant tanker with a petrochemical process plant on top of it.

For marine systems, Safety Of Life At Sea (SOLAS), a set of standards established within the International Maritime Organization, takes primacy; this is reflected in Classification Society Rules for the construction of ships and also in the International Safety Guide for Oil Tankers and Terminals (ISGOTT). However, once an FPSO process plant has been fitted on its main deck, the objective is to get the maximum physical separation (vertical and horizontal distance) between vent outlet and potential sources of ignition. So, knowing and understanding the physical properties of the vented gases, it is important to assist the decision-making process for locating the cold vents.

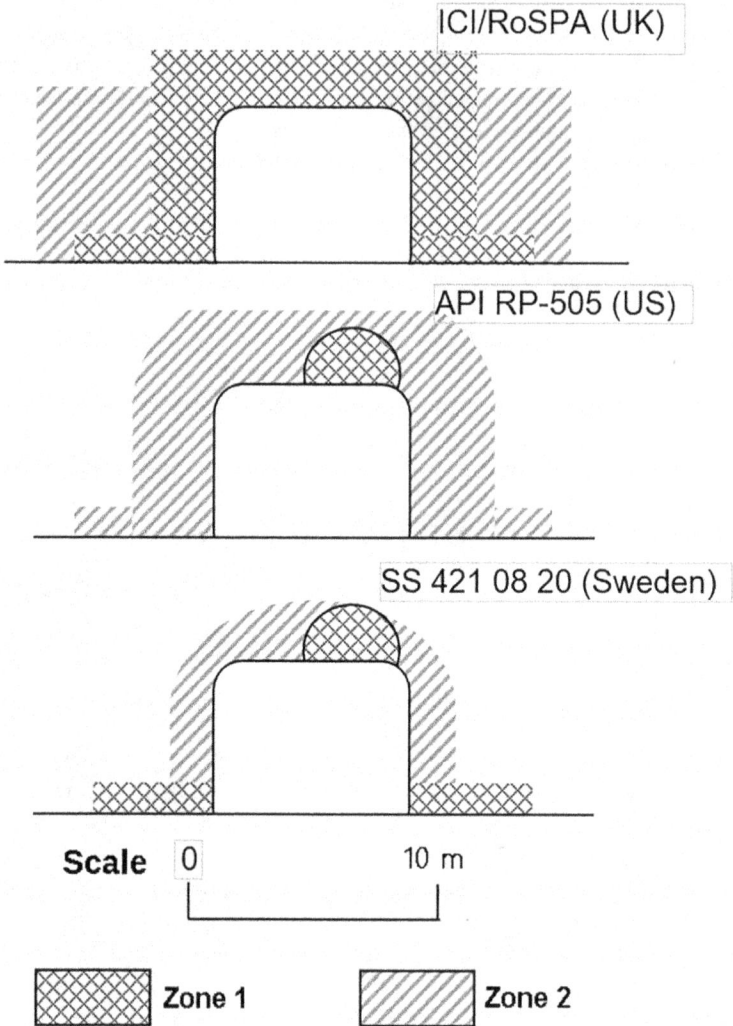

FIGURE 3.6 Area classification for the same situation, treated by different technical documents. Scale is valid for hatched areas only.

But, when considering the guidance contained with SOLAS, as well as other documents, some conflicts appear, because different basis and characteristics were considered, but not documented, as shown in Table 3.5.

It is clear that for the same process, different area extents are imposed by those documents, and no information is given about how such distances were defined, nor about what safety factors were used. Therefore, without indicating the justifications for the stipulated distances, there is no guarantee that the figures shown in those documents apply to all installations.

"Classifying areas" that are done only by mechanically copying figures is not engineering, and may even compromise the safety of the plant. For the correct

TABLE 3.5

Comparison of the Dimensions of Hazardous Areas Derived from Different Area Classification Documents for the Same Cargo Tank Vent

Code Reference	Directions
IP-15 *	10 m Hazard radius Zone 1 at tank vent tip
	10 m Hazard column Zone 2 extended down to grade
IEC 60092–502	6 m radius Zone 1 sphere around the cargo vent tip
	6 m radius column extended indefinitely upwards (?!)
	4 m radius Zone 2 column beyond the 6 m radius Zone 1 column, the cargo vent tip extending down to tank top/deck
DNV OS A101	Hazardous area defined by "recognized codes and standards"
	Large releases should consider the extent of the Zone larger than the boundary of 50 % LEL concentration.
SOLAS	Vent not located <6 m above cargo deck <2 m (Hi-jet types)
	Vent not be located <10 m horizontally from nearest intake or ignition source
ISGOTT	Vents to be sited in locations to prevent the accumulation of a flammable atmosphere and should be located as high as practical, unobstructed, and away from potential ignition sources. Maximum flow rate 36 m/s. SOLAS to be complied with for the vessel.

Note: * IP-15 (currently EI-15) in Appendix C provides a series of tables to enable the calculation of hazardous areas based on fluid composition and release rate vent size for specific applications, which is the recommended procedure. The distances shown in Table 3.5 for IP-15 are an example for a cargo tank vent based on values in that document.

dimensioning of the classified area extents, the responsible team must have considerable knowledge and experience.

3.3.4.2   The "By Source of Release" Method

In this method, the contribution of each source of release is assessed, taking into account the local ventilation and the process parameters at that point. Computational fluid dynamics (CFD) is a helpful tool to predict the behaviour of explosive atmospheres under the influence of local ventilation.

NBR IEC 60079–10–1 is considered as one attempt on presenting a mathematical model to define the extents of hazardous areas due to flammable gases. It is important to say that, although, apparently, NBR IEC 60079–10–1 gave the permission for "other standards and codes to be used for classifying areas" as said on its item 5.3.1, but, its Annex K establishes that if a study is conducted on the basis of a standard other than NBR IEC 60079–10–1, then it cannot be cited in such document as having been the basis for the study. In this way, Annex K corrects an error found in several audits, when the study indicates IEC 60079–10 as a reference, but, in fact, the extents were just copies of typical figures from API RP-500/505.

It is a mistake to assume that "the different standards for classifying areas are complementary", because if a given document indicates a certain extent, another standard that indicates a smaller extension for a similar situation cannot be sought. A single document must be adopted as a reference throughout the study.

3.3.4.2.1 Flammable Gas Atmospheres

The simulations to define the area classification extents are based on mathematical models that need to take into account many parameters such as the gas mass flow rate and the air temperature density, which can lead to very complexed models.

In order to familiarize the reader with the principles that guide the development of a model, we will present one of the simplest models. To estimate the gas mass flow rate, the Equation 3.2 describes a simplified model to relate the volumetric concentrations densities of the gas mass flow with the axial distance, based on the Gaussian plume model in the simplest case: a subsonic jet issuing from a circular orifice. The fundamental assumptions are: Plan terrain; constant winds with unidirectional speed; and atmospheric turbulence as homogeneous and stationary.

$$\frac{Cx}{C_0} = 5\frac{d_0}{x}\sqrt{(\rho_A/\rho_0)} \tag{3.2}$$

Where:

Cx is the volumetric concentration at x metres $[m^3/m^3]$.
C_o is the volumetric concentration at the source of release orifice $[m^3/m^3]$.
d_0 is the diameter of the orifice [m].
x is the axial distance from the origin of the emission [m].
ρ_A is the density of the air at site $[kg. m^{-3}]$.
ρ_0 is the density of the gas at outlet $[kg. m^{-3}]$.

This model is limited, as many important characteristics are not taken into consideration. Calculation of the classified areas' extents is also dependent upon the velocity of air at the point of leakage and the quantity of air which is available. If sufficient air is not available, the leak will steadily increase the amount of flammable gas or vapour in air at any point and progressively increase the extent of the classified area throughout the period of leakage until equilibrium is reached.

Therefore, the simplified calculative methods are generally only valid in areas which are very well ventilated and where a large amount of unconfined air is available. For calculating the extent of classified areas outdoors, there are some CFD software available in the market, which take additional parameters into consideration.

Given the complexity of the phenomenon, currently the simulation and analysis by CFD, in which the governing equations of mass, momentum, and energy as well as the turbulence models are solved in controlled volumes in space and time, show a more reliable alternative to the prescriptive models or using simplified equations. NBR IEC 60079–10–1 ed. 3.0 on its Annex D.3 presents the logarithmic diagram, shown in Figure 3.7, based on CFD, for different ventilation speeds of the site. The abscissa shows the release flow rate, and the ordinate indicates the extent of the explosive atmosphere for three specific types of release. Using the diagram, we note that, for distances greater than 1 m, the results are consistent for methane gas, while in the cases of propane and hydrogen, the distances shown are smaller than those obtained by dispersion models already consolidated in commercial software.

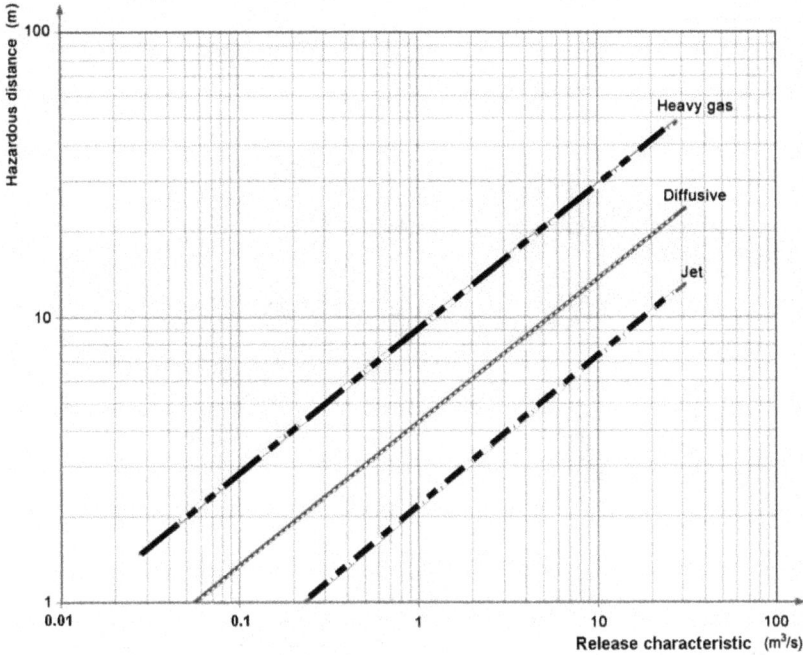

FIGURE 3.7 Chart for estimating hazardous distances.

This indicates that the line representing the distances for jet occurrence was obtained with dispersion simulations using natural gas as the representative substance. Thus, it is necessary to alert that the diagram does not seem to be applicable to most situations encountered in industries.

3.3.4.2.2 Comparing "by Source of Release" and "By Example" Results

The method "by example" has been the most widely used in studies of classified areas worldwide. The most used technical documents for this method are the American Recommended Practices NFPA 497 (for chemical industries) and API RP-500/505 (for oil and gas installations). There is a "myth" that these documents lead to "conservative" extents because the application of their figures would result in more expensive but "safer" installations. However, as already highlighted, as the conditions of pressure, temperature, flow, and ambient ventilation considered in the preparation of such figures are not available, it is not possible to state that they are "conservative" nor that they can be used in "similar" situations.

To illustrate how applying the Recommended Practice figures can result in an underestimated risk, we will compare the distances from a CFD simulation with those shown in NFPA 497.

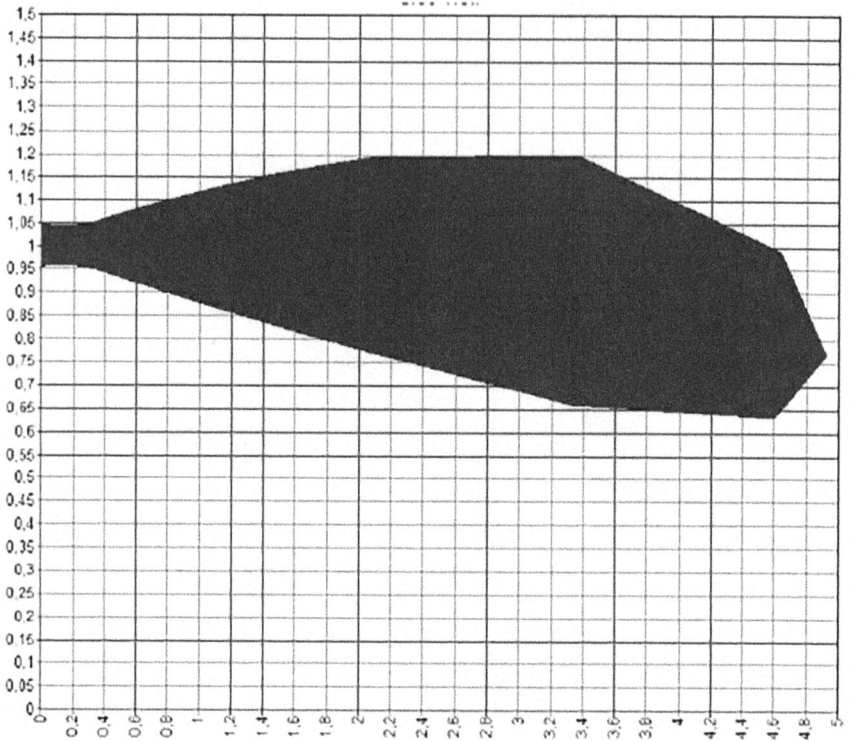

FIGURE 3.8 A given isopentane release at 30°C.

In Figure 3.8, where height [m] is on y axis and horizontal distance [m] is on x axis, are shown the extents and shape of a simulation using CFD for the explosive atmosphere due to isopentane at 30°C, with the source of release at 1 m height. Figures 3.9 and 3.10 show other simulations with isopentane at the same pressure, but at 90°C and 120°C.

The comparison is summarized in Table 3.6.

Another difference brought about by the "source of release" approach is the clear identification for the first time that Zone 1 is not automatically surrounded by Zone 2. While this was also true in the "by example" method, this fact tended to be obscured by the generality, and it became common belief that Zone 1 was always surrounded by Zone 2.

The source of release method automatically dispersed that misunderstanding as to have a hazardous area, it is first necessary to identify a source of release which creates it. Therefore, a leak would have to be identified as possibly behaving in two distinctly different ways to produce both a Zone 1 and a Zone 2. This would lead to another fact that was not before clearly identified, that is a single source of release can behave in different ways, depending on the circumstances.

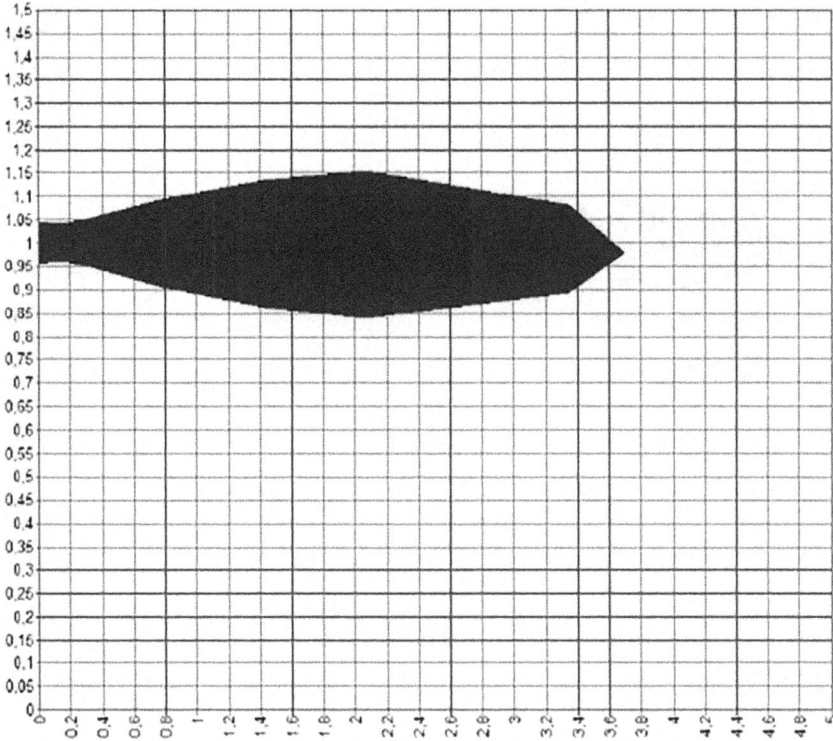

FIGURE 3.9   The isopentane release of Figure 3.8 at 90°C.

It can be seen that the shapes and extents of the explosive atmosphere given by CFD simulations are very different from those given by recommended practices, which "typically" define the extent of the classified locations as invariably 3 m, for all situations without a clear justification. CFD models offer the most complete and detailed description of the flow physics and provide a wealth of information about the flow in terms of flow velocities, temperature, fuel concentrations, and so on. CFD modelling is a knowledge-based process and requires the user to understand the fluid flow behaviour.

3.3.5   Interior of Equipment Containing Flammable Substances

The interior of various equipment containing flammable substances (also known as the generic term "process equipment"), such as atmospheric tanks, may be considered a classified area, although an explosive gas atmosphere cannot normally be present, considering the possibility of air entering the equipment.

When specific controls are used, such as inerting, the interior of equipment containing flammable substances may not need to be considered a classified area or

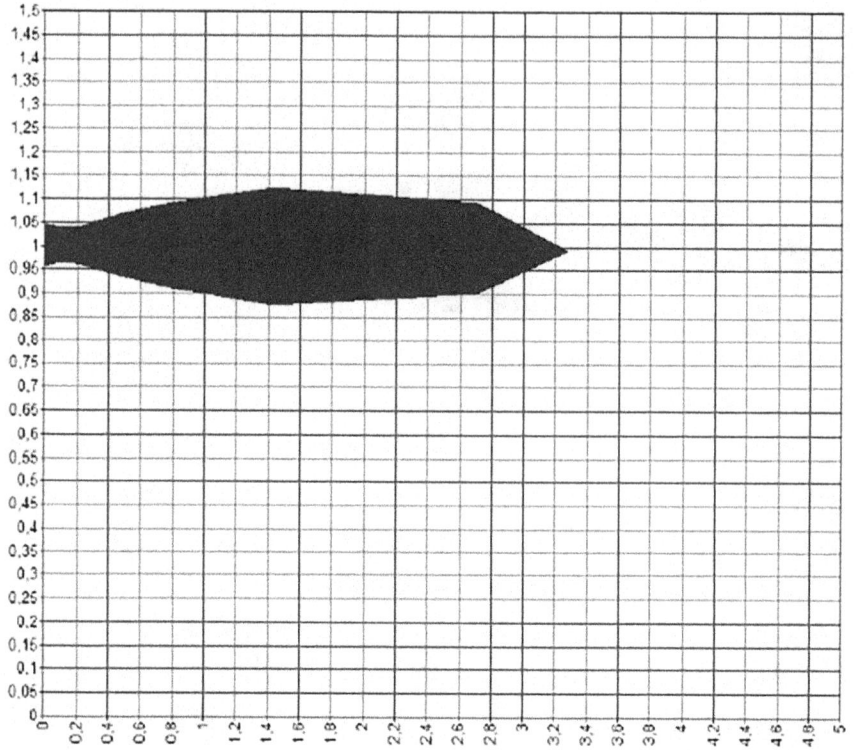

FIGURE 3.10   The isopentane release of Figure 3.8 at 120°C.

TABLE 3.6

Comparing Results between "By Example" and "By Source of Release" Methods with Isopentane

Method "By Source of Release"		Method "By Example"
Temperature [°C]	LEL Extents [m]	NFPA 497 [m]
30	4.9	3.0
90	3.7	3.0
120	3.3	3.0

---

may be designated as a lower risk Zone. In such cases, it is recommended that the reliability of the control measures be quantified to reduce the area classification that is determined for the interior of the equipment. For example, the control measures may be evaluated using an appropriate study, such as a Safety Integrity Level (SIL) assessment, in accordance with IEC 61511.

Inerting (the replacement of atmospheric oxygen in a system with a non-reactive or non-flammable gas) aims to render the atmosphere within the system incapable of propagating an ignition. The addition of flammable gases to ensure that the environment is always outside the flammability limits may also prevent a classified area inside the equipment.

3.3.6 CONCLUSIONS

CFD modelling is the most appropriate technique to investigate the dispersion of a flammable vapour cloud on industrial plants as the dispersion will be strongly impacted by the interaction of the wind with the topography of the location. This is why area classification studies made by just copying the "typical figures" shown in the API RP-500/505 and NFPA 497 are not reliable!

3.4 AREA CLASSIFICATION FOR COMBUSTIBLE DUSTS

For the correct specification of electrical and electronic equipment used in industries that process combustible dusts, it is essential to consult the unit's area classification plan. Only in this way will the electrical installation designer be able to correctly specify safe electrical and electronic equipment for each region of the industrial plant. The area classification for dust atmospheres needs to consider not only the work areas but also the interior of the equipment. This chapter lists the basic steps of an area classification study for dust atmospheres.

3.4.1 GENERAL DEFINITIONS

Hybrid Mixture

Mixture of a flammable gas, or vapour, with a dust.

Combustible Dust

Small solid particles, nominally 500 µm in size or smaller, which can form explosive mixtures with air under normal conditions of temperature and pressure.
  Notes:

- This definition includes dusts and particles as defined in ISO 4225.
- The term "solid particles" is intended to specify particles in the solid phase, and not in the gas, vapour, or liquid phases, but does not exclude hollow particles.
- Materials passing a no. 40 sieve as defined in ASTM E 11–04 are considered to meet the 500 µm criterion.
- Test methods for combustible dusts can be found in ISO/IEC 80079–20–2.

Conductive Dust

Combustible dust with electrical resistivity equal to or less than 10^3 Ω.m, which is classified as IEC Group IIIC.

Non-Conductive Dust

Combustible dust with electrical resistivity greater than 10^3 Ω.m, which is classified as IEC Group IIIB.

Combustible Flyings

Solid particles, including fibres, with a nominal size greater than 500 μm, which may form explosive mixtures with air under normal conditions of temperature and pressure. Examples of combustible flyings include rayon, cotton (also cotton waste and cotton scraps), sisal, jute, hemp, coir, tow, and kapok scraps. They are classified as IEC Group IIIA.

Hazardous Area (Dust)

The area in which combustible dust is present in the form of a cloud, or can be expected to be present, in quantities that require special precautions for construction, installation, and use of equipment. The potential for the formation of a combustible dust cloud from a layer of dust also needs to be considered.

Dust Containment

Compartments of process equipment, designed to handle, process, transport, or store materials inside them, in order to minimize the risk of releasing dust into the surrounding atmosphere.

Normal Operation

Operation of the equipment in accordance with its electrical and mechanical design specification and used within the limits specified by the manufacturer.

Ignition Temperature of a Dust Layer

Lowest temperature of a surface at which the ignition of a dust layer on that surface occurs. It can be determined by the test method given in ISO/IEC 80079–20–2.

Ignition Temperature of a Dust Cloud

The lowest temperature of the hot inner wall of a furnace at which the ignition of a dust cloud occurs in the air contained within the furnace. It may be determined by the test method given in ISO/IEC 80079–20–2.

Minimum Ignition Energy

The Minimum Ignition Energy (MIE) of a dust is defined as the lowest amount of energy at which the dust cloud will ignite. It is influenced by the following dust characteristics:

- Physical and chemical properties of the dust (particle size distribution and chemical composition)
- Concentration of dust in the dust–air mixture
- Homogeneity and turbulence of the dust–air mixture
- Geometry of the vessel in which the explosion is occurring.

Determination of MIE must be made over a range of dust concentrations.

3.4.2 CLASSIFICATION FRAMEWORK FOR COMBUSTIBLE DUST ATMOSPHERES

The procedure for the dust area classification has some particular characteristics. In the case of dust layers, the data are impacted by the thickness of the layer, the temperature of the surface on which the dust rests, and that of the immediate surroundings. For example, the ignition temperature may fall as the thickness of the layer increases.

It will vary should the dust be exposed to hot surfaces for a period and although in many cases the ignition temperature will rise when the dust degrades, there are other instances when it will fall. In the case of dust clouds, the data may be impacted by solvent content, the presence of additives, and by the particle size distribution in the cloud. For example, for a given material, a cloud of dust with a higher proportion of smaller particles is likely to have a lower ignition temperature than one with a predominance of large particles. It is therefore essential to obtain expert advice to provide such data and to ensure that it is relevant to the particular dust in the form in which it will occur in the plant.

Often, large parts of the work areas are classified as Zone 21, which leads to major problems when assessing the conformity of existing material and when purchasing new material. The argument given to justify such a classification is the presence of "excessive accumulations" of dust.

A layer of dust leads to a Zone 21 classification if, for some reason, this layer of dust can be suspended during the normal operation of the plant. An example of suspension during normal operation would be cleaning work areas with compressed air, which is considered a bad practice.

Using compressed air for cleaning would cause the layers of accumulated dust to be lifted, generating dust clouds. The formation of such dust clouds would occur every time cleaning is carried out, presumably during the normal operation. In this case, the classification of the work areas in Zone 21 would be justified. If the layers of dust accumulated in the work areas can only be suspended as a result of malfunctions, the work areas must be classified in Zone 22.

A possible malfunction could be the breakage of a pressurized air pipe, such as those used to operate cutters, for example, causing the accumulated dust in the work areas to be suspended. Or, when a pressurized air pipe, that feeds an Ex p panel, has a leakage and disturbs layers of dust, promoting the formation of dust clouds. So, although allowed by standards, the use of Ex p control panels in Zones 21 and 22 areas should be avoided.

Taking the situation in respect of dusts, the following steps are important to perform the area classification study:

3.4.2.1 Step 1: Identify the Dust

Identify if the material is combustible, according to its own characteristics, such as:

- The particle size
- The moisture content
- The Minimum Ignition Temperature (MIT) of the cloud
- The electrical resistivity.

The particle size significantly impacts the explosiveness of a dust cloud. Smaller particles have a larger surface area relative to their volume, which can lead to a higher rate of combustion. And fine particles can remain suspended in the air longer, increasing the likelihood of forming an explosive dust cloud. They also have a lower MIE, making them easier to ignite.

The moisture content in dust particles can influence their combustibility, reducing the risk of ignition and explosion, because it increases the energy required to ignite the dust. On the opposite side, very dry dust can be highly explosive.

The MIT of the cloud is the lowest temperature at which a dust cloud will ignite. Knowing the MIT helps in selecting appropriate equipment for hazardous areas. Lower MIT values indicate a higher risk of ignition, necessitating stricter control measures.

The electrical resistivity of dust reveals its ability to accumulate static charge. High values mean that the dust can accumulate static electricity, which can be a potential ignition source. Proper earthing and bonding are essential to mitigate this risk. These factors are integral to the classification and management of areas where combustible dust is present.

A comparison between explosive parameters of gas and dust is shown in Table 3.7. It is important to say that the minimum ignition energies can be considerably lower than these values when long spark discharge times are used.

TABLE 3.7
Comparison of Ignition Properties of Some Gases and Dusts

	Gases and Vapours				Dust Clouds			
Ignition Property	Methane	Propane	Hydrogen	Acetylene	Wheat Starch	Rice	Safflower Meal	Sugar
MIE [mJ]	0.30	0.25	0.019	0.017	20	40	20	30
Maximum explosion pressure [kg/cm²]	7.17	8.58	7.39	10.3	7.38	6.54	5.90	6.40
Time to peak pressure [s]	70	46	7	14	—	—	—	—
Explosion pressure rate of rise [kg/cm²/s]	—	—	—	—	597.6	253.1	203.9	351.5
Lower explosive limit [% volume]	5.0	2.1	4.0	2.5	—	—	—	—
Minimum explosion concentration [mg/L]	—	—	—	—	25	45	55	35
Auto-ignition temperature [°C]	435	198	470	272	—	—	—	—
Layer surface ignition temperature [°C]	—	—	—	—	—	220	210	400
Cloud/surface ignition temperature [°C]	—	—	—	—	420	440	460	370

3.4.2.2   Step 2: Identify the Sources of Release

The second step consists in identifying where the containment of dust, or dust emission sources, may be present. But the assessment needs to be done also inside equipment, as explosive atmospheres can be formed, as an example, inside pneumatic transportation ducts. The shape, size, and the mass of the particles greatly influence the combustion of dust atmospheres. In any case, the ignition energy for the combustion of combustible dust is much higher than that required for gases, shown in Table 3.7.

IEC defined three Groups for dusts:

- Group IIIA for combustible fibres and flyings
- Group IIIB for non-conductive dusts
- Group IIIC for metal dusts.

NEC also has three Groups of dust in the Class/Division System:
- Group E is defined as combustible metal dust atmospheres, including aluminium, magnesium and their commercial alloys, or other combustible dusts whose particle size, abrasiveness, and conductivity present similar hazards in the use of electrical equipment.
- Group F is defined as combustible dust other than combustible metal dust atmospheres, containing combustible carbonaceous dusts that have more than 8% total entrapped volatiles, or that have been sensitized by other materials, so that they present an explosion hazard. Representative combustible dusts that fall into this grouping are coal, carbon black, charcoal and coke.
- Group G is defined as fibres and flyings atmospheres containing other combustible dusts, including flour, grain, wood flour, plastic, and chemicals.

All three groups are combustible. Table 3.8 shows the relationship between IEC and NEC classifications.

3.4.2.3   Step 3: Define the Zones

It is not uncommon to find out in audits that the designations of Zones were determined "intuitively", that is to say, if there is a possibility of a dust cloud be formed during normal operation, it is considered Zone 21; otherwise, if the possibility is only

TABLE 3.8
NEC and IEC Groups for Dusts

NEC	IEC	Definitions
G	IIIA	Combustible fibres and flyings
F	IIIB	Non-conductive dusts
E	IIIC	Metal dusts

in abnormal operations, it is considered Zone 22. A Zone 21 in the work area can be illustrated by a bag-emptying hopper without an aspiration system. Every time a bag is emptied into the hopper, a dust cloud is generated outside the hopper.

And, a Zone 21 is usually defined around the hopper (depending on the product characteristics, it is "defined" as the distance of 1 m around the hopper). If the hopper has an aspiration system, the dust is retained. Under normal circumstances, no dust cloud is formed due to the aspiration system (if it is well designed).

Only if the aspiration system fails can an explosive atmosphere form outside the hopper. Therefore, a Zone 22 (in case of aspiration failure) can be defined around the hopper, which, depending on the process characteristics, can reach 1 m around the hopper.

3.4.2.3.1 Dust Layer Thickness

In practice, it is very difficult to indicate the maximum dust layer thickness that must not be exceeded in order to avoid the area classification, as it is difficult to monitor it in the field.

There is some information, from unknown origin, that is used as a rule-of-thumb guideline in determining dust layer accumulations versus the required area classification, as shown in Table 3.9. It is said that the dust accumulations in Table 3.9 are based upon a 24-hour build-up on horizontal surfaces; however, without knowing the characteristics of the combustible dust considered or the process parameters, there is no way to ensure that such classifications can be used in a project in a reliable way. Without informing the kind of dust and other parameters as the ambient characteristics, which were considered to create Table 3.9, if the responsible team for the area classification study adopted it, they just "threw the dice", because the safety of the plant is compromised.

In audits carried out in some companies, it was found that the value of 1 mm had also been used as a criterion, based on a popular belief, of unknown origin, that "layers of dust with a thickness of less than 1 mm are tolerable". However, it must be taken into account that a thickness of 1 mm is well above the tolerable limit!

Theoretically, a 1 mm thick dust layer can form a cloud 5 m high with a dust concentration in air of 100 g/m^3 (assuming a dust density of 500 kg/m^3). The resulting dust concentration in air is well within the explosive range (the explosive range for most substances usually is from a lower explosive limit of 30 g/m^3 to an UEL of 5 kg/m^3). So, there are cases when it is not feasible to meet these "1 mm criteria".

TABLE 3.9
Proposal for Area Classification Due to the Dust Layer Accumulation

Dust Layer Thickness	Area Classification
Greater than 3 mm	Zone 21
Less than 3 mm, but surface colour not discernible	Zone 22
Surface colour discernible under the dust layer	Unclassified

Experience shows that many older factories in the food industry, for example, have enormous difficulties in maintaining cleanliness in their work areas. In some cases, work areas are covered with a fine layer of dust again shortly after cleaning. In these cases, a Zone 22 classification can be opted for, if the operational team could keep the cleaning procedures in order to limit the thickness of dust layers. A Zone 22 classification covers the presence of layers of dust in the work areas and also the possibility of leaks.

3.4.2.3.2   Interior of Equipment

The rate of dust cloud dissipation is relevant to the decision whether the inside of the vessel should be classified as Zone 20, 21, or 22. Clearly, while any powder is being charged to the vessel, a potential explosive atmosphere is likely to exist. If the powder is charged continuously or frequently, then the inside of the vessel should be classified as Zone 20.

If the powder is not charged frequently, then the inside of the vessel can be classified as Zone 21. Although some experiments indicate that the dust cloud is likely to dissipate rapidly, it is not possible to classify the inside of the vessel as Zone 22 because the current indications are that an explosive dust atmosphere is likely to always occur in normal operations and not as a result of rarely occurring abnormal operations alone.

To allow the inside of vessel to be classified as Zone 22 requires more studies to be carried out, possibly using CFD, to gain a greater understanding of the generation of dust clouds and whether any situation exists where the assumption that an explosive dust atmosphere will be created in normal operation can be replaced by confirmation that an explosive dust atmosphere would only be created by a rarely occurring mal-operation.

As seen, there are some aspects that must be taken into account when classifying the interior of equipment. Inside some equipment, the dust concentration will be above the UEL. This is the case for some dense pneumatic conveyors.

If 40 tonnes of dust per hour are conveyed with 40 m^3/min of air, the average concentration in the conveying pipe is 16.7 kg/m^3. This concentration is very above the average UEL, which is around 5 kg/m^3 for many dusts.

Other examples where the UEL is exceeded are certain mills, sieves, and mixers. During the process, the dust is evenly distributed inside the equipment, reaching concentrations of over 10 kg/m^3. Usually, during the start-up and stop phases, its concentration occasionally enters the explosive range, and, in this case, the interior of this equipment usually is classified as Zone 20.

Zone 22 can also exist inside the equipment. An example is the clean side of bag filters. The clean side can contain an explosive dust/air atmosphere if the filter elements (bags) fail. It is important to emphasize the concept of failure. An explosive dust/air atmosphere will only exist in the event of a failure (there will be none during normal operation), and this is why a Zone 22 classification is usually chosen. Such classification will impact the fan that is usually installed on the clean side of the filters, as it must meet the requirements of a Zone 22 equipment. The requirements for the fan motor will be defined by the Zone classification carried out in the work areas.

3.4.2.3.3   Sack Tip Stations

NBR IEC 60079–10–2 recommends that the interior of a sack tip station (a simple method for the emptying of sacks, also known as bag dumping station) be classified as Zone 20. The simplest assumption is that this Zone 20 extends all the way through the dust extraction system to the dust collection filter, but it is possible that CFD could demonstrate that this is unnecessarily cautious.

NBR IEC 60079–10–2 recommends that the exterior of a sack tip station be classified as Zone 22. However, it does not define their extents. This sizing needs to be based on the process and the dustiness of the powders handled; therefore, it is not a case of "one size fits all". This is an example where the level of ventilation is important.

In the case of a sack tip station with no extract, a Zone 21 is recommended by the standard; again without any sizing. Dustiness and CFD studies will help to size such a Zone.

3.4.2.3.4   Dust Characteristics

The type of product being handled must be taken into account. The area classification of the inside of a silo, probably, will not be the same if flour or wheat is stored, for example, because the larger the grain size, the more difficult it will be to form explosive atmospheres.

Finally, it should not be forgotten that dust behaves very differently from gases and vapours. Zone classifications for dusts can often be found that followed (mistakenly) the scheme of classification that is usually made for gases and vapours: Zone 20 inside the equipment, surrounded by a Zone 21 which in turn is surrounded by a Zone 22.

This is the "typical" methodology suggested by recommended practices such as API RP-505 for hazardous areas due to flammable gases and vapours (remember that recommended practices are not standards). But, just mechanically applying this to the dust environment is conceptually wrong! Gases and vapours that are released from a continuous source of leakage are dispersed in the atmosphere forming a cloud that is characterized by a certain concentration gradient; dust that escapes from equipment due to a leak tends to be deposited on the floor, forming a layer of dust.

The formation of an explosive dust atmosphere is possible if the layer of dust is suspended, which usually only occurs if a malfunction occurs in the process. Therefore, a Zone 20 can be surrounded by a Zone 22, and not by a Zone 21.

3.4.2.3.5   Ventilation

Some people take ventilation into account when zoning areas where the risk is due to the presence of combustible dust. Ventilation plays an important role in reducing the zoning for gases and vapours, but this is not true for dusts. Unlike for gases and vapours, ventilation can even be counterproductive, as it can cause the accumulated dust in the work areas to be suspended, forming explosive dust clouds and increasing the classified area extent.

3.4.2.4   Step 4: Define the Hazardous Area Extents

The extent of an explosive dust atmosphere Zone is defined as the distance in any direction from the edge of a dust release source to the point where the hazard associated with that Zone is considered to be non-existent, that is to say, where the concentration is below its LEL.

The explosive dust atmosphere generated by a dust cloud can normally be considered negligible if the dust concentration is within an adequate safety margin in relation to the lower explosive limit required to form an explosive dust atmosphere. The responsible team for the area classification study is tasked with defining the value of the safety margin.

It is recommended that allowance be made for the fact that fine dust may be carried from a release source by the movement of air within a room.

As shown in Step 3, determining the extent of the hazardous areas by "intuition" does not guarantee reliable results. Everything being done by "guesswork", without due justification regarding the mass of the particles, their temperature, ventilation, and the ambient temperature of the location, does not offer reliability, nor safety, to the users of the plant. Therefore, the typical figures that illustrate the technical literature on the area classification for combustible dusts, in the same way as occurs with generic figures that illustrate the hazardous areas due to flammable gas atmospheres, cannot be copied mechanically!

As the characteristics of how such "typical figures" were deduced are not disclosed, there is no way to guarantee that they have the same characteristics of the dust and processes of the plant under study. Therefore, an "area classification" done in this way is not an engineering service, and the owner of the industrial plant will be left to "roll the dice" and hope that an explosive atmosphere does not occur in a location which such a study indicates as a "non-hazardous area".

Because the physical and chemical characteristics of a dust cloud are different compared to a gas cloud, the mathematical models for dust cloud dispersion will be slightly different. Dust often disperses horizontally from its source, and the extent of dispersion depends on:

- The initial horizontal air velocity
- The release height
- The particle settling time.

Combustible dust atmospheres are formed by sources of dust release that are points, or locations, at which dust can be released or raised so that an explosive dust atmosphere can be formed. This definition includes layers of dust capable of being dispersed to form a dust cloud. Depending on the circumstances, not every release source produces an explosive dust atmosphere. On the other hand, a continuous, dilute, or small release source can produce a dust layer.

For spherical particles, the settling time can be estimated from Stokes's Law, as given in Equation 3.3:

$$V_s = \frac{h}{t} = \frac{g\rho D^2 10^{-8}}{18\mu} \qquad (3.3)$$

Where:

V_s is the still air settling velocity [cm/s]
h is the release height [cm]
t is the settling time [s]
g is the gravitational constant (9.80 m/s²)
ρ is the particle density [g/cm³] (dry flour has density of 1.44 g/cm³)
D is the particle size (diameter) [µm]
μ is the air viscosity [P] (poises). Note: The viscosity of air depends mostly
on the temperature. At 15 °C is 18.1 µPa·s or 1.81 × 10⁻⁵ Pa·s.

Equation 3.3 is applicable for particles from 1 to 100 µm in diameter (in free fall), with approximately a 10% error at low concentrations. The larger particles will fall nearby; the smaller particles will fall some distance from the release point.

As seen on area classification with flammable gases, there are some documents suggesting the use of "typical" figures on classifying locations due to combustible dusts, as shown in Figure 3.11 taken from NBR IEC 60079–10–2, and, unfortunately, we detected many area classification studies just reproducing them. It is necessary to highlight that, without a clear indication on what product, what particle size, and what humidity content were considered to draw such figures, they cannot be applied in area classification studies, as there are no warranties that they are corresponding to what is being processed in the plant.

It is also necessary to assess the resulting dust cloud that is formed when a dust layer is perturbed, and Figure 3.12 gives an idea of the behaviours of corn starch, sugar, and wood flour.

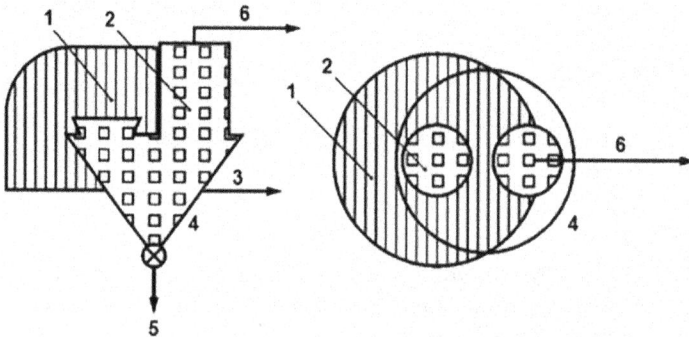

FIGURE 3.11 "Typical figure" (without any consideration on the real product's characteristics) that can only be considered as a suggestion for the area classification of hoppers.

Where:

1 represents Zone 22.
2 represents Zone 20.
3 represents floor.
4 represents bag discharge hopper.
5 represents to process via a rotary valve.
6 represents to extract within containment.

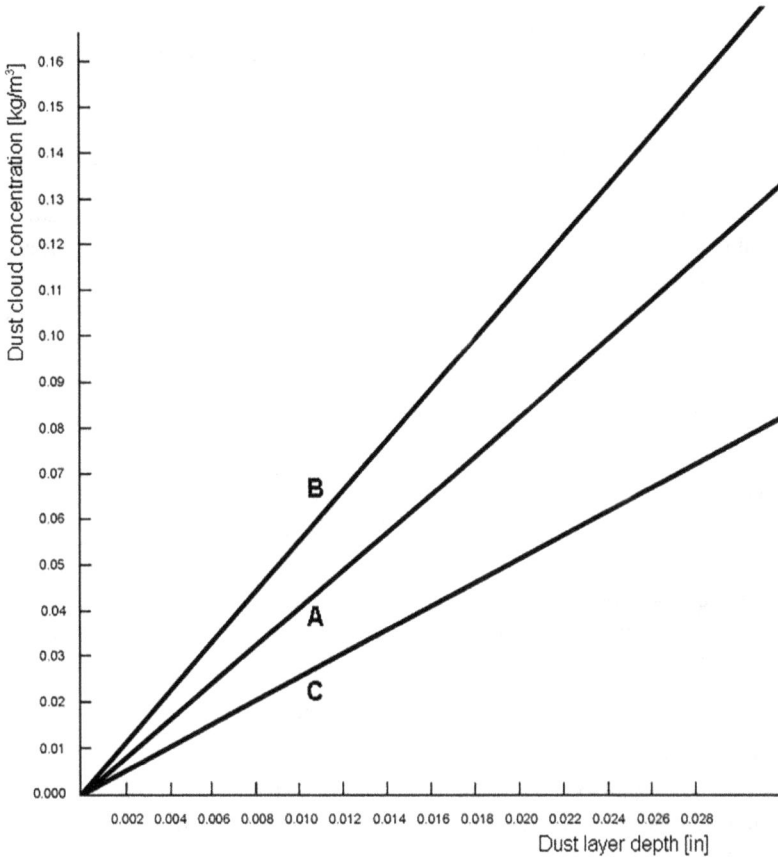

FIGURE 3.12 Dust cloud concentration versus dust layer depth, for a height of 3 m assuming a uniform dust distribution.

Where:
   A represents corn starch.
   – Bulk density is 490 kg/m³.
   – Minimum Explosive Concentration (MEC) is 0.04 kg/m³.
   – Minimum dust depth to create a cloud within its LEL is 0.3048 mm.
   B represents sugar.
   – Bulk density is 650 kg/m³.
   – MEC is 0.045 kg/m³.
   – Minimum dust depth to create a cloud within its LEL is 0.1727 mm.
   C represents wood flour.
   – Bulk density is 320 kg/m³.
   – MEC is 0.035 kg/m³.
   – Minimum dust depth to create a cloud within its LEL is 0.3048 mm.

There are some commercial software to estimate the classified area extents for dust atmospheres, but, before using them, it is necessary to check which ones are validated.

3.4.2.5   Step 5: Report

Issue a detailed assessment report that documents, at least, the following information:

- The critical information on the combustible dusts, obtained from the safety data sheets or, preferably, obtained from tests carried out on the products existing in the plant.
- A detailed list of all point sources of dust emissions indicated on the drawings
- The characteristics of the processes considered
- Any decisions taken on exceptional situations
- The results, or findings, obtained from the site assessment audit
- The justifications for the distances defined in the area classification study
- The safety signage plan, indicating on the drawings the location of the signs containing the Zone, Group, and Temperature Class.

All area classification documentation should be placed under the protection of the facilities management of change process control. As modifications are made to the facility, these documents should be reviewed to verify the impact of these modifications.

3.4.2.6   Step 6: Signage of the Hazardous Areas

It is very important to place signs to alert workers that an area is hazardous. There are differences between what a classified area is, and what an area with explosive atmosphere is. A classified area is a location where there is a probability of an explosive atmosphere occurring, that is to say it can happen sometimes.

Regarding combustible dusts atmospheres, which are visible, flammable gas atmospheres can happen invisibly. So, it is necessary to alert the persons in the industrial area about the classified locations with alert signs, indicating the Zone, Group, and the maximum surface temperature allowable for equipment to be used there.

Figure 3.13 shows an example that follows the ISO recommendations for alert signs in areas under fire risks, for a classified location due to the probable occurrence of combustible dust clouds above its LEL. The sign identifies a Zone 22 location due to an atmosphere of combustible fibres and flyings.

3.5   HYBRID MIXTURES

Among the events of particularly devastating hybrid mixture explosions, we can cite two examples that occurred in polyethylene and polypropylene manufacturing facilities.

1) On March 14, 2013, an explosion occurred in the High Density Polyethylene (HDPE) plant of the Yeosu Industrial Complex, South Korea. During a scheduled maintenance shutdown, the team was tasked with installing

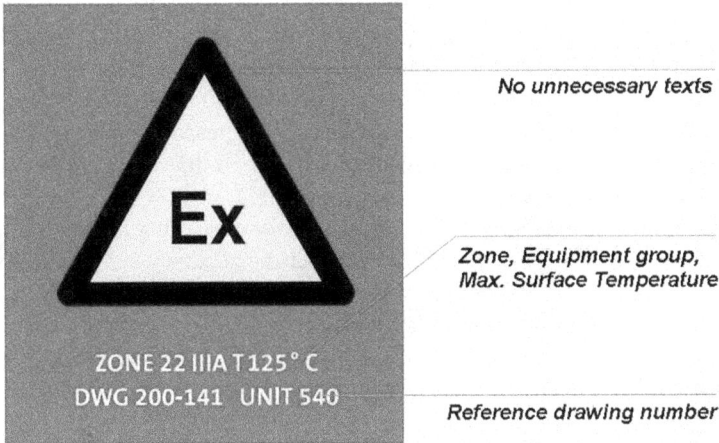

FIGURE 3.13   Alert sign for classified areas due to combustible dusts.

a manhole on the side of an aluminium silo, which was used to store HDPE. The operation involved cutting and welding on the outside of the silo, approximately 30 m above the ground. An explosion occurred, and reports indicated that it occurred during the welding operation, causing workers on scaffolding to fall and the fire to spread to two other silos and neighbouring companies.

The investigation found that there was HDPE dust inside the silo walls and that the tools used to create the opening in the silo likely disturbed it and created clouds inside the silo. HDPE dust has an explosion classification of St1, and the cutting and welding processes involve ignition sources. This "typical" explosion scenario is not yet complete, as stored HDPE releases monomeric hydrocarbon vapours (e.g. ethylene) at a decreasing rate after manufacture, leading to safety recommendations for a "rest period" to allow degassing and exhaustion of these flammable gases from the silos. However, in this case, the HDPE had only been produced three days before the explosion.

Considering this detail, it can be concluded that the explosion was due to a hybrid mixture of hydrocarbon gas and HDPE dust. In this case, the explosion is stronger than the explosive characteristics of the individual components. Recent studies have identified that the MIE of the dust cloud is generally reduced by the presence of a flammable gas, even if the concentration of this gas is below its LEL.

2) On July 10, 2017, an explosion occurred inside the polypropylene (PP) silo at the Yeosu Industrial Complex, which involved a pneumatic conveyor of polypropylene (PP) pellets with high levels of static electricity. Although this is a common phenomenon, in this specific case, the electrostatic charge caused an explosion inside a PP silo.

The particular characteristics of polypropylene can, in the case of electrostatic charging, lead to various types of discharges, such as the "brush" type. PP does not produce large amounts of dust when transported pneumatically, but its MIE can vary, depending on the particle size, from 50 mJ to 1,000 mJ; and the MIE values of propylene and ethylene gases are 0.28 mJ and 0.07 mJ, respectively. Therefore, when there is a cloud of combustible dust with an MIE of 50 mJ, and a flammable gas is added, the MIE of the hybrid mixture begins to fall.

Although the plant had silos with instruments for measuring the concentration of flammable gases, there were uncertainties about the accuracy and reliability of the readings, as the periodic calibration was not up to date. These events show that the risks of explosion of hybrid mixtures need to be properly assessed by a specialized consultancy. A reliable area classification study should guide the creation of safety procedures and the implementation of appropriate control measures.

3.5.1 FLAMMABILITY AND EXPLOSION BEHAVIOUR OF HYBRID MIXTURES

Analysis and study of the flammability and explosion parameters of several hybrid mixtures used in mining, nuclear, pharmaceutical, polymer, biomass, food processes, and textile manufacturing have been performed. Some examples of the studied hybrid mixtures as related to different industries are given in Table 3.10.

Nevertheless, during the risk assessment phase or explosion hazard assessment, especially during the application of ATEX Directives, the case of these mixtures is not frequently appreciated at its true value.

3.6 MYTHS ABOUT AREA CLASSIFICATION

During Ex safety audits on industrial plants with hazardous areas, some misconceptions were detected when talking with plant managers. Some of them are described next.

TABLE 3.10
Industries Where Hybrid Flammable Mixtures Can Be Found

Hybrid Mixture	Industry Reference
Methane and coal dust	Mining
Hydrogen gas with the addition of graphite/ carbon, tungsten, and aluminium dusts	Nuclear
Methanol, ethanol, propanol, acetone, di-isopropyl ether and nicotinic acid, lactose, cellulose, magnesium stearate, vitamins, antibiotics	Pharmaceutical
Solid polymers (e.g. polyethylene) and flammable gas (e.g. ethylene)	Polymer
Biomass dusts and ethane, hydrogen, carbon monoxide	Biomass
Cork and methane, corn starch and propane, wheat dusts, and fermentation gases	Food

a) "To determine whether or not an area is classified, it is sufficient to measure the concentration of the flammable substance present in the atmosphere."

Myth! The area classification study is not based on concentration measurement but on the assessment of situations that may lead to the formation of an explosive atmosphere. An example that illustrates the independence of measurement is the execution of an area classification study for an industrial plant design, when the unit is still on paper, without the physical conditions to make such "concentration measurements".

b) "When using an explosimeter, if it indicates 'zero', we know that the region is not classified."

Misconception! A classified area is a region where an explosive atmosphere may occur, under normal or abnormal operating conditions, occasionally, and for a short, or long period of time. It is not defined by measuring instruments.

c) "Since in our factory, we have a unit as a classified area Zone 1, in another 'similar' factory, we can adopt the same hazardous area classification study."

Myth! Although they are "similar" units in functional terms, the characteristics of the process, the products used, and the local ventilation may lead to a different classification.

d) "A company made an area classification drawing of our factory some years ago and classified a certain area as Zone 22. We consider it that way until today."

Myth! Area classification drawings are prepared in the basis of the conditions considered for the study, which must be periodically assessed again. Changes in process procedures, equipment, and products handled may change the classification determined years ago.

e) "When an area is classified, it is the warning that it will explode someday."

Myth! The area classification indicates the locations where there is a risk of explosive atmospheres forming. An explosion will only occur if the appropriate control measures are not adopted and an ignition source enters the environment.

f) "Our company does not have area classification drawings because this would lead employees to ask for a salary increase, because it would legally characterize the dangerousness of the work."

Myth! The purpose of area classification drawings is to indicate regions where explosive atmospheres may form. This indication will allow for the correct specification of electrical and electronic equipment, as well as provide important data for the preparation of procedures for the safe execution of services in these regions. If the activity performed by employees requires the payment of an additional salary, this is an issue to be addressed within the legislation sphere, therefore, independently of the area classification drawings.

g) "We want our industrial plant to be classified as Zone 2 because we hear that equipment for such area are cheaper than those for Zone 1."

Myth! The classification of areas cannot be previously imposed by the client. It will be the result of the evaluation of the professionals who will conduct the study, considering, among other things, the characteristics of the processes, products, and ventilation of the environment.

h) "Our European plants have an 'ATEX document', so we should also have a similar document here."

ATEX is a legal Directive applicable to member countries of the European Union, which mandates to employers, the obligation to elaborate an Explosion Protection Document for plants with hazardous locations. In other countries, there are other legal requirements to be met. However, it is important to check whether the headquarters establishes that all subsidiaries must follow the same procedures.

i) "Our company purchased the area classification standard and distributed it to employees so that they could prepare the area classification drawings for the facility."

Myth! This is a serious mistake! First, because a standard is not a "cake recipe" that can be used by laymen; it requires a lot of prior knowledge to be used. In addition, it is not enough to read the standard; it is necessary to interpret it and to know its application limitations, which only specialists are capable of.

j) "Our employees participated in a course on electrical installations in explosive atmospheres, so they can now classify the areas of our facility."

Myth! The Ex installation course only covers the requirements needed for selecting, installing, and maintaining Ex equipment, i.e., steps subsequent to the area classification study. To conduct an area classification study, the professional must have solid knowledge of the processes, participate in specific refresher training, and have the necessary experience to analyse situations that arise in practice and that are not in the standard.

k) "An area classification study must show the same figures that appear in the standard."

Myth! It is important to consider that there are differences between a "Standard" and a "Recommended Practice". The Standard presents the minimum requirements for performing a service or product; the recommended practice may present illustrations of certain situations, but it is up to the person responsible for carrying out the study to verify whether such illustrations are applicable to the reality of the plant under study.

There is no "one size fits all" here!

l) "The responsible person for elaborating the area classification plan is the electrical engineer."

Myth! In a given audit, it was found an incorrect procedure that said:

It will be the responsibility of the electrical engineer, as designated by project management, or area management, to interpret the rules and regulations as set forth in the NFPA codes, standards, and recommended practices for establishing the classification of an area.

This myth emerged because the classification of areas began in technical documents issued by organizations focusing on the requirements for electrical installations. These documents are used by electrical engineers and include the NFPA 70 (NEC) and the IEC 79–10.

As the mathematical models of gas and dust dispersions are not included in the electrical engineering courses, the electrical engineers were compelled to use the recommended practices such as NFPA 497 (note that "recommended practices" are not standards).

The results, as they did not know the chemical properties nor the physical concepts of the formation of explosive atmospheres, were just a mere copy of the generic figures from those documents, including them in their projects.

The consequences, as we can imagine, were that the most diverse industrial units were "classified" in the same way, with the same "figures", what, in fact, created a "fake safety sensation" in the industrial units that received such "area classification studies".

m) "Over zoning is better."

Myth! Audits have found many powder handling facilities where large blanket areas were classified as hazardous as ones where insufficient provision has been made. This approach is often exaggerated and hence acceptable as long as the company is willing to also accept the costs associated with it.

Minimizing dust releases and avoiding dust accumulations should therefore be the primary objective, rather than accepting large hazardous areas.

n) "The inside of the equipment is Zone 22, as it is used only 9 h/yr".

Myth! The "duration factor" is a common cause of wrong attitudes due to misinterpretations.

The important definition of a Zone 22 in this context is that a flammable dust cloud should not be expected as part of normal operation. So, the equipment running can actually meets the requirements for Zone 20.

o) "We have the hazardous area drawings; so, we don't need to know the justifications."

Myth! This is very common to find out in audits. The report is essential for, as a minimum, to know the justifications made when arriving at the zoning, including flammability and process parameters data.

3.7   THE DURATION FACTOR

We have observed that on several occasions, when the company responsible for area classification study is asked on what they consider to be "frequently", "occasionally", and "short or long period", they do not provide convincing answers. Some argue that IEC should include a table containing the number of hours that characterize each term, as published in other publications, like:

Zone 0: Explosive atmosphere for more than 1,000 hours/year
Zone 1: Explosive atmosphere for more than 10 but less than 1,000 hours/year
Zone 2: Explosive atmosphere for less than 10 hours/year

However, criticisms of these periods soon arose, because they were, apparently, launched without due basis. Other proposals emerged with the aim of reconciling the definitions with the application of the Ex protection types, but they did not reach consensus. One of the main criticisms was how could a lower degree of risk be attributed to a plant subject to an explosive atmosphere for 999 hours/year (maximum for Zone 1) compared to the one that presented 1,000 hours/year (minimum for Zone 0)?

It is clear that classifying areas based solely on such "numbers" presents major problems, as in many situations, the plant operator considers leaks to be "normal". This can largely explain the absurdities we find in area classification studies, which show a lack of knowledge of the process interfaces on the part of the person responsible for the study. As a result, "surreal" specifications of Ex equipment for the plant arise, which lead to unnecessary expenses.

The IEC did not include tables with "hours/year" because the consensus reached was that classifying areas by simply adopting such tables would be too simplistic and, therefore, inaccurate. Although these numbers are no longer published in standards, they continue to be cited in commercial booklets, which have caused confusion.

An example that illustrates the conflict with the criterion of pre-defined hour bands would be the application of the restricted-breathing protection type (Ex nR). It can be applied in Zone 2 according to the qualitative description of the standard. If the number of hours were defined and the plant presented a possibility of the formation of an explosive atmosphere for 10 hours/year, every year, this type of protection could not offer a guarantee of non-ignition. Therefore, this is a clash between qualitative versus quantitative assessments, that is not a matter of choosing numbers, but which requires the discernment of an experienced professional for proper application.

3.8   BATTERY ROOMS

It is known that batteries, such as lead-acid batteries, release hydrogen when charged, which requires the development of safety procedures, both for operation and maintenance activities. And the existence of several examples of explosions in battery rooms, especially those that supply Data Processing Centres, typically the bigger ones, shows the importance of a properly based analysis aimed at correctly specifying the electrical and electronic equipment that can be safely installed in them.

In one of our audits, we found a document issued by the company's project department, which was actually a survey of several foreign standards, reproducing all the recommendations mentioned in them regarding battery rooms. It revealed some conflicts between the various standards consulted, but in its conclusions, the report limited itself to saying that "the provisions of the standards should be complied". This means that divergent recommendations were identified, but they were not analysed to determine which guidelines the company should follow. That report was considered ineffective, which led to the hiring of a consultancy.

In our audits, it is noted to raise questions like this:

It is easy to say that all equipment intended for use in classified areas must be Ex; however, overdoing it on safety, may not be the best solution.

So, for rooms with batteries and UPSs, some preventive measures are usually applied, as example:

- Exhaust fan sized for adequate air changes per hour, aiming at keeping H_2 concentration levels below 1%.
- If the H_2 level exceeds 1%, all power to the room will be cut off.
- The exhaust is interlocked with the charger, turning it off if the exhaust stops.
- The light switch is located outside the room.
- The gas detector and the exhaust fan motor have been selected appropriately, taking into account the H_2 gas and the EPL Gb (IIC, T1).

Under these conditions, do the room's lighting fixtures need to be Ex? And if someone guarantees that the H_2 concentration will always be below 1%, should the room be kept as classified?

The questions are not simple. It is necessary to know what type of battery and capacity we are dealing with. Impacting factors are:

- The hydrogen emission rate (which depends on the charging profile programmed in the charger and the ventilation conditions of the room)
- The location of gas sensors and their characteristics
- The topography of the ceiling (as some "air pockets" probably can be formed).

Providing a high rate of air exchanges per hour does not guarantee the absence of an explosive atmosphere if there is no correct interlock with the charger or, due to operational needs, such interlock will be asked for setting on "bypass". In addition, if the battery room is for an offshore oil platform, there are some guidelines previously established by certifying entities that may impose alternatives like interlocks not accepted.

Considering all the implications mentioned before, in order to answer in a safe way, a comprehensive analysis of the system and the ventilation characteristics of the room needs to be done.

3.9   CLOSING REMARKS

It is said that the typical figures shown in recommended practices lead to a "conservative area classification", which causes higher costs on purchasing the required equipment; but, based on what reference people stated that?

If just considering the distance from the source of release, it is possible to find situations involving flammable gases where 1 m is already extremely oversized. In other situations, 3 m can be not "conservative", but an under-evaluated practice, which compromises the plant safety. A misinterpretation of API RP-500/505 is responsible for such "labelling" that the "standardized distance" for flammable gases area classification will be always "3 m", for any case, independently from the product, the ventilation, the altitude, the process pressure, etc.

And, as a result, we can say in a "conservative" way that 90% of area classification plans are not reliable, because they were made under the "copy-and-paste" of API RP-500/505 "typical figures", without any consideration of the texts of those two recommended practices (it is important to say that both are not "standards"). It is therefore necessary to turn to specialists in explosion protection to carry out the area classification.

It is worth noting that in NBR IEC 60079–10–1, it is emphasized that the knowledge necessary to perform area classification studies does not belong to the electricity discipline: "The area classification should be carried out by those who understand the relevance and significance of the properties of the flammable substances, principles of gas/vapour dispersion and those who are familiar with the process and the equipment". This is very important to clarify and stop the myth that "electrical engineers are the responsible persons to execute the area classification study" of a plant. Area classification requires knowledge of the flammable gases' properties, gas dispersion, thermal properties of dusts, process pressures and flows, and a lot of disciplines not familiar to electrical engineers.

It seems that the origin of such myth was caused by the information about hazardous locations that have been included in NEC, i.e. a document written to be used by electrical engineers. And, years after, the IEC contributed to the myth when issued the standard 79–10 on area classification. Because the IEC issues standards for electrical equipment and installations, when 79–10 was issued, it led many managers, and professionals, to believe that area classification was the responsibility of electrical engineers, which unfortunately has led to completely erroneous and unsafe studies.

Nowadays, there are commercial software available for assessing area classification, but, if such computational tool is mechanically manipulated by people without the necessary theoretical background, the results may even compromise the safety of the unit, as they are dissociated from the operational reality of the plant.

As there is no single dispersion model applicable to all situations, something "simple and quick", it is questionable that IEC issued 60079–10–1 and –2 as "standards". It would be more appropriate to issue them as Technical Reports (TR), especially as their texts have been confusing, even presenting conflicts with their own definitions.

As fluid dynamics is not a discipline in the day-to-day life of electricity professionals, IEC 60079–10–1 and –2 should not be issued by IEC, but by ISO, because as it acts in a broader scenario, receiving contributions from professionals of various technological areas, it is the appropriate forum for such documents. As an example NBR IEC 60079–10–2 requires that the dispersion of the powder is taken into consideration as part of the hazardous area classification exercise but without defining the parameters to be considered when calculating or modelling this.

A standard has to define the minimum requirements only; imposing a model, as done in IEC 60079–10–1, may result in inaccuracies, as in engineering, a "one size fits all" solution rarely is feasible.

The lack of scientifically derived data has led to frequent conflicts between designers and clients. Designers are rightly cautious and usually classify the inside of vessels as Zone 21 or Zone 20. Users tend to argue that they have been operating these areas as "safe" for many years "without problems". Are the designers being overcautious, or are the users taking unnecessary risks?

In conclusion, area classification is an engineering study, specific to each plant, and cannot be obtained through generic figures but rather through a comprehensive analysis and simulations made with an experienced team. Given that the area classification study contributes decisively to the safety of the unit, its documents must be prepared on a solid basis, as they will be consulted both for the specification of equipment and for the issuance of operational procedures. Such a study should only be conducted by specialists who are experienced and technically up to date, through participation in specific courses and conferences. The aim is to produce reliable and safe documents and procedures.

BIBLIOGRAPHY

Amyotte, P. and Kahn, F.—*Methods in chemical process safety, vol. 3—Dust explosions*. Elsevier, 2019.

API RP-500—*Recommended practice for classification of locations for electrical installations at petroleum facilities classified as Class 1, Division 1, and Division 2*. American Petroleum Institute, 2023.

API RP-505—*Recommended practice for classification of locations for electrical installations at petroleum facilities classified as Class I, Zone 0, Zone 1, and Zone 2*. American Petroleum Institute, 2018.

ASTM E11-04—*Standard specification for wire cloth and sieves for testing purposes*. ASTM International, 2010.

Bozek, A.—Application of IEC 60079-10-1 edition 2.0 for hazardous area classification. In: *Petroleum and Chemical Industry Conference*. Calgary, CA, 2017. Available at: https://ieeexplore.ieee.org/document/8188767

Cole, M., Driscoll, T. and Martin, K.—Installation, operation and maintenance considerations for electrical equipment in hazardous location. In: *IEEE IAS Electrical Safety Workshop*. IEEE, 2010.

Cox, A. W., Lees, F. P. and Ang, M. L.—*Classification of hazardous locations*. Institute of Chemical Engineers, 1990.

De Souza, A. O., Luiz, A M, Pereira Neto, A. T., De Araujo, A. C. B., Da Silva, H. B., Da Silva, S. K. and Alves, J. J. N.—CFD predictions for hazardous area classification. *Chinese Journal of Chemical Engineering*, vol. 27, 2019, pp. 21–31.

Di Benedetto, A., Russo, P., Di Sarlia, V. and Sanchirico, R.—CFD Simulations of the effect of dust diameter on the dispersion in the 20 L bomb. *Chemical Engineering Transactions*, vol. 31, AIDC, 2013, pp. 727–732.

DNV OS A101—*Offshore standard—Safety principles and arrangements*. Det Norske Veritas, 2008.

Dong, L., Zuo, H., Hu, L., Yang, B., Li, L. and Wu, L.—Simulation of heavy gas dispersion in a large indoor space using CFD model. *Journal of Loss Prevention in the Process Industries*, vol. 46, 2017, pp. 1–12.

EI 15 (former IP-15)—*Model Code of Safe Practice, Part 15: area classification for installations handling flammable fluids*. Energy Institute, 2015.

Gant, S. E., Jones, A., Ivings, M. J. and Santon, R. C.—Hazardous area classification of low-pressure natural gas systems using CFD predictions. In: *XIX Hazards*. Institution of Chemical Engineers, Manchester, UK, 2006.

IEC 60079-10-1—*Explosive atmospheres—Part 10–1: Classification of areas—Explosive gas atmospheres*. International Electrotechnical Commission, 2020.

IEC 60092-502—*Electrical installations in ships—Part 502: Tankers—Special features*. International Electrotechnical Commission, 1999.

IEC 61511-3—*Functional safety—Safety instrumented systems for the process industry sector—Part 3: Guidance for the determination of the required safety integrity levels*. International Electrotechnical Commission, 2016.

IGEM/SR/25—*Hazardous area classification of natural gas installations*. Institution of Gas Engineers and Managers, 2013.

ISO/IEC 80079-20-1—*Explosive atmospheres—Part 20–1: Material characteristics for gas and vapour classification—Test methods and data*. International Organization for Standardization/International Electrotechnical Commission, 2017.

ISO/IEC 80079-20-2—*Explosive atmospheres—Part 20–2: Material characteristics—Combustible dusts test methods*. International Organization for Standardization/International Electrotechnical Commission, 2015.

Ivings, M. J., et al.—*Area classification for secondary releases from low-pressure natural gas systems*. Research Report RR630, Health and Safety Executive, UK, 2008.

Jordão, D. M.—A nova IEC 60079-10-1: 2011—Os impactos nos futuros trabalhos de classificação de áreas. In: *VI Encontro Petrobrás sobre Instalações Elétricas em Atmosferas Explosivas*. Petrobras/Universidade Corporativa, Rio de Janeiro, Brasil, 2010.

Keane, B., Schwarz, G. and Thurnherr, P.—Cables and cable glands for hazardous locations. In: *LXV IEEE Petroleum and Chemical Industry Committee Conference*. Institute of Electrical and Electronics Engineers, 2018.

Leblanc, N., Berner, W and Ogilvie, P.—Conduit seals in hazardous locations—An ongoing safety issue: the past and the future. *IEEE Transactions on Industry Applications*, vol. 52, no. 1, IEEE, 2016.

Lorente, V. and Daniel, L.—*Simulación de atmósferas explosivas mediante CFD*. Universitat Politécnica de Catalunya, 2007.

Marigo, Marzio—*Rischio atmosfere esplosive ATEX: Classificazione, valutazione, prevenzione e protezione*. Wolters Kluwer, 2017.

NBR IEC 60079-10-1—*Atmosferas explosivas. Parte 10–1: Classificação de áreas—Atmosferas explosivas de gás*. Associação Brasileira de Normas Técnicas, 2022.

NBR IEC 60079-10-2—*Explosive atmospheres—Part 10–2: Classification of areas—Explosive dust atmospheres*. International Electrotechnical Commission, 2015.

NFPA RP-497—*Recommended practice for the classification of flammable liquids, gases, or vapours and of hazardous (classified) locations for electrical installations in chemical process areas*. National Fire Protection Association, 2017.

Rangel Jr., Estellito—A definição das zonas. *Section EMEx, Eletricidade Moderna Magazine*, São Paulo, ed. 483, Jun. 2014, p. 132.

Rangel Jr., Estellito—A nova IEC 60079-10-1. *Section EMEx, Eletricidade Moderna Magazine*, São Paulo, ed. 513, Dec. 2016, pp. 72–73.

Rangel Jr., Estellito—Análise de risco na classificação de áreas. *Section EMEx, Eletricidade Moderna Magazine,* São Paulo, ed. 436, Jul. 2010, p. 176.

Rangel Jr., Estellito—Análise de riscos. *Section EMEx, Eletricidade Moderna Magazine*, São Paulo, ed. 510, Sep. 2016, p. 64.

Rangel Jr., Estellito—Análise de riscos na classificação de áreas. *Section EMEx, Eletricidade Moderna Magazine*, São Paulo, ed. 373, Apr. 2005, p. 296.

Rangel JR., Estellito—Brasil moves from divisions to zones. In: *XLIX Petroleum and Chemical Industry Conference*. New Orleans, US, 2002. Available at: https://ieeexplore.ieee.org/document/1044981

Rangel Jr., Estellito—Classificação de áreas. *Section EMEx, Eletricidade Moderna Magazine*, São Paulo, ed. 391, Oct. 2006, p. 204–206.

Rangel Jr., Estellito—Classificação de áreas. *Section EMEx, Eletricidade Moderna Magazine*, São Paulo, ed. 504, Mar. 2016, p. 58.

Rangel JR., Estellito—Classification of hazardous areas: Standard, theory and practice. *Ex Magazine*, vol. 42, no. 1, Germany, 2010, pp. 15–21. Available at: https://bit.ly/3qz4udi

Rangel Jr., Estellito—Considerations for the prevention of combustible dust explosions. *Fire-and-explosion Magazine*, year I, no. 1, May 2019, Bulkmedia, Germany, pp. 48–54. Available at: https://bit.ly/3Ki3rio.

Rangel Jr., Estellito—Contratando a classificação de áreas. *Section EMEx, Eletricidade Moderna Magazine,* São Paulo, ed. 459, Jun. 2012, p. 164–165.

Rangel Jr., Estellito—Cuidados na classificação de áreas. *Section EMEx, Eletricidade Moderna Magazine,* São Paulo, ed. 428, Nov. 2009, p. 164.

Rangel Jr., Estellito—Definição de áreas classificadas por modelos de dispersão de gases. In: *CONTECC 2021—Congresso Tecnico Cientifico da Engenharia e da Agronomia, 76a SOEA CONNECT,* on line. EXP-54, GO, Confea, Sep. 15 to Sep. 17, 2021.

Rangel Jr., Estellito—Dúvidas na classificação. *Section EMEx, Eletricidade Moderna Magazine,* São Paulo, ed. 566, Jul/Aug. 2022, p. 60.

Rangel Jr., Estellito—Folhas de dados. *Section EMEx, Eletricidade Moderna Magazine,* São Paulo, ed. 493, Apr. 2015, p. 92.

Rangel Jr., Estellito—IEC 60079-10-1:2020 (Part I). *Section EMEx, Eletricidade Moderna Magazine,* São Paulo, ed. 562, Nov/Dec. 2021, p. 60.

Rangel Jr., Estellito—IEC 60079-10-1:2020 (Part II). *Section EMEx, Eletricidade Moderna Magazine,* São Paulo, ed. 563, Jan/Feb. 2022, p. 59.

Rangel Jr., Estellito—ISO 80079-36 e 37. Section EMEx, Eletricidade Moderna Magazine, São Paulo, ed. 552, Mar/Apr. 2020, p. 58.

Rangel Jr., Estellito—Misturas híbridas. *Section EMEx, Eletricidade Moderna Magazine,* São Paulo, ed. 559, May/Jun. 2021, p. 58.

Rangel Jr., Estellito—Normas complementares. *Section EMEx, Eletricidade Moderna Magazine,* São Paulo, ed. 498, Sep. 2015, p. 94–95.

Rangel Jr., Estellito—O que é classificação de áreas? *Section EMEx, Eletricidade Moderna Magazine,* São Paulo, ed. 411, Jun. 2008, p. 204, 206.

Rangel Jr., Estellito—Reviews on the area classification in the oil and gas industry. *Bulletin of Petroleum Production Technology,* vol. 10, no. 1 & 2, Petrobras, 2015, pp. 25–36. Available at: https://bit.ly/3yrTLed.

Rangel Jr., Estellito—Safety in hazardous locations is paramount! *LinkedIn,* @Estellito, Mar. 18, 2018. Available at: https://bit.ly/2S2THMz

Rangel Jr., Estellito—Sala de baterias. *Section EMEx, Eletricidade Moderna Magazine,* São Paulo, ed. 557, Jan/Feb. 2021, p. 60.

Rangel Jr., Estellito—Sinalização de áreas classificadas. *Section EMEx, Eletricidade Moderna Magazine,* São Paulo, ed. 372, Mar. 2005, p. 128.

Rangel Jr., Estellito, Luiz, Aurélio M. and Madureira Jr., Hilton L. P.—Area classification is not a copy and paste process. In: *LXI Petroleum and Chemical Industry Conference.* San Francisco, US, 2014. Available at: http://ieeexplore.ieee.org/document/6961903/

Rangel, J. R. Estellito and Naegeli, G. S. T.—Métodos alternativos para classificação de áreas: o uso da API-RP-505. In: *VI Encontro de Engenharia Elétrica Petrobras.* Petrobras/ Universidade Corporativa, Rio de Janeiro, Brasil, vol. 1, pp. 108–119, 2001.

Schlegel, C., Pfuetzenreiter, A. A. e Costa, J. A.—Confiabilidade em estudos de classificação de áreas. In: *XIII Brasil Automation.* Instrument Society of America -ISA, Distrito 4. Sao Paulo, Brasil, vol. 1, pp. 56–69, 2009.

Seonggyu, P., Seongho, J., Changhyun, R. and Chankyu, K.—*Case studies for dangerous dust explosions in South Korea during recent years.* MDPI, Sep. 2019.

SOLAS, Classifications Register Rules and Regulations—*Rules and regulations for the classification of offshore units, part 7: Safety systems, Hazardous areas and fire, chapter 2: Hazardous areas and ventilation, section 2: Classification of hazardous areas.* Safety of Life at Sea, 2022.

TNO Yellow Book. *Methods for the calculation of the physical effects due to releases of hazardous materials (liquids and gases).* Publication Series on Dangerous Substances, PGS 2, The Netherlands Organization of Applied Scientific Research. The Hague, 2005.

Tommasini, R., Pons, E. e Palamara, F.—Area classification for explosive atmosphere: Comparison between European and North American approaches. In: *LX Petroleum and Chemical Industry Conference.* Chicago, US, 2013. Available at: https://ieeexplore.ieee.org/document/6746223

4 Designing Installations in Hazardous Areas

4.1 INTRODUCTION

It is interesting to know that the objective of an IEC international standard is to facilitate the trade of electrical equipment. To do so, each standard is developed by its respective Maintenance Team (MT), which is formed by experts chosen from the countries that demonstrated interest on participating in that MT. The average number of countries participating on each MT is around nine. After a consensus is achieved by the MT members, the text is sent for appreciation and vote of the countries that demonstrated interest on that standard. If the votes against are lower than 25% of the total, the text is considered approved, and the standard is issued as is.

Despite joint efforts and consideration of regional, cultural, and technological issues, the IEC emphasises that the full adoption of an international standard may not be possible. This is clearly described in the Foreword of IEC standards at, a) item 3)

> IEC publications are in the form of recommendations for international use and are accepted by the IEC National Committees in this regard.

In this way, IEC standards are voluntary and could be treated as recommendations for the elaboration of national standards of each country member. National standards, on the other hand, are supported by their countries' legislation. In the electricity segment, in the European Union (EU), for example, countries are required to follow the CENELEC—Comité Européen de Normalisation Electrotechnique standards—not the IEC ones. Regarding the design of installations in hazardous areas, in addition of taking into consideration the technical requirements given by IEC standards (voluntary) and the National Electrical Code (NEC - compulsory in US), there are also administrative requirements to follow.

4.2 LEGISLATION

The success of a safe electrical system design in hazardous areas, in addition to adhering to the current technical standards, depends also on complying with the regulations issued by the country's government entities, that is to say, the legal requirements.

The following items are commonly required by legislations in order to authorize the erection of electrical installations in hazardous areas:

DOI: 10.1201/9781003500001-4

4.2.1 Worker Competence

Electrical installations in classified areas are the key for safety, because equipment inadequately specified, or installed, or maintained, may behave as sources of ignition, which could lead to an explosion with heavy consequences to the environment and lives. So, the competency of workers plays an important role here, not only for designers but also for installers and maintenance teams.

All activities related with electrical installations in classified areas can only be performed by skilled workers, from the people involved in the design of the electrical installations, to the people responsible for Ex equipment repairs, because not only the design, but also the installation and maintenance must follow the standards' requirements. In a brief, the required knowledge is dependent on the task that the work is designed, as given next.

a) **Designers**

Designers must have detailed knowledge of the principles of explosion protection, of types of protection, of the techniques to be used in the selection of Ex equipment, and of the requirements established in IEC 60079–14 for electrical systems in classified areas.

b) **Installers**

Installers shall possess knowledge of the requirements for each type of explosion protection, of the work permit system, and of manufacturers' orientations for installing equipment and should be aware of the legal regulation on conducting electrical services.

c) **Inspectors**

Inspectors give the final word about the correctness of an installation, which is a great responsibility to safety. They shall be able to provide evidence of having achieved the knowledge about the types of protection and respective installation requirements, of the work permit system, of the use of the installation documentation, to read and interpret the conformity certificates, and at last, but not least, to have skills for the production of written inspection reports.

d) **Maintainers**

Maintenance personnel must be familiar with the requirements for each type of explosion protection method, the properties of Ex marking, the work permit system and manufacturers' guidelines for maintaining equipment. They must also be aware of the local legal regulations governing electrical services.

Based on these descriptions, it can be understood that hiring workers for services in classified areas requires analysing the competence of individuals on specific tasks. And, if the volume of services will require hiring not a single person but a whole team from a human resources company, the care must be doubled, as audits have revealed that the competencies of employees from outsourced companies are usually in a very low level.

4.2.2 PERSONAL EX CERTIFICATION

When the IEC 60079–14 mentions that "Ex workers should have ability to demonstrate competence", it suggests that a competence verification system may be used by contractors to hire the workers. It should be noted that a legal requirement in many countries establishes that a worker might only perform the services after receiving a formal authorization granted by the employer, based on the technical knowledge he possesses or that was given to him. Personal certification, therefore, is voluntary.

There are some Ex personal certification schemes in the market, and they offer flexibility when defining "units of competence", related with the tasks that will be executed by the worker such as inspection, design, maintenance, etc. After an assessment, usually with practical tests, an "Ex personal certificate" is issued, related with the units the worker achieved to pass the exams.

So, a worker may have a certificate in a given task, as example, to install Ex d equipment but has no certificate on, as example, installing Ex i systems. Even if the worker has personal Ex certificates, it is recommended to apply a simplified test before contracting, because some entities give Ex certificates without practical exams, only making assessment using written questions, and in this way, the "certified" worker did not demonstrate their skills to perform the tasks as supposed by the employer.

4.3 TECHNICAL REQUIREMENTS

The technical requirements for explosion protection include:

- Prevention of hazardous explosive atmospheres
- Mitigation of the effects of explosions
- Application of Process Control Engineering (PCE)
- Requirements for work equipment.

In this chapter, the requirements for work equipment that do not act as a source of ignition will be described, focusing that all installation materials are suitable for use in hazardous areas, under the possible ambient conditions at the workplace in question.

4.3.1 ELECTRICAL SYSTEM DESIGN

The IEC 60079–0 defined standard atmospheric conditions (relating to the explosion characteristics of the atmosphere) under which it may be assumed that electrical Ex equipment can be operated, and these are:

- Temperature: $-20°C$ to $+60°C$
- Pressure: 80 kPa (0.8 bar) to 110 kPa (1.1 bar)
- Air with normal oxygen content, typically 21% v/v.

The design, first speaking, needs to comply with the technical standards of the client's country related to the electrical systems in non-hazardous areas, supposing that the installation may be inspected before the plant come into operation by inspectors of the National Agency, Security officials, and the client inspectors. All parts of the electrical system parts need special attention, as they have a high impact on the safety of the plant.

4.3.1.1 Documentation

The detailed design for electrical installations in classified areas must include:

- Single-line and multi-line diagrams, showing the distribution of electrical systems
- Technical specifications detailing the equipment and materials to be used
- Installation plans including cable routing, equipment location, and earthing points.

The requirements for the design and assembly of electrical installations in hazardous areas are described in IEC 60079–14. We will describe them in two sections: Requirements for the components installations and for the Ex equipment.

All documentation related to the selection, installation, inspection, and maintenance of equipment must be kept in an organized and accessible manner. This includes certificates of conformity, inspection reports, maintenance records, and risk analysis considered in the safety procedures.

4.3.1.2 Earthing Systems

For protection against dangerous sparks, possible accidental contact with live bare parts, other than intrinsically safe parts, shall be prevented. Furthermore, the fundamental principle on which safety depends is the limitation of earth-fault currents in frames or enclosures and the prevention of high potentials on equipotential bonding conductors.

There are three main earthing systems, as described next:

TN type

If a type of system earthing TN is used, it shall be type TN-S (with separate neutral N and protective conductor PE) in the hazardous area, i.e. the neutral and the protective conductor shall not be connected together, or combined in a single conductor, in the hazardous area. At any point of transition from TN-C to TN-S, the protective conductor shall be connected to the equipotential bonding system in the non-hazardous area.

TT type

If a type of system earthing TT (separate earths for power system and exposed conductive parts) is used, then it shall be protected by a residual current device.

Note: Where the earth resistivity is high, such a system may not be acceptable.

IT type

If a type of system earthing IT (neutral isolated from earth or earthed through sufficiently high impedance) is used, an insulation monitoring device shall be provided to indicate the first earth fault.

Note 1: If the first fault is not removed, a subsequent fault on the same phase will not be detected, possibly leading to a dangerous situation.

Note 2: Local bonding, known as supplementary equipotential bonding, can be necessary.

It is highlighted that for TN, TT, and IT systems, all exposed and external conductive parts must be connected to the equipotential bonding system. This connection need not be made for intrinsically safe enclosures and for cathodic protection systems, unless specifically requested.

4.3.1.3 Earthing and Bonding

In industries, in addition to preventing accidents caused by electrical or atmospheric discharges, an adequate earthing system prevents energy disturbances that harm the efficiency of equipment. And in hazardous areas, such as those used for the storage of flammable products, chemical products, and silos for storing agricultural products, it serves to prevent explosions and fires caused by the accumulation of static energy.

Without proper earthing and equipotential bonding, these environments can suffer serious accidents with the potential risk of death and material damage. All metallic equipment must be earthed to prevent the accumulation of electrostatic charges.

The following topics need to be considered on the design of the earthing system.

a) **Earth resistance**

Earth resistance is a crucial factor in the design of earthing systems, especially in hazardous areas, where safety against electrical discharges and potential differences is very important. In the 20th century, it was observed among several normative bodies, there was a requirement of a value of 10 ohms for earth resistance. This requirement was reproduced in several documents, and for a long time, it was considered as a goal to be achieved in any situation. However, the traditional value of 10 ohms may not be adequate for all situations, as resistivity varies significantly with the type of soil.

b) **Soil resistivity**

The first step in designing an earthing grid is soil resistivity survey. This process involves measuring soil resistivity at different depths and locations to determine soil stratification. Methods such as Vertical Electrical Sounding (VES) and the use of earth meters are common for this purpose.

c) **Soil stratification**

Soil stratification is essential to understanding how electrical current will disperse in the soil during a fault. The resistivity of the soil layers can vary, impacting the effectiveness of the earthing. Stratification analysis helps determine the ideal earthing grid configuration.

d) **Conductor sizing**

Earthing grid conductors must be sized on the basis of maximum expected short-circuit current. This includes selecting suitable materials (such as copper or galvanized steel) and determining the cross-sections and lengths of the conductors.

e) **Calculation of maximum allowable potentials**

Touch and step potentials are critical for the safety of people. The design must ensure that these potentials are within the safe limits.

f) **Integration with other systems**

The earthing grid must be integrated with other protection systems, such as the Lightning Protection System (LPS) and equipotential bonding systems. This ensures comprehensive protection against electrical surges and lightning strikes.

g) **Earth resistance measurement**

Figure 4.1, which is about measuring the impedance of a earthing grid using a 25 kHz earthing tester, under the 3 pole method, is very didactic, because it shows that the equipment only "sees" part of the grid (the black portion) due to the high frequency of its signal. Another point of attention is the fact that the conductors of the grid have longitudinal inductance, which is exactly what limits the circulation of the signal injected by the equipment throughout the grid, resulting in it being restricted to an area close to the current injection point (in black). This is why measurements of "earthing resistance" (which actually measure earthing impedance), touch and step voltages, performed with equipment of this type, are questionable, especially in medium-sized and larger grids (with a diagonal of more than 70 m). It is worth remembering that the so-called "high-frequency earthing meter" was developed specifically to measure earthing of transmission line towers, with the assumption that the high frequency would reduce the leakage of the measurement current through the lightning arrester wires of the HV transmission line.

4.3.1.4 Overload and Short-Circuit Protection

Protection devices are essential for the safety of the electrical installation. Therefore, power supply circuits and electrical equipment must be protected against overload,

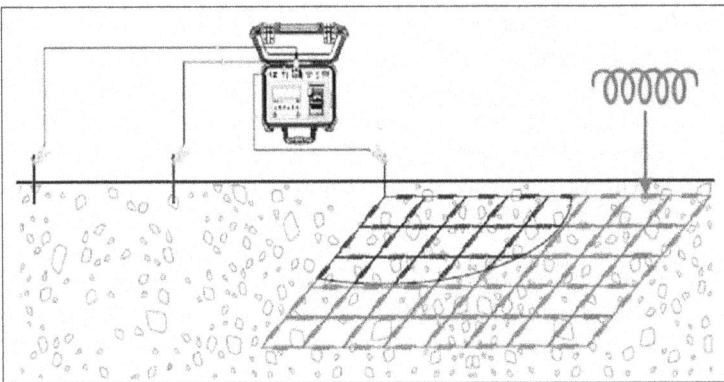

FIGURE 4.1 Limitations on the measurement of the earth grid impedance.

short circuits, and earth faults, preventing excessive currents from causing damage to the equipment and generating dangerous conditions. Short-circuit and earth-fault protection must be such that automatic closing under fault conditions is avoided.

Another important point is to provide protection against the loss of one or more phases of polyphase electrical equipment (e.g. three-phase motors) which causes overheating. Although IEC 60079–14 allows an alarm to be used as an alternative to automatic shutdown (in situations where an automatic shutdown of electrical equipment poses a greater safety risk than the risk of ignition), we do not recommend this alternative, as it is difficult to guarantee the conditions established by the standard (such an alarm being readily perceptible and immediate corrective action being taken) in practice.

Overload protection devices should be set to operate at an overload current slightly higher than the rated current of the circuits. Verify if your national standard defines a given setpoint to operate, as it is around 110% and 125% of the rated current, depending on the equipment specification and environmental conditions. Precautions shall be taken to prevent the operation of electrical equipment in the absence of one of the supply phases to avoid dangerous overheating of their surfaces.

For short-circuit protection, circuit breakers should be selected on the basis of the maximum current interruption capacity expected at the point of installation. If installed in hazardous areas, it is essential that these devices are certified, ensuring that they are not capable of interrupting the fault current generating arcs or sparks that could cause ignition. And, protective devices shall not be self-reclosing.

4.3.1.5   Routes for Electrical Feeders

The routing of electrical cables and feeders must be carefully planned to minimize hazards and facilitate maintenance. Routs above pump and compressor skids are to be avoided, as in case of fire, the flames will burn the cables, stopping control signals to operate. Cable routing should avoid locations where they may be subject to mechanical damage.

The use of metal ducts and raceways is recommended to protect cables against mechanical damage. These ducts must be properly earthed to prevent the accumulation of electrostatic charges.

Power, control, and instrumentation cables must be segregated to prevent interference and signal crossover. This segregation is particularly important in hazardous areas to ensure the integrity of control signals and the overall safety of the installation.

With regard to the triggering by accumulated electrostatic discharges and hot surfaces, it is recommended to focus attention on the laying of the ducts. In fact, the cable routes and their laying method must be arranged in such a way as not to expose them to friction and accumulation of electrostatic charges due to the passage of dust, which, moreover, by settling, could lead to the reaching of dangerous surface temperatures. All cable entry and exit points in enclosures must be properly sealed with cable glands certified for use in hazardous areas, ensuring that there is no possibility of ingress of liquids or dust.

1.3.1.6   Cables

Under IEC 60079–14, cables used for fixed installations in hazardous areas shall be sheathed with thermoplastic, thermosetting, or elastomeric material, circular and compact. Annex C requires a pressure test to evaluate the gas permeability in cables. It requires a piece of cable with a length of 0.5 m, installed into a sealed enclosure of 5 L under constant temperature conditions. The cable is considered acceptable if the time interval required for an internal overpressure of at least 0.3 kPa to drop by 0.15 kPa is not less than 5 s.

Cables, including flexible ones, to be used in hazardous areas, shall have one of the following characteristics:

- Ordinary tough rubber sheathed
- Ordinary polychloroprene sheathed
- Heavy tough rubber sheathed
- Heavy polychloroprene sheathed
- Plastic insulated and of equally robust construction to heavy tough rubber sheathed and flexible.

They must be suitable for the type of protection required by the equipment and must be installed in a manner that prevents mechanical damage and exposure to chemical substances. Connections should be minimized and, when necessary, should be made in junction boxes certified for hazardous areas.

The selection of appropriate electrical cables is critical to ensure the safety and functionality of installations in hazardous areas. The type needs to be chosen considering the regulations (especially if the use is offshore units), the intended use, and the environmental characteristics of the location. The most used types are described next.

a) **Flame-retardant jacket**

Cables must be of the flame-resistant type to prevent them from contributing to the spread of fires. Cables with Low Smoke Zero Halogen (LSZH) sheaths are recommended.

b) **Shielded cables**

Shielded cables are preferred in hazardous areas to provide additional mechanical protection and minimize Electromagnetic Interference (EMI).

4.3.1.6.1 Voltage Drop

The energy loss involved in distributing electrical power takes the form of a reduction in voltage at the receiving end of each cable run. This reduction is dependent on the impedance of the cable, values of which can be calculated.

For cable sizes up to about 35 mm², the impedance at 50 Hz is practically equal to the d.c. resistance of the conductors, but above this size, especially with single-core cables, the inductance becomes increasingly important and, at the largest sizes, is the dominant component. As the inductance decreases only logarithmically with conductor diameter, it becomes more and more difficult at these large sizes to reduce the voltage drop by increasing the size of conductor, and splitting the circuit into two or more independent parallel limbs may be necessary.

Circuit inductance increases as conductors are separated, so that from the point of view of voltage drop, it is best to keep single-core cables as close together as possible. Although it is a general practice to use tabulated values of voltage drop directly without any adjustment, they are, in reality, vectorial values, and in a.c. circuits neglecting the effect of their phase angles amounts to taking the worst case. Actual circuits may have a lower voltage drop than is given by such a simple approach. The effective voltage drop depends on both the power factor of the load and the phase angle of the cable impedance.

Single-core cables in trefoil have slightly lower ratings than those laid in flat formation but have the advantage of presenting lower impedance, which is important at the load currents for which single-core cables are commonly used; in addition, this impedance is balanced between the lines. As a general rule for 400 V circuits, it is usually better to obtain a high current-carrying capacity by paralleling more than one run of cables in trefoil than to employ larger cables in flat formation.

For three-phase circuits having single-core cables installed in flat formation, the apparent impedances of the three conductors are unequal, and the impedance between the two outer conductors is tabulated, this being the one having the greatest magnitude. This will yield the highest value of voltage drop for all sizes of armoured cable with all values of load power factor. However, when using the voltage drop equations, it is not a matter of the largest magnitude of impedance voltage but of the largest component of that voltage in phase with the line-to-line load end voltage.

For non-armoured cables, the impedance across the outer conductors does not yield the highest voltage drop, when the load power factor is higher than about 0.9 and may be seriously low for the larger single-core conductors. Under these conditions, the highest voltage drop occurs between the centre and an outer conductor. The matter is somewhat complicated, and designers of installations with very large unarmoured cables are expected to verify carefully the manufacturers' recommendations.

The unbalanced nature of the impedances of single-core cables in flat formation introduces considerable difficulties when cables are connected in parallel. Of all the available combinations of phase positions, only few offer the possibility of reasonable current share between cables in parallel in the same phase. With three cables per phase, there is no arrangement of cables which will give equal current sharing; however, with a few arrangements, balance within about 5% can be obtained.

It should be noted that the problem is not confined to the line conductors; neutral and protective conductors run with the phase conductors may be subject to considerable circulating currents.

Especially with shorter runs, the arrangement at the terminations where cables break formation in order to connect to equipment terminals can seriously impact the current balance between parallel conductors. Suitable terminal positions can reduce this difficulty, and it may be advisable to use larger conductor sizes to provide for inevitable poor current sharing.

If materials which either surround or are close to cables (e.g. conduits, trunking, trays, structural items) are magnetic, the effect on current ratings is small, but the voltage drop can be increased. Unfortunately, the general situation is too complicated to deal with in a simple manner as it is critically dependent on the disposition of the cables in relation to the steel. However, variation in cable disposition in conduits is limited, and it is possible to tabulate values of voltage drop figures in rating tables for "enclosed cables" which are based on experimental data.

The impedance of both multi-core and single-core cables up to 25 mm² is unimpacted by running them in contact with steel.

Where single-core cables are used to feed large discharge lamp loads, the neutral conductor carries a considerable third harmonic current which results in a significant voltage drop unless the neutral conductor is laid close to the phase conductors. The loaded neutral must then be taken into account when considering the effect of grouping on the rating of the cables.

4.3.1.7 Lighting

Lighting in hazardous areas must be designed to provide a safe environment and adequate visibility without introducing additional ignition risks. So, all luminaires must be Ex certified.

The lighting level must be appropriate for the activities carried out in the area, considering ergonomic and safety standards. For work areas, a minimum level of 200 to 300 lux is recommended, but higher levels can be required by local or specific regulation.

Fixed luminaires (lighting fittings) are subject to additional requirements to reduce the risk of sparking, and low-pressure sodium lamps are not permitted, as is the case with Ex e type of protection. Portable luminaires must comply with the same requirements. In addition, the luminaire must be capable of being dropped four times from a height of 1 m onto a smooth concrete floor without sustaining any damage other than superficial damage and without the lamp envelope breaking. Each type of luminaire requires special precautions as described next.

a) **LED luminaires**

When the first edition of IEC 60079–28 was published in 2007, it was believed that it was applicable only to high-power laser systems and fibre-optic communications systems, as it only made a brief reference to LEDs. As the word "luminaire" was not mentioned in the document, many manufacturers and certification bodies ignored the document's existence when designing and certifying LED luminaires.

In the second edition of IEC 60079–28, published in 2015, the Maintenance Team reinforced that luminaires containing LEDs had to be in the scope. Why was it necessary to apply this standard to luminaires? Because, when the light beam of an LED is concentrated—for example, in a laser pointer—its radiation can heat a dust particle to the point of making it a source of ignition of the explosive atmosphere of flammable gas that may be present, or even of an atmosphere of combustible dust above the LEL.

Between the first and the start of work on the second edition of the standard, the price and configuration of LEDs changed significantly, as the market began to demand a large volume of high-power LED luminaires. With the increase in the power of LEDs used in lighting equipment, the concentration of energy from the light beam close to the LED chip may be sufficient to ignite a dust particle, even if the LED is intended for lighting.

LEDs are becoming increasingly powerful, with relative power levels reaching double-digit percentage increases per year. Despite all the advantages in terms of lifespan and efficiency, experts have been discussing the safety of this technology in hazardous areas.

There is a concern that under certain conditions, high-power LEDs can act as a possible ignition source in explosive atmospheres. Several certificates of conformity have been issued by IECEx without assessing the risks of a concentrated optical radiation, which could compromise the user's safety.

One of the standards applicable to luminaires in classified areas is IEC 60079–28, the first edition of which, although published in 2006, had not yet been consistently implemented until 2016. A report showed, among other things, that out of approximately 1,000 IECEx certificates, less than 10% considered the risk of optical radiation based on the requirements of this standard for the type Ex op is.

There are four factors that must be considered to assess whether a light source can be an ignition source:
- The energy of the light source
- The focal point of the emission
- The distance from the light source
- The presence of an energy receiver (absorber).

Just as a magnifying glass can concentrate sunlight on a small point, which can burn the nearby paper, LEDs can be sources of ignition if certain precautions are not taken of when designing and/or assembling luminaires for classified areas. Figure 4.2 illustrates the situation with an IP 6X luminaire.

To illustrate the risk, consider a toaster with P = 750 W, which, on average, provides 5 mW/mm^2 of radiation, enough to toast two slices of sliced

FIGURE 4.2 An IP 6 X luminaire, where several adjacent light beams may overlap and create a region of intense radiation, especially in explosive atmospheres of combustible dusts.

bread or even burn them after a long time. This energy level is precisely one of the limits established in IEC 60079–28. The wavelength of the radiation from the toaster is different, but the example aims to show its impact.

The risk is subtle: Various substances can adhere to the surface of the lenses of the luminaires, absorb the optical radiation, and start to heat up. As the luminaires tested by the certification bodies are new and clean, many certificates were issued without a complete and safe assessment, is to say, without considering the "op is" type of protection.

As companies do not necessarily have experts who understand the differences between Ex product certificates, trust in these documents is seen as absolute. Thus, the buyer only checks whether the Ex certificate exists and is unable to critically analyse what the document actually covers.

Luminaires that have an "op is" (intrinsically safe optical radiation) marking may be more expensive, the reasons for which, given the current inconsistencies in the interpretation of the standard, are unknown to the buyer. It should also be noted that, for certification bodies, the assessment of optical radiation requires investment in personnel training and laboratory training, which generates costs.

There is still much work to be done, as the risk of optical radiation remains high until the implementation of the standard is consistent across all Ex certification bodies.

b) **Fluorescent luminaires**

They are still widely used despite losing market share to LED luminaires. They are made of non-metallic material or stainless steel, with one or two-pin fluorescent lamps and high-power factor ballast. As the price difference between plastic and stainless steel can be considerable, it is up to the user to properly define the type of housing desired. In the case of Ex ballasts, there are other requirements such as protection against the end of life (EOL) of the fluorescent lamp.

4.4   SPECIAL COMPONENTS

4.4.1   CABLE GLANDS IEC

One of the functions of the cable gland is to maintain the IP rating of the enclosure; in certain cases, it may be recommended to use washers to ensure sealing, which usually are not provided by cable gland manufacturers; so, their specification and acquisition are the installer's duty. At this point, doubts arise about the recommended thickness and the material used to manufacture the washers. In addition, another concern is the possibility of their damage due to excessive tightening during the assembly of the cable gland.

Aiming to minimize inadequate assemblies, IEC 60079–0 determines that cable glands must be tested with the washers installed and that these must be properly identified in the certificates of conformity. In other words, the washers are now components of the certified cable gland that are subjected to durability testing, need to be provided by the manufacturers, and purchased by the installer. To meet the ingress protection requirement, it is necessary to check the sealing condition between cable glands, adapter, blanking elements, and the enclosure (e.g. by means of a sealing washer or thread sealant).

Please note that, to meet the IP54 minimum requirement, there is no need for additional sealing between threaded cable entry devices and threaded cable entry plates or enclosures of 6 mm or greater thickness, provided the axis of the cable entry device is perpendicular to the external surface of the entry plate or enclosure.

Another question is the torque to be applied to cable glands, an information that is not usually included in installation manuals or certificates of conformity. In light of this issue, IEC 60079–0 requires manufacturers to indicate in their manuals the torque to be used in assembly.

Traditionally, Ex training centres have always developed practical activities for assembling cable glands without the use of torque wrenches. Based on the aforementioned, the new requirement of the IEC 60079–0 will impact many stakeholders, not just manufacturers. In practice, the torque value to be included in the manuals will be that which is used by the test laboratory and that led to the approval of the component.

The cable gland shall be selected to match the cable diameter. The use of sealing tape, heat shrink tube, or other materials is not permitted to make the cable fit to the cable gland.

Cable glands and/or cables shall be selected to reduce the effects of "coldflow characteristics", which can be described as the movement of the cable sheath under the compressive forces created by the displacement of seals in cable glands, where the compressive force applied by the seal is greater than the resistance of the cable sheath to deformation. Coldflow could give rise to a reduction in the insulation resistance of the cable. Low-smoke and/or fire-resistant cables usually exhibit significant coldflow characteristics.

Cable glands shall be in accordance with IEC 60079–0 and shall be selected to maintain the requirements of the protection technique, as shown in Table 4.1.

Ex t glands, adapters or blanking elements, having parallel threads, may be fitted with a sealing washer between the entry device and the "t" enclosure. If no washer is used, the thread engagement shall be at least five full threads. Tapered threaded

TABLE 4.1
Suitable Cable Glands for Each Protection Technique

Protection Technique of the Equipment	Glands, Adapters, and Blanking Element Protection Technique		
	Ex d	Ex e	Ex nR
Ex d	X		
Ex e	X	X	
Ex nR	X	X	X
Ex i—Group II	X	X	X
Ex p	X	X	X

TABLE 4.2
IP Requirements for Cable Glands in Group III Installations

Type of Protection	Group IIIC	Group IIIB	Group IIIA
ta	IP 6X	IP 6X	IP 6X
tb	IP 6X	IP 6X	IP 5X
tc	IP 6X	IP 5X	IP 5X

joints without an additional seal, or gasket, shall engage no less than 3½ threads. Table 4.2 shows the cable glands requirements for Ex t enclosures.

4.4.1.1   Barrier Gland—Ex d

The cable entry system shall comply with one of the following:

- Cable glands sealed with setting compound (barrier cable glands)
- Cables glands meeting all of the following:
  - Comply with IEC 60079–1 and are certified as equipment
  - Cables used are circular and compact; any bedding, or sheath, shall be extruded; and fillers, if any, shall be non-hygroscopic
  - The connected cable is at least 3 m in length
- Indirect cable entry using the combination of flameproof enclosure with a bushing and increased safety terminal box
- Mineral-insulated metal-sheathed cable, with or without plastic outer covering, with appropriate flameproof cable gland complying with IEC 60079–1
- Flameproof sealing device specified in the equipment documentation or complying with IEC 60079–1 and employing a cable gland appropriate to the cables used. The sealing device shall incorporate compound or other appropriate seals which permit stopping around individual cores. The sealing device shall be fitted at the point of entry of cables to the equipment.

Barrier glands should also be used:

- In Ex e applications when there is a risk of gas migrating down a cable
- In Ex nR applications where the cable is not sealed.

4.4.2   Cable Glands—NEC

Requirements for cable glands for use in hazardous locations are given in UL 2225, which includes construction and testing requirements for cable glands being marked with Class I approvals and also for cable glands being marked AEx. As the permissible cable types for hazardous areas are TC-ER-HL and MC-HL, cable glands approved by UL 2225 are designed to work with these cables. Additionally, as flexible cord is permitted, subject to certain restrictions, cable glands can be also evaluated for use with this type of cable.

UL 2225 tests glands with the specific cables they are designed and approved for. Therefore, the glands' usage is limited to those cable types that they have been evaluated with.

Part II of UL 2225 addresses the approval of cable glands for Class I Division 1. This section provides the construction and testing requirements for explosion-proof and dust-ignition-proof cable sealing fittings. Tests included are:

- Torque
- Resistance to impact tests
- Explosion tests
- Hydrostatic pressure tests.

4.4.3 Conduit Seals

One of the most widely used types of protection in industrial electrical installations that process flammable products is flameproof (IEC) or explosion-proof (NEC). This type of protection has special construction characteristics, and its installation can only be entrusted to professionals who have completed the required technical training.

One of the safety requirements for these installations using conduits is the use of sealing units. These units prevent explosions from travelling throughout the conduit system and minimize the movement of gases or vapours from hazardous to non-hazardous areas through connecting raceways or enclosures. They also stop pressure piling, which is the build-up of pressure inside conduit lines caused by precompression as the explosion travels through the conduit.

Proper installation is crucial for it to function well. In other words, the presence of the sealing unit alone does not guarantee that the installation complies with standards and it is safe for users. Sealing fittings must be certified, cannot contain splices, and have to be installed in accessible locations, without exception. Conduit seals perform as intended if the seal is dammed and poured in an acceptable fashion, as shown in Figure 4.3. When a sealing process is not complete, it is a risk to property.

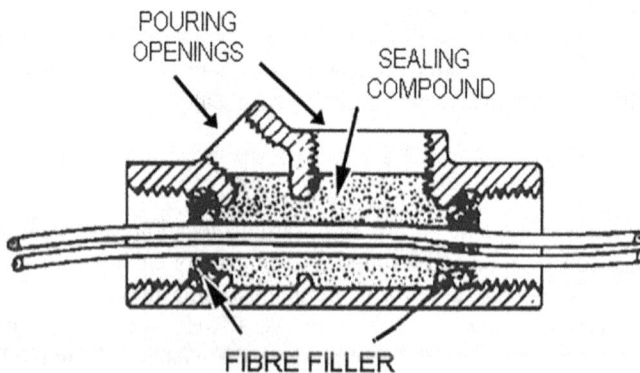

FIGURE 4.3 Sealing unit for Ex d installations using conduits.

Inferior sealing methods, such as filling a conduit seal with silicone or electrical sealing putty, are unacceptable. These remedial methods fall short of standards' compliance and achieving functional objectives. The seal fitting installation is not complete if the seal is not dammed and poured.

Under the NEC, completing an effective conduit seal requires installing damming fibre and sealing compound specific to the certified sealing fitting. This compound must be durable for the surrounding atmosphere or liquids, having at least 16mm thickness and not less than the trade size of the conduit in which it is installed.

The standard type of conduit seals are not intended to prevent the passage of liquids, gases, or vapours at pressures continuously above atmospheric. Extreme temperature and highly corrosive liquids and vapours may impact the ability of seals to perform their intended functions.

The melting point of the sealing compound cannot be less than 93°C to withstand flames and heat from explosions. And its thickness needs to be at least equal to the conduit diameter. For example, if the seal is installed in a 2 inch conduit, the compound in a completed seal cannot be less than 2 inches thick.

Conductor fill is restricted in conduit runs where sealing fittings are installed. Generally, it is permitted for up to 40% fill for conduits containing more than two conductors, and limiting the fill to 25% of the conduit cross-sectional area of the raceway provides adequate room in the fitting for separating the conductors and achieving an effective seal around each conductor as it passes through the fitting.

If the fill exceeds 25% of the conductors, it can bunch up in the middle of the seal, creating spaces between them through which gases or vapours could migrate. There are oversized certified sealing fittings that can accommodate up to 40% fill requirements.

Where multiconductor cables pass through conduit seals, there are two methods of achieving an effective seal, which depend on whether the cable jacket can transmit gases or vapours through the cable's core. If the cable jacket, or sheath, allows passage of gases or vapours, the jacket must be removed so the conductors in the cable assembly can be separated when installing the sealing compound. For cables that do not transmit gases or vapours through the sheath to the cable core, the cable can be dammed and sealed without removing the sheath. Multiconductor cable manufacturers should be able to provide the evidence of sheath suitability.

On the IEC side, the installation requirements for sealing units are

the distance from the nearest surface of the sealing to the housing (or the housing provided for final use), and the outer wall of the housing (or the housing provided for final use), must be as small as possible, but in no case be greater than the size of conduit, or 50 mm, whichever is the smaller of the two values.

IEC 60079–1 defines that

A sealing device is considered to be inserted immediately on entry of the flameproof housing when the device is fixed to the same directly, or by means of a coupling accessory.

4.5 SPECIAL EQUIPMENT

4.5.1 Variable Frequency Drive (VFD)

Three-phase asynchronous motors with variable frequency drives (VFD) are increasingly being used in the chemical and petrochemical industries, allowing for the appropriate control of various processes. Although they offer several advantages, the use of asynchronous motors controlled by converters in classified areas needs to be properly evaluated.

A drive system by converter can be said to consist of three units: Frequency converter, motor and logic control unit. Although manufactured separately, these three units need to be evaluated together in order to ensure explosion safety.

The system can be described: First, the 60 Hz AC of the industrial system is converted to DC. Then, fast-switching solid-state devices such as Insulated Gate Bipolar Transistor (IGBT) convert the direct voltage back to alternate voltage, as required to power the motor. The AC voltage, with adjustable voltage and frequency characteristics, drives the three-phase induction motor within the desired rotation and torque range for the industrial process.

The VFD is generally installed in the electrical panel room, outside the hazardous area. The motor is installed in the field in a hazardous area.

It is recommended to install temperature sensors in the motor to provide information to a protection device, which must shut down the machine if the temperature rises considerably. Figure 4.4 illustrates the installation in a simplified manner.

When specifying the motors, it is necessary to take notice of the fact that the winding insulation class must be defined on the basis of the estimated temperature for continuous operation. When using converters, attention must be paid to the thermal stresses resulting from the specific duty cycle of the application, and also to the temperature rise in the stator and internal connections, due to the high-frequency currents.

Any plastic components installed in the machine (e.g. auxiliary connection boxes and fans) are also subjected to additional tests to verify their thermal stability.

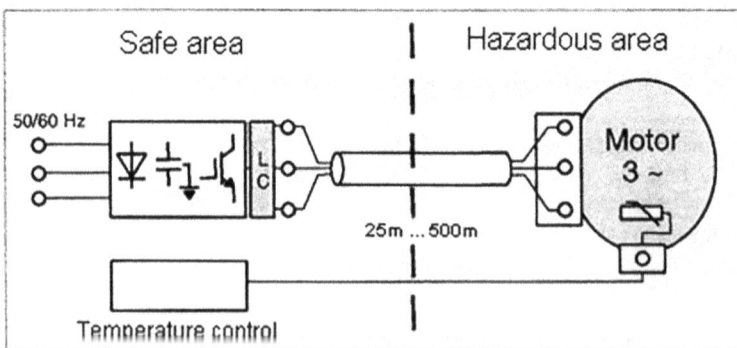

FIGURE 4.4   Installation diagram of a motor with VFD in hazardous location.

Regarding the protection, it is important to know that earth-fault relays at 60 Hz have performance limitations when the power system includes VFD. They will not accurately detect an earth fault when the VFD output frequency is significantly below 60 Hz. A total inability to detect an earth fault can be expected at frequencies below 12 Hz.

Neither a direct current (DC) earth fault internal to the VFD nor an earth fault in the motor or cable, can be detected using conventional ground fault relays (GFR) when the VFD is operating at a low speed. A typical current transformer (CT) can only detect alternate current (AC), and, therefore, DC fault currents are not detected.

Some drives are equipped with their own internal scheme to detect earth faults and protect themselves against high-current AC faults. Often, drives are manufactured to operate on solidly earthed systems where there is no limitation to the earth-fault current. The earth-fault current sensitivity of the drive should be checked to determine if it is compatible with resistance-earthed systems and lower-current earth faults. Early warning or personnel protection quite likely requires a supplemental GFR.

Other issues with VFD are as follows:

- VFDs often include power quality and EMI filters. Filters and surge protective devices provide leakage paths to earth and add to the overall system leakage current.
- Filter components may be rated for line-to-neutral voltage. When installed on an unearthed or resistance-earthed system, catastrophic component and drive failure can occur during an earth fault.
- VFDs switch at a kHz carrier frequency. Capacitive reactance decreases as frequency increases, therefore, higher frequencies generate higher levels of leakage. Carrier frequencies across insulation, which is a distributed capacitance, add further to the leakage inherent to the device.
- Harmonic frequencies that are a result of the switching operations further add to the impact of these leakage paths.
- Transient voltage spikes are a known issue on drives, in particular, when the manufacturer's recommended separation distance between the drive and the motor is exceeded.

4.5.2 AIR COMPRESSORS

Air compressors are extremely important equipment, as they supply the media required by devices such as control valves and also that required by pressurized Ex p panels. Usually mounted on metallic skids, they are assembled as a unit with the compressor, the electric motor, filters, and the control panel. All components need to be verified on the earthing and bonding requirements, in order to avoid electrical charge accumulation.

Considering the greater operational versatility, the option of driving the motor with a VFD has been preferred, and, in this case, the precautions expressed in Section 4.5.1 must be taken. The prices of certified VFD for use in hazardous locations can vary up to 100 times, depending on the technology used and the taxes applied.

FIGURE 4.5   Air compressor skid with a certified VFD for Zone 2.

For the control panel, the cheaper options include the Ex d type due to the ease of selection of internal industrial electrical and electronics components. Figure 4.5 shows an air compressor for Zone 2, with its Ex d control panel.

4.6   NON-ELECTRICAL EQUIPMENT

In 2016, the ISO 80079–36 and –37 standards were published with requirements for non-electrical equipment in hazardous locations, and they have been adopted as European Normative (EN) standards and supersede the EN 13463 series of standards which have been withdrawn. Therefore, it is important to become familiar with the new standards.

In 2008, the IECEx decided the transition to the ISO 80079 series, bringing non-electrical requirements to the IECEx scheme. As consequence, products that complied with the basic requirements of EN 13463-1 should meet the general requirements of ISO 80079–36 with a few updates in the design documentation. The changes included:

- Introduction of Equipment Protection Levels (EPLs) Ga/Gb/Gc and Da/Db/Dc, which align with IECEx requirements where EPLs are used instead of categories or Zones
- Introduction Dust Subdivisions IIIA/IIIB/IIIC to align with the requirements of EN IEC 60079–0. It is now required to be categorized depending on the equipment's suitability for combustible flyings (IIIA), combustible

flyings and non-conductive dust (IIIB), or combustible flyings, and non-conductive dusts and conductive dusts (IIIC)

- Manufacturers of ATEX Category 2 and 3 non-electrical equipment became subjected to quality system audits with a Quality Assessment Report (QAR) if interested in the IECEx certification scheme.

The main protection concepts for assessing non-electrical equipment have been combined into ISO 80079–37. They cover construction safety (Ex c), control of ignition source (Ex b), and liquid immersion (Ex k).

In addition to combining these considerations, the ISO 80079–36 includes:

- An additional method of ensuring lubrication, which should be present in moving parts
- Additional information for brakes and their use
- Expanded requirements for maintaining correct belt tension
- Further clarity on when the control of ignition source "Ex b" is to be applied
- Requirement to mark equipment with "Ex h" rather than "Ex b", "Ex c", or "Ex k"
- Manual details/information about which protection concept has been applied, along with the necessary information for each concept
- The marking requirement has changed from "Ex b" (control of ignition source), "Ex c" (constructional safety) and "Ex k" (liquid immersion) to "Ex h", for any or all of these considerations.

ISO 80079–36 and ISO 80079–37 are now harmonized for ATEX and IECEx, giving manufacturers the option to obtain dual certification without requiring additional testing. Since October 31, 2019, the EN 13463 series of standards no longer provide a presumption of conformity to the ATEX Directive. Manufacturers needed to conduct a gap analysis on the ISO 80079–36 and –37 standards and update their Declarations of Conformity accordingly.

At a minimum, revised ignition hazard assessment, instructions manual, and schedule drawings were needed for updating the standards from EN 13463 to the ISO 80079–36 and ISO 80079–37 standards. If there were any changes in the product design, the certification body would need to reanalyse and/or retest the equipment based on the scope of revision.

Under the ISO 80079–36 standard, ATEX certification requirements for non-electrical equipment vary, depending on the Zone the intended equipment will be used in. The variations include options for self-declaration, requirements for a Quality Assessment Notification (QAN), and more. They can be broken down in the following way:

Zone 0: Manufacturers are not permitted to self-declare their non-electrical equipment. An EU Type-Exam Certificate and a QAN is required.

Zone 1: Manufacturers are permitted to self-declare non-electrical equipment, provided they:

- Assess the new standard
- Complete an ignition hazard assessment
- Mitigate potential ignition sources that may be present during normal operation
- Construct a technical dossier with minimum criteria outlined, lodge with a notified body for a period of 10 years after the manufacture date, and keep an identical copy of the file in their quality system.

Zone 2: Manufacturers are permitted to self-declare non-electrical equipment provided they:

- Assess the equipment on the basis of the ISO 80079–36 standard
- Complete an ignition hazard assessment
- Mitigate potential ignition sources that may be present during normal operation
- Construct a technical dossier with the minimum criteria outlined in the standard and retain a copy of the file for a period of 10 years after the manufacture date.

It should be noted that the IECEx scheme requires a QAR issued for all certificates (except unit verifications) issued, regardless of intended area of installation.

4.7   INITIAL INSPECTION

During commissioning, and before the installation comes into operation, a detailed initial inspection must be done, in accordance with IEC 60079–14. This includes:

Detailed inspection:
To verify compliance with the design and the absence of damages, also.
Functional tests:
To ensure that all systems are operating correctly.

The initial inspection includes a detailed review of the equipment approvals and the special conditions contained in them. It is important to verify the conformity certificates' sections "Description of the Equipment", and "Special Conditions", because a closer examination of these sections can reveal if the selected equipment has restrictions to be used in specific applications. If the "Special Conditions" state that the equipment must not be exposed to ultraviolet (UV) rays, it cannot, under any circumstances, be installed outdoors, without additional protection.

Experience shows that if these points are not taken seriously enough, as a consequence, strictly speaking, the equipment should be replaced. The inspection will check if the equipment has been properly selected in accordance with the standards' requirements, and if all the relevant documentation is available.

The IEC 60079–14 contains in its Annex C, inspection schedules for the initial inspection related with the type of protection of the equipment. Originally, these schedules were in IEC 60079-17, but their inclusion in the installation standard was well received by installers.

4.7.1 Verification of the Hazardous Areas' Marking

Directive 1999/92/EC stipulates that the points of entry to hazardous areas be marked by the employer with a warning sign that is a simple and efficient safety measure for preventing risks. The signage plan for hazardous areas must use the icons and colours standardized by ISO and should be included in the area classification drawings, which will clarify if the whole space concerned is classified, or only part of it, and also will give details of the height of the estimated Zones.

Such marking is required, as example, for rooms (such as battery rooms, shown in Figure 4.6) or fenced enclosures where flammable liquids are stored. The yellow backgrounded triangle with "Ex" inside is the standardized icon used for warning on hazardous areas, but it is necessary to add other details in order to effectively guide workers on the safe requirements for carrying out services in that area. The sign plate must include information about the Zone, Equipment Group, and Temperature Class to enable workers to check that the marking on the Ex instruments they are going to use confirms their safe use in that location.

It is worthwhile to say that if the sign plate has only the "Ex triangle" on it, important information is missing, as an example if the Zone is 1 or 2. Without such information, workers could be led to believe that a tool marked "Ex n IIA T3" with a "yellow Ex triangle" sticker on it would be appropriate, as example, for use in a battery room, as shown in Figure 4.6.

FIGURE 4.6 A warning sign of hazardous location at the entrance door of a battery room. In addition to the Ex icon, it provides the necessary information for the safety of workers: The Zone, Group, Temperature Class, and the area classification drawing number.

It is not recommended to add text to Ex warning signs (such as "do not smoke", etc.), not only because then they will need to be written in other languages if there are workers from other countries on site, but also because warning signs need to quickly reach the brain and generate an immediate reaction. Forcing workers to read texts in risky environments, especially when written in a foreign language, has a significant potential to cause accidents.

Note that warning signs do not replace the required safety training. Training will inform workers about the risks of hazardous areas, the meaning of the "yellow backgrounded Ex triangle", and what to do when one of these signs is seen.

4.8 CLOSING REMARKS

Designing electrical installations in potentially explosive atmospheres requires a meticulous approach based on technical standards and also on legal regulations.

The combination of detailed risk analysis, area classification, appropriate equipment selection, careful assembly, and initial inspection is essential to ensure safe and efficient operations in hazardous areas.

BIBLIOGRAPHY

ATEX 1999/92/EC—Directive of the European Parliament and of the Council of 16 December 1999 on minimum requirements for improving the safety and health protection of workers potentially at risk from explosive atmospheres.
ATEX 2014/34/EU—Directive of the European Parliament and of the Council of 26 February 2014 on the harmonisation of the laws of the Member States relating to equipment and protective systems intended for use in potentially explosive atmospheres.
Borges, Giovanni Hummel—*Manual de segurança intrínseca: do projeto à instalação*. GHB, 1997.
CEPEL Report 20080915—*Spark risks in HV Ex e motors*. Centro de Pesquisas de Energia Elétrica, Brasil, 2008.
Cole, M., Driscoll, T. and Martin, K.—Installation, operation and maintenance considerations for electrical equipment in hazardous location. In: *IEEE IAS electrical safety workshop*. IEEE, 2010.
DSEAR—*Dangerous substances and explosive atmospheres regulations*. HSE, UK, 2002.
IEC 60079-0—*Explosive atmospheres—Part 0: Equipment—General requirements*. International Electrotechnical Commission, 2017.
IEC 60079-14—*Explosive atmospheres—Part 14: Electrical installation design, selection and installation of equipment, including initial inspection*. International Electrotechnical Commission, 2024.
IEC 60079-19—*Explosive atmospheres—Part 19: Equipment repair, overhaul and reclamation*, International Electrotechnical Commission, 2019.
IEC 60079-28—*Explosive atmospheres—Part 28: Protection of equipment and transmission systems using optical radiation*. International Electrotechnical Commission, 2015.
IEC 62305 SER—*Protection against lightning—All parts*. International Electrotechnical Commission, 2024.
ISO 80079-36—*Explosive atmospheres—Part 36: Non-electrical equipment for explosive atmospheres—Basic method and requirements*. International Organization for Standardization, 2016.

ISO 80079-37—*Explosive atmospheres—Part 37: Non-electrical equipment for explosive atmospheres—Non-electrical type of protection constructional safety "c", control of ignition sources "b", liquid immersion "k"*. International Organization for Standardization, 2016.

Jordão, Dácio Miranda—*Manual de instalações elétricas em indústrias químicas, petroquímicas e de petróleo*. Editora Qualitymark, 2008.

Keane, B., Schwarz, G. and Thurnherr, P.—Cables and cable glands for hazardous locations. In: *LXV IEEE Petroleum and Chemical Industry Committee Conference*. Institute of Electrical and Electronic Engineers, 2018.

Leblanc, N., Berner, W. and Ogilvie, P.—Conduit seals in hazardous locations—An ongoing safety issue: The past and the future. *IEEE Transactions on Industry Applications*, vol. 52, no. 1, IEEE, 2016.

NFPA 70—*National Electrical Code (NEC)*. National Fire Protection Association, 2023.

Rad, Marcel D., Sălăsan, Diana, Fotau, Dragos and Zsido, Sorin—Considerations regarding the choice of cable glands for electrical equipment used in potentially explosive atmospheres. In: *10th International Symposium on Occupational Health and Safety*. SESAM. 2021.

Rangel Jr., Estellito—Certificação e legislação. *Section EMEx, Eletricidade Moderna Magazine*, São Paulo, ed. 474, Sep. 2013, p. 146.

Rangel Jr., Estellito—Instalação Ex d. *Section EMEx, Eletricidade Moderna Magazine*, São Paulo, ed. 487, Oct. 2014, p. 136.

Rangel Jr., Estellito—ISO 80079-36 e 37. *Section EMEx, Eletricidade Moderna Magazine*, São Paulo, ed. 552, Mar/Apr. 2020, p. 58.

Rangel Jr., Estellito—Luminárias Ex led. *Section EMEx, Eletricidade Moderna Magazine*, São Paulo, ed. 512, Nov. 2016, p. 93.

Rangel Jr., Estellito—Motores Ex acionados por conversores. *Section EMEx, Eletricidade Moderna Magazine*, São Paulo, ed. 402, Sep. 2007, p. 190.

Rangel Jr., Estellito—Nova ATEX. *Section EMEx, Eletricidade Moderna Magazine*, São Paulo, ed. 506, May 2016, p. 86.

Rangel Jr., Estellito—Prensas-cabos. *Section EMEx, Eletricidade Moderna Magazine*, São Paulo, ed. 485, Aug. 2014, p. 146.

Rangel Jr., Estellito and Sanguedo, C. A.—Considerations on the new requirements for electrical installations in hazardous locations. *IEEE Transactions on Industry Applications*, vol. 55, no. 1, IEEE, 2019. Available at: https://ieeexplore.ieee.org/document/8451968

Rintala, Tarmo—Is certification of LED luminaires causing higher risks on Ex areas? *Atexor*, Jul. 2017.

Schram, P. J., Benedetti, R. and Earley, M. W.—*Electrical installations in hazardous locations*. J&B Learning, 2009.

UL 2225—*Cables and cable fittings for use in hazardous (classified) locations*. Underwriters Laboratories, 2022.

5 Ex Equipment Selection

5.1 THE NEED FOR SPECIAL EQUIPMENT

In Chapter 3, the concepts of area classification were discussed, that is an evaluation of locations where may exist the presence of flammable gases, vapours, dust or fibres in sufficient quantity to form explosive mixtures. Once the area is "classified", it will be necessary to choose the suitable electrical and electronic equipment considered safe for the application, that is to say, that would not act as sources of ignition. These special electrical and electronic equipment are also known as "Ex equipment".

The classified areas are divided into Zones where, probably, flammable gases can be present:

- Zone 0: Area where an explosive atmosphere of flammable gases or vapours is present continuously or for long periods
- Zone 1: Area where an explosive atmosphere of flammable gases or vapours is likely to occur under normal operating conditions
- Zone 2: Area where an explosive atmosphere of flammable gases or vapours is unlikely to occur under normal operating conditions and, if it does occur, it is for a short period

And where, probably, combustible dusts are present:

- Zone 20: Area where an explosive atmosphere of combustible dust is present continuously or for long periods
- Zone 21: Area where an explosive atmosphere of combustible dust is likely to occur under normal operating conditions
- Zone 22: Area where an explosive atmosphere of combustible dust is unlikely to occur under normal operating conditions and, if it does occur, it is for a short period.

Traditionally, the selection of Ex types of protection was made, choosing the types of protection approved for use in certain classified areas, as shown in Table 5.1.

Nowadays, a second option is available: The selection of Ex types of protection by the Ex equipment categories, also known as Equipment Protection Level (EPL), according to IEC 60079–0, shown in Table 5.2. The IEC 60079–0 standard clarified that EPLs were introduced as an alternative way for selecting Ex equipment, but it can only be used in engineering projects that have carried out a risk analysis of the consequences of explosions in the plant. In facilities where this analysis has not been carried out, equipment selection is maintained by linking it to the types of protection adequate for each Zone.

DOI: 10.1201/9781003500001-5

TABLE 5.1
Selection of Ex Types of Protection Related to Zones

Zone	Ex Types Approved For Use
0	Intrinsic safety "ia"
	Special certified equipment "s"
1	All above, and:
	Flameproof (explosion-proof) "d"
	Intrinsic safety "ib"
	Pressurized "p"
	Increased safety "e"
	Liquid immersion "o"
	Powder filling "q"
	Encapsulation "m"
2	All above, and:
	Non-sparking "n"

TABLE 5.2
The EPL Selection Based on Zones

EPL	Zone	Characteristic
Ga	Zone 0	Very high level of protection
Gb	Zone 1	High level of protection
Gc	Zone 2	Enhanced level of protection
Da	Zone 20	Very high level of protection
Db	Zone 21	High level of protection
Dc	Zone 22	Enhanced level of protection

It is important to note that not all types of protection can be used in all applications. Part of the equipment selection process needs the definition of which type of protection is the most suitable for a given application in a given set of circumstances.

5.2 HAZARDOUS (CLASSIFIED) AREAS

Potentially explosive atmospheres are found in a wide variety of industries. Although the principles for operating safely in these facilities are well known, defining the best way to meet the requirements for specifying electrical equipment suitable for such use still causes some confusion.

Such selection requires extreme care, as this kind of equipment is specially designed to prevent it from being a source of ignition due to the possible occurrence of sparks or hot spots during its operation. A gas station is an example easily found in everyday urban life for this type of installation, and industries such as petrochemicals and pharmaceuticals are also classic examples of locations with hazardous areas.

Potentially explosive atmospheres can also be composed of combustible dusts, examples of which are revealed in sugar refineries, coal mining, metals' processing such as aluminium and magnesium, and many other locations.

5.3 SELECTION REQUIREMENTS

Equipment selection must, in addition to meeting the requirements of the driven loads, consider the area classification study. Electrical equipment suitable for explosive atmospheres should be selected ensuring that:

- It has adequate type of protection.
- It can withstand environmental conditions (temperature, humidity).
- It is compatible with the types of substances present in the atmosphere.

In order to make the selection of the Ex equipment, it is necessary to know the environmental characteristics of the location, because they can impact the safe properties of the equipment and the installation. Some key points that define the equipment specification are:

- Equipment Group
- Temperature Class
- EPL/categories
- IP—ingress protection.

The importance of each of these items needs to be shown in detail.

5.3.1 EQUIPMENT GROUP

Identifying the Equipment Group is important to the design of electrical installations in hazardous areas, as certain types of protection depend on this information to be correctly specified. Often, the elements present in industrial processes mean that the user does not know which Gas Subdivision the installation belongs to. How should the electrical system designer proceed in these cases?

The recommendation is to request tests to be carried out to correctly identify the Gas Subdivision. Two methods are available:

- Determination of the Maximum Experimental Safe Gap (MESG)
- Determination of the Minimum Ignition Current (MIC).

5.3.1.1 Maximum Experimental Safe Gap

Under ISO/IEC 80079-20-1, to identify the MESG, two ignition tests must be carried out in a series of gaps, at intervals of 0.02 mm, covering the range from a safe gap to an unsafe gap. From the results, the highest value of the gap, g_0 (whose ignition probability is 0%), and the lowest value, g_{100} (which presents a 100% ignition probability), are determined.

The series of tests are repeated for a range of concentrations of the mixture, and the variations of the gaps g_0 and g_{100} are then obtained. The concentration of the mixture that presents minimum values of g_0 and g_{100} is the one of greatest risk. This method for determining the MESG uses a special vessel; however, experiments and determinations carried out only in an 8-litre spherical flask with ignition close to the gap of the flange can be accepted, in principle.

The MESG is used to classify flammable gases into groups based on their flame propagation characteristics, and, for example, gases with an MESG of less than 0.45 mm are classified as Group I, while gases with an MESG of 0.45 mm or greater are classified as Group II. The National Electrical Code (NEC) also uses MESG to classify flammable gases into Class I Groups A, B, C, and D. For instance, gases with an MESG of 0.25 mm or less are classified as Group A, while gases with an MESG greater than 0.75 mm are classified as Group D. So, the specific MESG values used to define the groups can differ slightly between IEC and the NEC.

For IEC, the groups of equipment for explosive atmospheres are:

- Group I: Equipment for coal mines subject to the presence of firedamp
- Group II: Equipment for locations with a gas atmosphere different from that found in coal mines subject to the presence of firedamp.

5.3.1.2   Minimum Ignition Current

Considering that each flammable gas has a particular Minimum Ignition Energy (MIE), it is necessary to know the energy emission levels of the electrical equipment, in order to know on which gases it would be put on work without risk of acting as a source of ignition.

By measuring their Minimum Ignition Current (MIC), gases and vapours can be classified according to the ratio of their MIC with the ignition current of laboratorial methane. The standardized method for determining MIC relationships uses the equipment described in IEC 60079–11. However, experiments performed for determinations obtained with other equipment may be accepted, exceptionally.

Some standards present such gas data, but it is important to emphasize that those are obtained by experimental tests, being influenced by variations in laboratory instruments, procedures, and precision of measuring devices. In certain cases, some data may have been determined at temperatures above ambient, so that the vapour would be in the explosive range. Furthermore, the data are subject to revision, and it is recommended that an up-to-date database be used in cases where more recent information is required.

Due to the nature of the environment, mining and surface industries are treated differently within the ATEX and IEC coding.

The IEC Groups characteristics are as follows:

Group I

Electrical equipment for underground mines susceptible to firedamp. Specifically for underground mines rather than surface mining, predominantly for coal mines, and has additional requirements over Group II, as coal dust generally must be considered.

Group II

For surface industries and excluded mines susceptible to firedamp. The equipment for Group II are split into three sub-groups: IIA, IIB, and IIC, based on the amount of energy required to ignite the gas, which is evaluated by the MESG and MIC tests.

Group IIA

It consists of the gases having either a maximum experiment safe gap (MESG) value greater than 0.90 mm or minimum ignition current ratio (MIC ratio) greater than 0.80. They require the highest energy to ignite, with a typical gas being methane, the main constituent of firedamp, which is a Group I gas.

Group IIB

It consists of gases having either a maximum experiment safe gap (MESG) value greater than 0.50 mm and less than or equal to 0.90 mm or minimum ignition current ratio (MIC ratio) greater than 0.45 and less than or equal to 0.80. This is best explained by considering the emission of energy from electrical equipment. This Group has the majority of flammable gases.

Group IIC

It includes only a small number of gases, which are the most easily ignited, having either a maximum experiment safe gap (MESG) value less than or equal to 0.50 mm or minimum ignition current ratio (MIC ratio) less than or equal to 0.45.

For the purposes of classifying gases and vapours, the MIC ratios are:

- Group IIA: MIC > 0.8
- Group IIB: 0.45 < MIC < 0.8
- Group IIC: MIC < 0.45

The IEC Groups Subdivisions are shown in Table 5.3.

Equipment to be installed in petrochemical plants is majority certified for Group IIB or, occasionally, for IIB + Hydrogen. Certification for IIB + Hydrogen can often be easier to achieve than a full IIC, particularly for Ex d equipment. For equipment selection, an increasing level of protection is required going from sub-group A to B to C (e.g. Group IIC equipment can be used in an IIB environment but not vice versa).

Group III

This Group is for dust atmospheres in surface industries. Dusts are split into Groups based on dust types rather than ignition energy, as generally dust

TABLE 5.3
IEC Gas Subdivisions

Subdivision	Characteristics	Examples
IIA	MESG > 0.9 mm MIC > 0.8	Acetone, ammonia, ethyl alcohol, gasoline, methane
IIB	0.50 mm <MESG <0.9 mm 0.45 < MIC < 0.8	Acetaldehyde, ethylene
IIC	MESG < 0.50 mm MIC < 0.45	Acetylene, hydrogen, carbon disulphide.

has approximately 1,000 times the MIE than gases do. Therefore, ignition energy becomes largely irrelevant in terms of equipment certification, as most protection techniques were devised around the lower ignition levels for gases with the type of dust rather than ignition energy being the important factor in the risk assessment. The primary type of protection for dust is Ex t (protection by enclosure), where in most cases, ignition is not an issue.

The term "Gas Groups" is not rarely (erroneously) used to include Dust Subdivision. Definitions for dust classification are as follows:

Non-conductive combustible dust: Finely divided solid particles, 500 μm or less in nominal size, which may be suspended in air, may settle out of the atmosphere under their own weight, may burn or glow in air, and may form explosive mixtures with air at atmospheric pressure and normal temperatures. Its electrical resistivity is higher than 103 Ωm.

Conductive dust: Combustible dust with electrical resistivity equal to or less than 103 Ωm.

Combustible flyings: Solid particles, including fibres, greater than 500 μm in nominal size, which may be suspended in air and could settle out of the atmosphere under their own weight.

The Subdivisions for dust are shown in Table 5.4.

Regarding the sub-group characteristics, Group IIIA dusts are considered to be lower risk and Group IIIB is by far the largest group, as it contains almost all carbon-based dusts, e.g. foodstuffs, which are explosive. Group IIIC is considered high risk as static electricity can be concentrated within the dust cloud, and shorting of exposed electrical connections is a real hazard. As with Gas Subdivisions, an increasing level of protection is required going from sub-division A to B to C (e.g. Group IIIC equipment can be used in an IIIB environment, but not vice versa).

TABLE 5.4
IEC Dust Subdivisions

Group	Characteristics	Examples
IIIA	Combustible flyings	Rayon, cotton, sisal, jute, hemp, cocoa fibre, oakum, and baled waste kapok.
IIIB	Non-conductive dusts	Foodstuffs (e.g. sugar, flour, grain, & additives); paper; and wood.
IIIC	Conductive dusts	Aluminium, bronze, zinc, magnesium

TABLE 5.5
Gas Categories under NEC and IEC

Typical Gas	NEC	IEC
Methane	D	IIA
Propane	C	IIB
Hydrogen	B	IIC
Acetylene	A	

The grouping was originally referred to, in older standards, as "gas group" because the gases were initially tested to form the groups. Now it is known as "equipment group".

5.3.1.3   The NEC Grouping

While IEC categorizes gases and dusts in "subdivisions", NEC categorizes them in "groups". The Gas Groups in the United States are also designed by letters, but there are differences between NEC and IEC designations, as shown in Table 5.5. It should be noted that gas subdivisions are different from the Class and Division grouping in ANSI/UL1203. For Class and Division, Group A is the most onerous, whereas in IEC 60079–1, Group IIC is the most onerous. Most notably, Acetylene (North America Group A) and Hydrogen (North America Group B) are separated out in UL1203, whereas IEC 60079–1 put them together as Group IIC.

One of the most widely used types of protection in industrial electrical installations that process flammable products is Ex d—also known as "explosion-proof" (the United States) or "flameproof" (Europe).

The electrical equipment is tested for a given explosive atmosphere and marked accordingly with the applicable standard.

The Dust Groups in the NEC are shown in Table 5.6.

TABLE 5.6
NEC Dust Groups

Group	Characteristics	Examples
E	Metal dusts	Magnesium, aluminium, titanium
F	Carbon-based	Coal and charcoal
G	Non-conductive dusts	Flour, wood/sawdust, plastic dust

5.3.2 Temperature Class

The ignition temperature, i.e. the temperature at which an ignition could occur, for example due to a hot surface of the apparatus, is dependent on the type of existing gases or vapours. This ignition temperature is influenced by several factors and is thus dependent on the stipulated testing order. Depending on the measuring system, the results can thus differ in the various countries.

The ignition temperatures of gases and vapours are in no way related to the ease of ignition by energy. Ignition temperature has to be a completely separate consideration. In scientific circles, it is still not fully understood why this should be the case.

The maximum temperature of the exposed surface of electrical apparatus must always be lower than the ignition temperature of the dust, gas, or vapour mixture, where it is to be used. For all types of protection, the Temperature Classes' range is T1–T6, corresponding to the classification of electrical equipment according to its maximum surface temperature.

The temperature classification system requires that the maximum surface temperature of the apparatus is measured or assessed. The value must fall in between two Temperature Classes' T ratings in Table 5.7. The lower of the two is the rating given to the equipment.

The classification of mixtures of gases or vapours with air, according to their MESG, MIC, and ignition temperature of gases and vapours, is listed in ISO/IEC 80079–20–1. When an equipment is assigned to a given Temperature Class, the maximum surface temperature is defined that it can reach during working, allowing the designer to verify if it is suitable for the area where it is intended to be installed. The IEC established six classes, and the NEC established the same in its article 505, but in its article 500, there are sub-divisions in classes T2, T3, and T4.

It may be noted that aforementioned temperature classification is based on an ambient temperature of –20 to +40 °C. The T classes in IEC and NEC are shown in Table 5.7.

For example, a rating of T5 means that the maximum temperature of equipment can be up to 100°C, while a T4 rating would mean that the maximum surface temperature generated by equipment at room temperature cannot exceed 135°C.

TABLE 5.7
IEC and NEC Temperature Classes

IEC	NEC	Temperature [°C]	Safe for Use with (Gases with Higher Auto-Ignition Temperature)
T1	T1	450	Benzene, ammonia, carbon monoxide, coal gas, hydrogen, methane, acetone, ethylene, toluene
T2	T2	300	Acetone, ethane, isopropyl alcohol, xylene, methanol, naphthalene, propane, ethyl acetate, ethanol, isopentane, glycerol, acetic acid, butane, acetylene
	T2A	280	Kerosene
	T2B	260	
	T2C	230	Diesel, cyclohexane, triethylamine
T3	T3	200	Acetylene, crude oil, n-hexane,
	T3A	180	Heptane, benzaldehyde, dimethyl sulphate
	T3B	165	Acetaldehyde
	T3C	160	
T4	T4	135	Carbon bisulphide, trichlorosilane, ethyl glycol, Diethyl ether
	T4A	120	
T5	T5	100	No gas or vapour specified as yet
T6	T6	85	Carbon disulphide, ethyl nitrate

The Temperature Class is related to the auto-ignition temperature of the hazardous material. The auto-ignition temperature of a substance is the lowest temperature at which it spontaneously ignites. There are many documents, as manufacturers' catalogues and standards, that present the auto-ignition temperature data for many substances, but it is important to emphasize that those could be obtained through different tests and are influenced by variations in laboratory procedures and precision of the measuring devices.

In certain cases, some data may have been determined at temperatures above ambient, so that the vapour would be in the explosive range. As these data are subject to review, it is recommended to use an updated database in cases where more recent information is required.

5.3.3 EQUIPMENT PROTECTION LEVEL

The latest revisions of the IEC standards introduce the concept of Equipment Protection Level (EPL), to be aligned with the ATEX categories. IEC 60079–0 established, strangely, that EPL also takes into consideration, the "potential consequences" of a possible explosion. For Zones 0/20, the EPL required would be "a"; for Zones 1/21, the level would be "b"; and for Zones 2/22, the level would be "c".

5.3.4   ATEX CATEGORIES

Equipment categories are used in ATEX. The category indicates which safety level of product must be used in each Zone. In Zones 0/20, category 1 devices must be used; in Zones 1/21, category 2 devices; and in Zones 2/22, category 3 devices.

Classification into categories is of particular importance, because all the inspection, maintenance, and repair duties of the end user will depend on the category of the product/equipment and not on the Zone where it is installed.

5.3.5   IP—INGRESS PROTECTION

The IEC 60529 defines the degrees of protection provided by enclosures against the ingress of liquids and solids, that is important to describe the sealing effectiveness of enclosures of electrical equipment against the intrusion of foreign bodies and liquids into it (e.g. tools, dust, etc.). This classification system utilizes the letters "IP" ("ingress protection") usually followed by two digits, but a third character is sometimes used. If an "X" is used for one of the digits (e.g. IPX4), it means that the product was not tested for the protection related to the missed digit.

5.3.5.1   First Digit

The first digit of the IP code indicates the degree to which equipment is protected against solid foreign bodies entering an enclosure. This ranges from 0 to 6, as shown in Table 5.8.

TABLE 5.8
Meaning of the First IP Digit

Level	Protected against Solids	Description
X	Unnecessary	If one of the two digits is not given (because it is not known, nor necessary), this missing digit is replaced with an X (e.g. IPX6).
0	Not protected	No protection against contact and ingress of objects.
1	> 50 mm	Any large surface of the body, such as the back of a hand, but no protection against deliberate contact with a body part.
2	> 12.5 mm	Avoid penetration of fingers or similar objects.
3	> 2.5 mm	Avoid penetration of tools, thick wires, etc.
4	> 1 mm	Avoid penetration of most wires, slender screws, large ants, etc.
5	Dust protected	Ingress of dust is not entirely prevented, but it must not enter in sufficient quantity to interfere with the safe operation of the equipment.
5K	Dust protected	Protection against dust deposits inside (dust protected). *Note*: This is found in ISO 20653, and not in IEC 60529.
6	Dust-tight	No ingress of dust; complete protection against contact (dust-tight). A vacuum must be applied. Test duration of up to 8 hours based on airflow.
6K	Dust-tight	Protection against dust ingress (dust-tight). *Note*: This is found in ISO 20653 and not in IEC 60529.

5.3.5.2   Second Digit

The second digit indicates the degree of protection of the equipment inside the enclosure against the harmful entry of liquids (e.g. dripping, spraying, submersion). It varies from 0 to 9, as shown in Table 5.9.

TABLE 5.9
Meaning of the Second IP Digit

Level	Protected against	Description
X	Unnecessary	X means no data is available to specify a protection rating.
0	None	No protection against ingress of water.
1	Dripping water	Dripping water (vertically falling drops) shall have no unsafe effect on the specimen. Test duration: 10 minutes Water equivalent to 1 mm rainfall per minute.
2	Dripping water when tilted at 15°	Vertically dripping water shall have no harmful effect when the enclosure is tilted at an angle of 15° from its normal position. Test duration: 2.5 min for every direction of tilt (10 min total). Water equivalent to 3 mm rainfall per minute
3	Spraying water	Water falling as a spray at any angle up to 60° from the vertical shall have no harmful effect. Test duration: 1 min/m² for at least 5 minutes. Water volume: 10 L/min Pressure: 50–150 kPa
4	Splashing of water	Water splashing against the enclosure from any direction shall have no harmful effect. Duration: 10 minutes.
4K	Splashing of water	Protected against splash water with increased pressure on all sides. *Note*: This is found in ISO 20653 and not in IEC 60529.
5	Water jets	Water projected by a nozzle (Ø 6.3 mm) against enclosure from any direction shall have no harmful effects. Test duration: 1 minute per m² for at least 3 minutes. Water volume: 12.5 L/min Pressure: 30 kPa at distance of 3 m.
6	Powerful water jets	Water projected in powerful jets (Ø 12.5 mm) against the enclosure from any direction shall have no harmful effects. Test duration: 1 minute per m² for at least 3 minutes Water volume: 100 L/min Pressure: 100 kPa at distance of 3 m.
6K	Powerful water jets with increased pressure	Water projected in powerful jets (Ø 6.3 mm nozzle) against the enclosure from any direction, under elevated pressure, shall have no harmful effects. *Note*: This is found in ISO 20653, and not in IEC 60529.
7	Immersion (up to 1 metre depth)	Ingress of water in harmful quantity shall not be possible when the enclosure is immersed in water under defined conditions of pressure and time (up to 1 metre of submersion). Test duration: 30 minutes.

TABLE 5.9 (*Continued*)
Meaning of the Second IP Digit

Level	Protected against	Description
8	Immersion (1 metre or more depth)	The equipment is suitable for continuous immersion in water under conditions which the manufacturer shall specify. However, with certain types of equipment, it can mean that water can enter but only to an extent so that it produces no harmful effects. The test depth and duration are defined by agreement with the manufacturer.
9	Powerful high-temperature water jets	Protected against close-range high-pressure, high-temperature spray downs. Smaller specimens rotate slowly on a turntable from four specific angles. Test duration: 30 s in each of four angles (two minutes total). Pressure: 8–10 MPa at distance of 0.10–0.15 metres. Water temperature: 80 °C The specific requirements for the test nozzle are shown in figures 7, 8, and 9 of IEC 60529. This test is identified as IP X 9 in IEC 60529.

It is worth noting that the IEC 60529 introduced an additional digit to designate a new degree of protection against water ingress. It is the second digit 9, which characterizes an enclosure protected against high-pressure and high-temperature water jets. The test for which must verify the force distribution between the tolerance limits and the water temperature maintained between 80 ± 5 °C.

5.3.5.3 Supplementary Letter

Additional information about the IP rating, is shown in Table 5.10.

Note that in Table 5.9, the letter K is specified in ISO 20653, a standard that deals primarily with the degrees of protection for vehicles. In addition to the numbers, the IP codes of the ISO 20653 sometimes also include the letter "K", which indicates a difference in the test specification to IEC 60529. For example, IP X 9 (IEC 60529) and IPx9K stand for "protection against liquids during high-pressure cleaning and steam jet cleaning", but the specification is slightly different.

The letter W is often mistaken as a designation "for use in the rain" or even as a "more robust construction characteristic", and, consequently, presented by manufacturers as a more expensive option. In fact, the W designates a special condition of use, defined between the manufacturer and the buyer, resulting in the "W" for one supply being different from the "W" for another one. This explains some conflicts that exist when a purchase specification is made by "copying-and-pasting" from another.

Whether a higher protection class is required in a special case depends on further environmental conditions. In special climatic conditions, such as high humidity, ice, strong temperature fluctuations, or the formation of condensation, a higher protection rating may be required.

TABLE 5.10
Supplementary Letter (IEC 60529)

Letter	Meaning
F	Oil resistant
H	High-voltage apparatus
M	Motion during water test
S	Stationary during water test
W	Weather conditions

The IEC 60529 Ingress Protection code does not consider the ambient conditions, which are covered in NEMA 250.

5.3.6   NEMA ENCLOSURES

The National Electrical Manufacturers Association (NEMA) defines requirements for enclosures, in a different mode than IEC 60529. Table 5.11 shows the characteristics of the NEMA 250 standard for enclosures.

There is no direct correspondence between the IP code and the NEMA 250 for enclosures. IP code is only valid for ingress of solids and water, while NEMA code includes resistance to other factors. Table 5.12 can be used only to convert NEMA 250 to IP; the reverse is not valid.

A simplified two-way similarity, but non-official, is shown in Table 5.13.

5.3.7   THE PROTECTION CLASS

The protection class is covered in IEC 61140 and refers to the measures that are taken with electrical devices to protect against dangerous voltages, in a range from protection Class 0 to protection Class 3.

5.3.7.1   Protection Class 0

Class 0 equipment does not provide any special protection against electric shock, apart from basic insulation. There is no connection to a protective conductor system, and protection is provided only by the equipment surroundings. Protection Class 0 is not designated with its own symbol, as no labels are applied to it. Devices with protection Class 0 have the lowest protection, and in many countries, including Austria and Germany, operation of such unprotected equipment or systems is not permitted.

5.3.7.2   Protection Class I—Protective Conductor

As far as equipment with this protection class is concerned, all electric-conductive enclosure components are connected to a system of protective conductors of a fixed electric installation with the earth potential.

TABLE 5.11
NEMA 250—Types of Enclosures

Provides a Degree of Protection against the Following Conditions	Type of NEMA Enclosure									
	1*	2*	4	4X	5	6	6P	12	12K	13
Access to hazardous parts	X	X	X	X	X	X	X	X	X	X
Ingress of solid foreign objects (falling dirt)	X	X	X	X	X	X	X	X	X	X
Ingress of water (dripping and light splashing)	...	X	X	X	X	X	X	X	X	X
Ingress of solid foreign objects (circulating dust, lint, fibres, and flyings**)	...	...	X	X	...	X	X	X	X	X
Ingress of solid foreign objects (settling airborne dust, lint, fibres, and flyings**)	...	...	X	X	X	X	X	X	X	X
Ingress of water (hose down and splashing water)	...	...	X	X	...	X	X	...	...	...
Oil and coolant seepage	...	...	...	...	...	...	...	X	X	X
Oil or coolant spraying and splashing	...	...	...	...	...	...	...	...	...	X
Corrosive agents	...	...	...	X	...	...	X	...	...	...
Ingress of water (occasional temporary submersion)	...	...	...	...	...	X	X	...	...	...
Ingress of water (occasional prolonged submersion)	...	...	...	...	...	...	X	...	...	...

Notes:

* These enclosures may be ventilated.

** These fibres and flyings are non-hazardous materials and are not considered Class-III-type ignitable fibres or combustible flyings. For Class-III-type ignitable fibres or combustible flyings, the NEC should be consulted in the United States.

TABLE 5.12

Association of NEMA Enclosure Types with IEC IP

IP First Character	1 A	1 B	2 A	2 B	3, 3X, 3SX A	3, 3X, 3SX B	3R, 3RX A	3R, 3RX B	4, 4X A	4, 4X B	5 A	5 B	6 A	6 B	6P A	6P B	12, 12K, 13 A	12, 12K, 13 B	IP Second Character
IP0_	XX	XX	XX	XX	XX	XX	XX	XX	XX	XX	XX	XX	XX	XX	XX	XX	XX	XX	IP_0
IP1_	XX		XX	XX	XX	XX	XX	XX	XX	XX	XX	XX	XX	XX	XX	XX	XX	XX	IP_1
IP2_					XX	XX		XX	XX	XX	XX	XX	XX	XX	XX	XX	XX	XX	IP_2
IP3_					XX	XX		XX	XX	XX	XX		XX	XX	XX	XX	XX	XX	IP_3
IP4_					XX	XX		XX	XX	XX	XX		XX	XX	XX	XX	XX	XX	IP_4
IP5_					XX				XX	XX	XX		XX	XX	XX	XX	XX		IP_5
IP6_									XX	XX			XX	XX	XX	XX			IP_6
														XX		XX			IP_7
																XX			IP_8

Where:

A: A shaded block in the "A column" indicates that the NEMA enclosure type exceeds the requirements for the respective IEC 60529 IP first character designation. The IP first character designation is the protection against access to hazardous parts and solid foreign objects.

B: A "XX" block in the "B column" indicates that the NEMA enclosure type exceeds the requirements for the respective IEC 60529 IP second character designation. The IP second character is the protection against the ingress of water.

TABLE 5.13
A Simplified Correlation between IP × NEMA

NEMA Enclosure	IP Code
1	IP20
2	IP22
3, 3X, 3S, 3SX	IP55
3R, 3RX	IP24
4, 4X	IP44, IP66, IP65
5	IP53
6	IP67
6P	IP68
12, 12K, 13	IP54

Mobile equipment rated as protection Class I is equipped with a plug-in connector with a protective contact or a cable fitted with an additional protective conductor and a plug equipped with a protective contact. The protective conductor connection is designed as a leading contact, so that it is the first to be activated when a plug is inserted and is the last to be disconnected when it is removed.

In addition to the mechanical strain relief feature, a connection cable inserted into a device must be designed in such a way that, in case the cable is pulled out, the protective earth conductor gets pulled out last. If a live cable accidentally touches the enclosure connected to the protective conductor, a short-circuit occurs. The protective conductor enclosure connection is dimensioned in such a way that there is no permanently hazardous touch voltage on the enclosure, and a line breaker, fuse, or residual current switch-off device is triggered momentarily and disconnects the circuit power supply.

Classic neutral earthing with protective earth and neutral (PEN) conductors, still encountered in building systems, goes against the protection Class I concept. A PEN conductor combines a protective (PE) and neutral (N) conductor. To connect sockets, a neutral conductor is connected to socket protective conductor contacts, which results in neutralizing the required protective measure. PEN conductor interruptions may also result in hazardous voltage transfer onto enclosures of all protection Class I equipment connected to a corresponding circuit.

5.3.7.3 Protection Class II—Double Insulation

Protection Class II equipment comes with enhanced or double insulation at the rated insulation voltage between live parts and elements that can be touched. They are usually not connected to a protective conductor. If they come with an electrically conductive surface or conductive elements that can be touched, they are separated from the live parts with enhanced or double insulation, and their touch current value does not exceed 0.5 mA.

Mobile Class II equipment usually comes with plugs that are not fitted with a protective conductor connection or a protective conductor itself. It includes primarily small-sized white goods or tools, such as electric drills, or certain electronic appliances.

If a cable with a protective conductor is used, it must not be connected to an enclosure and must be treated as an active cable.

5.3.7.4   Protection Class III—Low-Voltage Protection

Protection Class III equipment operates at Safety Extra-Low Voltages (SELV) or Protective Extra-Low Voltages (PELV). Such devices can only be connected to SELV or PELV power supplies. They include:

- Safety transformers as per the applicable standard
- Electro-chemical power supplies (batteries and rechargeable batteries)
- Solar cells, generators, crank generators, and similar equipment supplying low voltage.

PELV equipment comes with double insulation between a mains connection and low-voltage carrying components, but the low-voltage circuits or enclosure can be earthed. The earthing is not provided for safety reasons, but in order to ensure electromagnetic compatibility (interference emissions, earthing loops, electrostatic discharge (ESD) protection).

Such solutions are treated as functional earthing. Laptop or audio equipment power supplies are examples of such devices.

Table 5.14 gives an overview of the protection class symbols and their meaning.

TABLE 5.14
Protection Classes

Protection Class	Meaning	Symbol
0	Basic insulation is given, but no other protection. There is no connection to a protective conductor system, and protection is provided only by the equipment surroundings.	
1	A protective earth conductor is present in the equipment. In the case of movable equipment, the protective earth conductor is installed in the plug.	
2	Double insulation is given as protection between the mains circuit and the output voltage. There is no connection to the protective conductor.	
3	The unit is protected by a SELV/PELV. A safety transformer according to EN 61558-2-6 is required.	

5.3.8 BALANCING COSTS

The selection of special electrical and electronic equipment for use in classified areas (Ex equipment) must be made considering not only their functional characteristics but also the process with which they will be installed and the material and personnel resources that will be available when the industrial unit starts operating.

Regarding the resources of the industrial unit that need to be assessed, we can cite:

a. Acquisition costs
b. Installation costs
c. Maintenance requirements
d. Flexibility

Let us break down what is involved in these items:

a) The acquisition costs of the equipment are the easiest to quantify. It is necessary to evaluate not only the sales price but also the warranty period, after-sales assistance (especially the ease of purchasing spare parts), and the delivery time.

b) There are specific requirements for each Ex type, which need to be evaluated in the design of the industrial unit. For example, pressurization requires the availability of an air compression system and respective ducts to the location where the Ex p electric panel will be. And in the event of a lower internal pressure, there is a possibility that the panel will have to be de-energized, interrupting the operation of the systems connected to it.

c) Some Ex types require more maintenance resources than others. For example, intrinsic safety has the advantage that maintenance can be performed without the need to turn off the power. Explosion-proof enclosures require that the power be turned off or that the absence of the flammable mixture be proven for the duration of the service. As Ex d enclosures are usually made of cast aluminium, they usually are heavier than other types, and they can also be impacted by aggressive atmospheres. The pressurized system involves the compressed air system and its accessories such as pressure switches and timers. All Ex types will require specialized training of the maintenance team.

d) Regarding flexibility, it is advisable to check whether there is a forecast of expansion of the industrial unit in the future, which could indicate the need to select an Ex type that could be used. For example, if a part of the existing plant is classified as Zone 2, and a future expansion is planned for that location, it could receive a certain process that is expected to generate Zones 1.

5.3.9 CERTIFICATION

Regarding the safety requirements for gas and vapour flammable atmospheres, the certification of conformity is required, as all equipment must follow the legislation of the country where they will be installed, regarding conformity certificates. In the European Union, compliance with ATEX Directive is mandatory; in Brasil,

INMETRO is the main body responsible for issuing the regulations for equipment to be installed in classified areas. It is important to verify these requirements before purchasing, as they can impact the time schedule of the enterprise.

For each Zone, there are recommended types of Ex protection, each one following its respective construction standard that can follow the IEC 60079 series or a national legislation. In the United States, the ANSI (American National Standards Institute) has adopted some IEC standards with deviations. Typically, the International Society of Automation (ISA, former Instrument Society of America), when it issues versions of the IEC standards, illustrates national differences from the IEC text through the use of legislative notation (strikeout and underlining).

The American certification body UL (Underwriters Laboratories) issues their IEC version highlighting the national differences immediately after the IEC text. National differences between the UL version and the ISA version should be checked word for word, except for editorial changes.

IEC has its own Ex certification scheme, IECEx—International Electrotechnical Commission Scheme for Certification to Standards Relating to Equipment for use in Hazardous Areas. The IECEx certification scheme is made up of three distinct certification sub-schemes, all voluntary and managed by a committee composed of member delegates:

IECEx Certified Equipment Scheme:
The scheme is aimed at certifying the conformity of Ex electrical and electronic equipment on the basis of IEC standards on Ex types of protection.
IECEx Certified Service Facilities Scheme:
The scheme is aimed at certifying the conformity of services providers and repair shops with the IEC 60079–19 standard.
IECEx Certification of Personnel Competence Scheme:
The scheme is aimed at certifying persons on their ability to apply their skills and experience to work safely in accordance with relevant IEC Ex Standards.

The certification requirements depend on the legislation of each country, so, if a given Ex equipment has a certification, that is to say "ATEX", it is not guaranteed that it will be approved in other certification system, as an example, the American.

5.3.10 MARKING

Equipment for use in potentially explosive atmospheres must have the appropriate marking, whose format is specified in IEC 60079-0. To clearly identify the equipment, the name of the manufacturer, model, and serial number must be indicated. The name of the certification body and the number of the certificate of conformity must also be included.

The basic information that allows the user to identify the application for which the equipment is approved to be used is:

* The Ex designation
* The types of protection that constitute the equipment
* The Gas Subdivision
* The Temperature Class or maximum surface temperature in °C (in case of hazardous areas due to combustible dusts).

Examples are given as follows:

> Ex d IIC T4
> Ex pD21 T120 °C.

Notes:
In the intrinsic safety type, associated equipment, installed in non-classified areas, requires a marking placed in square brackets. Example is: [Ex ia] IIC.

If the associated equipment needs to be installed in a classified area, it must be placed in a suitable enclosure. The intrinsic safety marking is placed in brackets. Example is: Ex-d [ib] IIC T6.

It is worth noting that this marking does not describe the level of protection of the equipment needed in relation to explosion protection.

The designations of the EPLs, which consist of two letters, the first one in capital letter and the second in lowercase, were included in IEC 60079–0. The first indicates the type of explosive atmosphere (G for gas and D for dust) and the second, the effective level of protection of the equipment, which can be a, b, or c. This information can be added to the "traditional" marking. Example is Ex de IIC T4 Gb.

In the case of associated equipment, the EPL can be mentioned after the type of protection. Example is: [Ex ia Ga] IIC.

When the equipment associated with an intrinsically safe "ia" circuit is mounted in an explosion-proof enclosure, the following marking can be found: Ex d [Ga ia] IIC T4 Gb.

An example is a temperature transmitter, intended to be assembled in Zone 1 with its sensor element inserted in Zone 0, certified, with the following data:

> Manufacturer A
> Model: TT672
> Serial number:44556677
> Certificate of Conformity: CC 1234/2006 X
> Ex ib [ia Ga] IIC T4 Gb IP 65
> U_i = 30 V
> I_i = 120 mA
> P_i = 750 mW
> U_o = 1 V
> I_o = 10 mA
> P_o = 2.5 mW
> C_o = 1,000 µF
> L_o = 400 mH
> $-20°C \leq T_{amb} \leq +70°C$

If the equipment is also approved for use in combustible dust atmospheres, Zones 20 and 21, an additional marking would be required:

> Ex ib D [iaD Da] III C Db T 80 °C

If the manufacturer A wishes to sell such a product in the EU, it must also comply with the ATEX Directives, which require the marking to include the year of manufacture, the Ex symbol, the category of the equipment, the CE mark, and the registration number of the certifying body.

With the publication of current IEC 63365, a new form of digital marking, using technologies such as the Quick Response Code (QR), RFID (Radio Frequency Identification), NFC (Near Field Communication), and even product firmware, became standardized.

All of these technologies were well known and used by civil society, such as QR code, which uses the internet; RFID, developed in the 1940s and used in special labels (tags), on packaging for inventory control, and on badges for access control to restricted areas uses radio waves; and NFC (used to pay bills by tapping a credit card or cell phone) exchanges information in the 13.56 MHz band among compatible devices that are nearby.

Digital marking will allow manufacturers to enable the placement of more information, where previously there was a limitation on the size of small labels, for example, using QR codes.

The long-term goal of IEC 63365 is to replace physical identification plates with electronically readable plates. However, the removal of conventional plates will have an impact on users.

The chemical industry also uses them to identify products packaged throughout the production process. Although it is possible to use barcodes for this function, RFID can provide more information, and the tags are also able to withstand changes in temperature, dust, and mechanical impacts.

RFID antennas are essential for the system to function correctly, as they communicate between the RFID tags and the reading devices and must be installed in a strategic location that allows a good coverage of the radio frequency signal. Although it is a technology with benefits, there are some problems involved with the exchange of information via radio frequency, such as:

- High cost when compared to the bar code system
- Interference of local magnetic field on metal products, restricting the operation of the readers. For this application, special labels are required, which are more expensive.

In classified areas, there are additional concerns, because, although there are passive labels, which can be considered as "simple" devices according to IEC 60079–14, depending on the application, active labels may be necessary, with batteries included, which must be evaluated for certification for classified areas. Furthermore, the RFID tag reader is an essential device for the correct functioning of the technology. It "reads" the tags and collects the available information. As it is an electronic device, when used in classified areas, it will need to be evaluated for the energy accumulated in its circuits and receive the respective certificate of conformity, which will indicate which Gas Subdivision it can be used for safely. These considerations should be taken into account by users, especially when contracting inspection services, which will be more expensive if it is necessary to use Ex-certified readers.

5.4   TYPES OF PROTECTION

IEC established the types of protection to be safely installed in hazardous locations, and their particular characteristics will be shown below.

5.4.1   Ex d

This type of protection has special construction characteristics, and its installation can only be entrusted to professionals who have undergone specific technical training and have thus received formal authorization. However, it is very common to find non-conformities in this type of installation, which indicates that it was entrusted to professionals without the necessary training.

Although often confused as if they were just one type, there are some differences between the "flameproof" (IEC 60079–1) and "explosionproof" (UL 1203).

5.4.1.1   IEC Standard

IEC 60079–1 contains specific requirements for the construction and testing of electrical equipment with type of protection "d", intended for use in explosive atmospheres for Group I (mining) or Group II (surface installations—IIA, IIB, and IIC). The requirements of this standard are complementary to the general requirements defined in IEC 60079–0.

One of the requirements for Ex d installation is the use of sealing units when using metallic conduits, which perform important functions for the safety of the plant, such as preventing the propagation of possible explosions along the conduits. However, for it to be fulfilling that function, it is essential that it is installed correctly.

As such, the presence of the sealing unit does not always guarantee that the installation complies with the standards and is safe for users. The installation of sealing units is required to be as close as possible to the explosion-proof box.

In corrosive environments, it is important that the Ex d enclosure, which is usually metallic, be protected against the effects of corrosion, with painting being one of the adopted measures. Ex d is frequently confused as "flameproof", and it is defined in IEC 60079–1 as

> enclosure in which the parts which can ignite an explosive gas atmosphere are placed and which can withstand the pressure developed during an internal explosion of an explosive mixture, and which prevents the transmission of the explosion to the explosive gas atmosphere surrounding the enclosure.

It is important to note that the term "flameproof" in Europe, despite conceptually corresponding to the term "explosion-proof" in the United States, is associated with slightly different technical standards. A flameproof enclosure is designed to withstand the pressure of an internal explosion; it is not necessary therefore to provide openings for pressure relief. Where there is a joint, however, or where a spindle or shaft passes through the enclosure, the products of the explosion can escape.

It should be understood that the aim of a flameproof enclosure is not necessarily the total avoidance of any gaps in an enclosure. The misconception that it should be "gas-tight" is misplaced.

The principle recognizes that some openings are unavoidable in practice and, so, restricts it to requiring that the size of such openings should not exceed the safe limit above which the nature of the escaping flame is such as to ignite a specified flammable atmosphere. On the other hand, it is not the aim to require joints to be deliberately spaced to give an opening.

The entry of cables or conduits requires careful specification to provide adequate physical protection and maintain the integrity of the protection. In certain processes, flammable gases may be emitted for a long time, which contraindicates this type of protection in Zone 0.

The Ex d has three EPLs:

- Ex da (EPL "Ma" or "Ga")—which can be used in Zone 0, but it is only applicable to catalytic sensors of portable combustible gas detectors, and with maximum free internal volume not exceeding 5 cm^3
- Ex db (EPL "Mb" or "Gb")—which can be used in Zones 1 and 2
- Ex dc (EPL "Gc")—which can be used only in Zone 2, but its free internal volume shall not exceed 20 cm^3. The requirements for Ex dc were copied and updated from the Ex nC part of IEC 60079-15 (equipment protection by type of protection "n"). Primarily, but not exclusively, used for components such as switches, the "C" part of Ex nC is "enclosed break", i.e. sparking parts are sealed to prevent ingress of gas.

The resultant of above is that an "Ex dc" assembly may not necessarily be "flameproof", but an Ex e style enclosure, which would have historically been certified Ex nC, may be suitable. Ex dc aside, the protection concept Ex d relies on a flamepath for its protection. One of the key criteria is that the flamepath should never be impeded. Dust hazards would almost certainly block that path, consequently conventional Ex db protection method, i.e. flameproof enclosures, and dusts are, as a rule, exclusive.

However, as Ex d protection is probably the first call for inclusion of uncertified equipment into the hazardous area prior to specific dust hazards protection methods (Ex t), it is widely used for category 2 (Zone 21) solutions. In this case, the flameproof enclosure essentially is being used as "protection by enclosure", ignoring the flamepath. This is why generally a flameproof enclosure can only be certified for gas or dust, not a hybrid atmosphere, i.e. gas and dust. Flameproof enclosures, recently certified, can show labels with separate lines for gas and dust, and sometimes, Ex t certification for dust. Flameproof enclosures can be used in Zone 1 and Zone 2. Only Ex da can be used in Zone 0.

5.4.1.2 NEC Standard

"Explosionproof" follows the standards issued by American certification bodies, as examples, FM 3600 and UL 1203, and it is one of the protection concepts accepted for Class 1 Division 1 in the United States. The explosion-proof type, which allows placing industrial electrical devices inside a special enclosure, requires full compliance certification also (including all internal components installed).

Division 1 is the equivalent of Zones 0 and 1, as, historically, the NEC did not differentiate between the two. However, since the NEC 1996 edition, the IEC three-Zone system for gases and vapours began to be accepted in the United States, but only in installations constructed after that date, and whose area classification and electrical installation specification were done using IEC standards.

5.4.1.3   Differences between IEC 60079–1 and ANSI/ISA 60079–1

UL 60079–1 included the following as national differences:

- It covers both Gas and Dust requirements, while IEC 60079–1 is Gas/Vapour specific.
- Taper-threaded joint requirements conform to the NEC thread engagement requirements.
- The copper content of the alloy shall be limited to 30%, whereas for IEC, the cooper content acceptable is 60%. UL has more stringent material requirement than IEC standard.
- All cable glands, whether integral or separate, must meet the requirements in UL 2225.

5.4.1.4   Applications

Examples of equipment that use Ex d are electric motors, lighting fixtures, instruments with small enclosures as flowmeters, and pressure switches.

5.4.2   Ex e

Increased safety protection applies only to electrical equipment where no part can produce sparks or arcs or exceed the temperature limit of the materials used in its construction.

It is a type of protection where, even though an explosive atmosphere can enter the enclosure, the equipment does not spark, arc, or overheat during normal operation. It also requires good-quality materials with well-defined insulation properties, and the mechanical and electrical design specifications are sufficiently reduced to make faults that could cause ignition, unlikely. This safety from fault is further increased by enclosure protection from its environment, reducing the risk of environmental conditions adversely impacting its operation.

Switches or switching mechanisms inside an Ex e enclosure are not permitted in this concept of protection, unless they are built with some Ex type of protection that prevents open sparks inside them. Sparks, therefore, cannot occur, and spark energy does not need to be considered.

Ex e motors need special winding design and the installation of temperature monitoring. They are also required to be further assessed for possible air-gap sparking and may require additional measures such as pre-start ventilation with air before start-up.

Electrical equipment with type of protection increased safety "e" are either:

a) Ex eb (EPL Gb), which can be used in Zones 1 and 2; or
b) Ex ec (EPL Gc), which can be used only in Zone 2.

The Ex eb applies to electrical equipment where the rated voltage does not exceed 11 kV rms, ac or dc; and Ex ec applies to electrical equipment where the rated voltage does not exceed 15 kV rms, ac or dc. Some of the requirements for an equipment to be certified and marked as Ex e are explained in the next subsections.

5.4.2.1 Enclosures

They must be designed to be weatherproof and to be impact-resistant. The minimum IP rating is IP 54 before and after a 7 Nm drop test has been performed, as specified in the IEC 60079–7. The enclosure is not designed to withstand external or internal explosion.

When applied to junction boxes, an Ex e enclosure is given a 'safety factor' when certified. This represents the highest number of "terminal-amps" permitted in the box.

Terminals mounted in the box must be component-approved. The total of terminal-amps must be calculated and must be equal to or less than the safety factor.

5.4.2.2 Terminations and Connections

They are made with a degree of security such that in the conditions of use, they are unlikely to become loose and spark. Electrical connections are oversized and of sufficient cross-section for required current-carrying capacity, in order to lower contact resistance and to help dissipate heat.

5.4.2.3 Clearance and Creepage

In Ex e terminals, bare live conductors must be separated by a minimum air clearance. The distance between the bare end of an insulated wire and the next terminal when separated by an insulating surface is called creepage. Both distances are illustrated in Figure 5.1 and tabulated in IEC 60079-7.

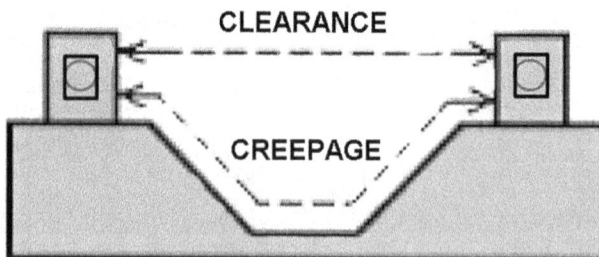

FIGURE 5.1   Clearance and creepage distances.

So, ordinary industrial terminals cannot be used in Ex e assemblies, as the insulating material needs also to be approved under the stipulated requirements for the Comparative Tracking Index (CTI), which is a measurement that evaluates the capability of the insulating surface to retain its properties in the presence of contaminants.

5.4.2.4   Differences between IEC 60079–7 and UL 60079–7

UL 60079–7 included the following as national differences:

- The electrical connection should be able to provide contact pressure that is not applied through the insulating material. However, IEC 60079–7 allows the contact pressure to be applied through the insulating material if the earth continuity test of IEC 60079–0 is accomplished.
- Terminals greater than 1,500 V be subjected to the tests in UL 1059 and UL 486E.
- Threaded connections can only be released, or removed, by use of a tool.
- Plugs and sockets shall be capable of being connected by wiring methods permitted in the NEC.
- Cable assemblies and associated plugs and sockets shall meet the requirements of UL 2238 and UL 2237.
- Arcing, or sparking, contacts are not permitted for level of protection "eb", and for level of protection "ec", manually operated arcing, or sparking, components located within an enclosure that are not accessible in normal operation without the use of a tool need only comply with the separation distances on the external connection points.

The requirements in IEC 60079–7 do not meet the requirements in UL 60079–7.

5.4.2.5   Applications

Examples of equipment constructed with the Ex e type are: Installation components such as junction boxes, control stations, and squirrel-cage induction motors.

5.4.3   Ex i

Intrinsically safe Ex i protection is defined in the IEC 60079–11 as a protection method in which electrical and thermal energies are limited to a level safe enough to prevent the ignition of explosive atmospheres. This is achieved through the use of electrical components and circuits that cannot generate enough energy to cause an ignition.

On the other hand, the IEC 60079–25 addresses the requirements for intrinsically safe systems, defined as a set of interconnected equipment that, as a whole, is intrinsically safe. IEC 60079–25 specifies the requirements for the design, installation, operation, and maintenance of these systems. In addition, there are several installation requirements for intrinsically safe circuits and supporting documentation to be carried out by users, according to IEC 60079–14.

Type of protection Ex i can be used in locations within any of Groups I, II or, III (Ex iD). The cables for Ex i systems can be installed with the protection of lower mechanical resistance, which helps to reduce the cost of installation, compared to other types of protection.

This type of protection is well known in Europe, and it is also accepted in the United States. Ex i has three EPL for gas atmospheres:

- Ex ia (EPL Ga)—which can be used in Zone 0
- Ex ib (EPL Gb)—which can be used in Zones 1 and 2
- Ex ic (EPL Gc)—which can be used only in Zone 2.

For dust atmospheres, there is Ex iD, as explained next.

Ex iD (intrinsic safety)

It has the same characteristic of limiting the available energy of the circuit to prevent ignition in a hazardous area due to combustible dust atmosphere.

The IEC 60079–0 has the procedure for defining the maximum surface temperature for intrinsically safe equipment of Group III. In these cases, the measurement must be made using the values of Ui and Ii specified for the intrinsically safe equipment without the 10% safety factor.

The temperature must be that of the surface of the intrinsically safe equipment that comes into contact with the dust. For example, for intrinsically safe equipment protected by an enclosure with a degree of protection of at least IP5X, the surface temperature of the enclosure must be measured.

Alternatively, the intrinsically safe equipment must be considered suitable for total immersion or covered by a layer of dust of undetermined thickness, provided that the combined power dissipation of any components complies with IEC 60079–0 requirements, and the continuous short-circuit current is less than 250 mA. Such intrinsically safe equipment must be marked with T value of 135°C.

Its EPL is available as Db.

5.4.3.1   Enclosures

When the intrinsic safety of intrinsically safe equipment may be compromised by the ingress of dust, or access to conductive parts, for example, if the circuits have infallible creepage distances, one of the following enclosures is required:

a) When the separation is obtained by meeting the separation and creepage distance requirements of IEC 60079–11, the enclosure must provide a degree of protection of at least IP5X, according to IEC 60529.

b) When the separation is obtained by meeting the requirements for distance under coating, encapsulating compound, or separation distances by means of solid insulation of IEC 60079–11, the enclosure must meet a degree of protection of at least IP2X.

5.4.3.2 Differences between IEC 60079–11andUL 60079–11

- The UL standard imposes stricter requirements than the IEC standard when it comes to clearances, creepage distances, and separations between conductive parts.
- UL 60079–11 requires additional condition to be met if "ia" apparatus uses series current limiters consisting of controllable, and non-controllable, semiconductor devices in Division 1. The conditions in the UL standard are that both the input and output circuits are to be intrinsically safe or it is to be demonstrated that the semiconductors, or controllable semiconductor devices, cannot be subjected to transients from the power supply network.

5.4.3.3 Applications

Examples of equipment that use Ex i protection include temperature sensors, pressure transmitters, intrinsic barriers, and measuring instruments used in industrial environments. Among the main benefits of this type of protection is that maintenance can be carried out with the circuits energized, as the level of energy available in these systems is insufficient to ignite an explosive atmosphere.

5.4.4 Ex M

Encapsulation is a type of protection where the electrical circuits are covered by a casting compound, following the requirements of IEC 60079–18. The standard permits a wide range of electrical and electronic components to be encapsulated, provided that the operating temperature of such components does not exceed their rating at the maximum ambient temperature envisaged for operation. Even components such as relays and switches are permitted, provided that they are themselves enclosed before being encapsulated.

The rating of the components used needs to be sufficient to ensure that they remain within it in case of a single fault elsewhere, or their failure in the worst possible way is assumed to be a part of that fault. The types of fault which are considered include, but are not limited to, short or open circuit of any component, or faults in printed circuit, such as open circuit failures or shorts between tracks.

The equipment is not permitted anywhere to exceed its specified operating temperature range in normal operation or fault conditions, so it needs to be protected from overheating as it might cause the encapsulation to degrade or cause excessive surface temperatures. Often thermal fuses are installed to ensure that the supply is isolated when there is an internal fault.

Equipment with type of protection encapsulation Ex m can be installed in all Groups (flammable gases, combustible dusts, and coal mines) and have the following EPL:

a) Ex ma (EPL Ma, Ga or Da), which can be used in Zone 0
b) Ex mb (EPL Mb, Gb or Db), which can be used in Zones 1 and 2
c) Ex mc (EPL Gc or Dc), which can be only used in Zone 2.

Equipment protected using this method are typically not repairable and are replaced in the field by new devices.

5.4.4.1   Differences between IEC 60079–18 andUL 60079–18

The requirements in IEC 60079–18 are considered to meet UL 60079–18 for equipment in explosive gas atmospheres.

5.4.4.2   Applications

The most common applications for the encapsulation type of protection are solenoid valves, relays, and ballasts for fluorescent luminaires.

5.4.5   Ex n

The IEC 60079–15 addresses specific considerations for the type of protection non-incendive, for electrical equipment in explosive atmospheres.

Ex n is fundamentally a non-sparking concept. Its limitation is for normal operating conditions only, and in general, fault conditions do not need to be taken into consideration. As there is no fault tolerance, Ex n has an EPL Gc (ATEX Category 3G), so, it is only suitable for use in Zone 2.

The perspective of the user is that Ex n is a less-costly approach than IS, because no barrier, or isolator, is required. But, the overall installation is less safe with Ex n than with Ex i, so much so that Ex n remains restricted to Zone 2 only. This raises the concern that area classification may be influenced, in order to accommodate Ex n equipment.

The IEC 60079–15: 2017 has the following types of protection:

- **Ex nR**

    The "nR" type of protection aims to limit the extent to which an explosive atmosphere can enter the enclosure, essentially making it almost gas-tight but not completely. This allows uncertified components to be used internally. However, components which go through the enclosure wall and therefore come in contact with any gas should be suitably certified. The sealing is not a hermetic seal per se, but there are stringent requirements requiring the enclosure to be pressure tested for leaks. Ex nR is based on a sealed enclosure; it does allow the potential for sparking internally but is designed to keep the gas out of the enclosure in normal use.
- **Ex nC**

    This type of protection encompasses the following techniques: Sealed devices "nC", hermetically sealed devices "nC", and non-incendive components "nC".

Among the changes introduced by the IEC 60079–15: 2017, compared with its previous edition, were the change of the Ex nA and Ex nL markings into Ex ec and Ex ic, respectively. Ex ec equipment is of increased safety and is designed to prevent the ignition of flammable gases in classified areas, just as the Ex nA safety principles were. Ex ic equipment is intrinsically safe, designed to limit electrical energy to safe levels, in the same way as the previous Ex nL type.

Some equipment can be found that was originally installed and certified as Ex nL. This concept was rarely used and cannot be used for new installations. Where practicable, it is recommended that older equipment be replaced.

The definitions of the types of protection previously included in IEC 60079–15: 2020, which were moved to other parts of IEC 60079, are described next for reference:

- **Ex nA** (since IEC 60079–7: 2015, it became Ex ec)
  Equipment with non-sparking protection is designed to prevent the generation of ignition sources. This is achieved through the use of non-incendive materials, current or voltage limitation, and component design that do not generate sparks or heat sufficiently to cause ignition.
- **Ex nC** (since IEC 60079–1: 2014, it became Ex dc)
  Similar concept to Ex nA, with the addition that sparking contacts are protected, meaning the sparking element is hermetically sealed or enclosed to prevent gas ingress (e.g. hermetically sealed relays). Components should be certified as for Ex nA, but it is less likely that a basic declaration would be sufficient as the hermetically sealed component aspect is not so obvious, and generally the technical file should include independent certification as justification for use. A hermetically sealed device is so constructed that the external atmosphere cannot gain access to the interior and in which any seal is made by fusion, for example, brazing, welding, or the fusion of glass to metal.

5.4.5.1   Differences between IEC 60079–15 and UL 60079–15

UL 60079–15 included additional requirements with regard to conduit entries, gasket seal, and cable glands for restricted-breathing enclosures protecting equipment producing arcs, sparks, or hot surfaces. So, IEC 60079–15 ed. 4 do not meet the requirements in UL 60079–15 ed. 4.

5.4.5.2   Applications

The most common applications for the encapsulation type of protection are solenoid valves, relays, and ballasts for fluorescent luminaires.

5.4.6   Ex o

It is a technique used to ensure the safety of electrical equipment in environments where there is a risk of the presence of flammable gases. In this method, the equipment is immersed in a tank containing insulating liquid, following the requirements of IEC 60079–6. The insulating liquid must provide a safe and reliable barrier, which not only electrically insulates the equipment but also helps to dissipate heat and to break electrical arcs.

Ex equipment and Ex components of type of protection liquid immersion "o" are either:

- Ex ob (EPL "Mb" or "Gb"); or
- Ex oc (EPL "Gc").

For Ex ob, IEC 60079–6 applies where the rated voltage does not exceed 11 kV rms ac or dc; for Ex oc, IEC 60079–6 applies where the rated voltage does not exceed 15 kV rms ac or dc.

5.4.6.1 Differences between IEC 60079–6 and UL 60079–6

In IEC 60079–6, it is required that switching devices protected by liquid immersion Protection Level "ob" be suitable for a prospective short-circuit current of 32 kA, unless marked with a lower value. UL 60079–6 has included a national difference to this requirement by adding a note stating that NEC limits the use of the increased safety termination to 10 kA short-circuit current.

And, considering that UL 60079–6 refers to the NEC for selection and installation of equipment, whereas IEC refers to IEC 60079–14, it is considered that the requirements in IEC 60079–6 do not meet the requirements in UL 60079–6.

5.4.6.2 Applications

Typical Ex o equipment includes transformers and switching devices. Ex o protection is widely used in environments such as refineries, offshore platforms, and petrochemical industries, where the presence of flammable gases requires strict safety measures to prevent possible ignitions.

5.4.7 Ex op

Optical radiation protection is a safety technique specified in IEC 60079–28, which also covers equipment located outside the explosive atmosphere or protected by an Ex type of protection but generates optical radiation that is intended to enter an explosive atmosphere. It covers Groups I, II, and III, and EPLs Ga, Gb, Gc, Da, Db, Dc, Ma, and Mb. In this method, the equipment is designed to prevent the risk of ignition due to optical radiation, as laser rays.

In principle, IEC 60079–28 discusses four potential ignition mechanisms:

- Optical radiation causes particles to heat up; they can, under certain conditions, attain a surface temperature that could ignite an explosive atmosphere.
- Thermal ignition of a gas volume because the optical wavelength matches an absorption band of the gas (a kind of resonance effect).
- Photochemical ignition due to photochemical dissociation of oxygen molecules by radiation in the ultraviolet range.
- Direct laser-induced breakdown of a gas at the focal point of a strong beam, producing plasma or a shock wave, both potentially acting as an ignition source.

There are three techniques of Ex op protection, as described next.

5.4.7.1 Ex op is

The Inherently Safe Optical Radiation (op is) is very similar to the "Ex I" electrical intrinsic safety concept. It is based on the idea of limiting the optical energy in a system—for example, in a fibre-optic cable, under normal operating conditions and under certain fault conditions. This limits the permissible optical radiated power for use in hazardous areas in Zone 1 and equipment Group IIB in Temperature Class T4 to a maximum of 35 mW.

5.4.7.2   Ex op pr

The Protected Optical Radiation (op pr) is designed with special enclosures to prevent ignition. Enclosures must be designed so that an explosion inside the enclosure cannot ignite the external atmosphere and so that no hazardous amount of light energy can reach the outside. As a result, they must not contain any inspection windows or similar features. This means that this method of protection largely corresponds to the "increased safety" and "flameproof enclosure" methods, both of which are electrical explosion protection methods.

5.4.7.3   Ex op sh

The Locking and Shutdown Principle (op sh) relies on broken fibres being detected immediately and the optical radiation being switched off safely as soon as this occurs. The protective principle underlying this method of protection is based on a risk assessment, and a recommended literature is the "functional safety" set of standards IEC 61508 and IEC 61511. Due to the extremely strict requirements for software and hardware, there are only a few products for this application on the market.

Of these three techniques above, only Ex op pr is valid to be installed in environments with combustible dusts; the other two can only be installed in environments with the eventual presence of an explosive atmosphere due to gases and vapours. However, equipment with this marking can be installed in a hazardous area due to combustible dusts, provided that it has another complementary protection technique, and it is certified for this type of installation.

5.4.7.4   Differences between IEC 60079–28 and UL 60079–28

UL 60079–28 references the National Electrical Code (NFPA 70), which is specific to the United States, while IEC 60079–28 follows IEC 60079–14 for installations in hazardous locations. Other differences include slight variations in testing procedures and in labelling, documentation, and certification requirements to meet U.S. regulatory standards. So, IEC 60079–28 does not meet UL 60079–28 requirements.

5.4.7.5   Applications

The "op is" type of protection is especially suited to industrial Ethernet installations, as fibre optics offer the same advantages as those already known from electrical intrinsic safety. The progress of digitalization is increasingly bringing Ethernet networking into the field and, therefore, into hazardous areas as well.

When bridging large distances in extended systems, in particular, and due to their excellent immunity, fibre optics are becoming an ever more popular choice for network technology. Depending on the optical fibre used, distances of 2 to 30 kilometres can be covered.

Examples of equipment that uses Ex op protection include luminaires, signalling devices, surveillance cameras, optical sensors and others using lasers, LED and similar components required in most systems for communication, monitoring, and measurement purposes.

5.4.8 Ex p

Pressurization is a type of protection established in IEC 60079–2, which involves the use of an enclosure maintained at a positive pressure (higher pressure) in relation to the external environment, using clean air or an inert gas, with the aim of preventing the entry of explosive atmospheres. It is relatively efficient, but the downside is the use of compressed air, which, although readily available, can come at a relatively high running cost.

Pressurized systems are better than some other protection methods for thermal management, particularly as larger enclosures can be used. An option is to use a vortex cooler, which is a small device using compressed air to create a spinning vortex within a cylindrical tube. This creates a rapid drop in pressure, causing a temperature drop which cools the enclosure.

The pressurization process of an Ex p enclosure has four stages:

- Purge the enclosure with clean compressed air to clear out any potential explosive gas or dust
- Pressurize the enclosure, then the outlet is closed
- Power on the enclosure
- Monitor and maintain the positive pressure.

Ex p equipment can be applied in Zones 1 and 2, and they are categorized into three types: px, py, and pz.

5.4.8.1 Type px

This type is used in environments where the hazardous areas are Zones 1. The enclosure is kept pressurized with decontaminated air or an inert gas, such as nitrogen, helium, or carbon dioxide, to prevent the entry of flammable gases or particles.

All equipment that interfaces with the classified area, i.e. buttons, switches, etc., must have the appropriate Ex protection.

Electrical components installed internally can be of industrial use, as this technique turns the internal pressurized environment into a non-classified area. An appropriate control system is used to maintain the internal pressure within the specified limits.

5.4.8.2 Type py

Used in classified areas Zones 1 and 2, the py enclosure is pressurized with air, or an inert gas, to prevent the entry of flammable gases. However, unlike the px type, this one does not "declassify" the pressurized environment; it only reduces its classification to Zone 2, and it is still necessary for all equipment internally installed to have at least EPL Gc.

5.4.8.3 Type pz

This type is used in classified areas Zone 2. The enclosure is pressurized with an inert gas to protect against gases penetration and makes the internal volume a non-classified area. Like the previous methods, all components that interface with the classified area, i.e. buttons, switches, etc. must be adequately Ex protected. An appropriate control system is employed to maintain the internal pressure within the specified limits.

5.4.8.4 Ex pD

The "pD" type of protection is a technique that consists of applying a protective gas to an enclosure in order to prevent the formation of an explosive atmosphere of combustible dust inside it. There are a series of requirements for this type of protection, from the capture of the protective gas to be supplied to the enclosure, which in case of air, must be done in a non-classified area, to the assembly of both the pressurization equipment and the inlet duct, which must prevent the entry or leakage of combustible dust into the system. The shielding gas exhaust is normally directed to a non-classified area.

This type of protection has the same precautions of the Ex p, regarding energization. For example, before connecting the power supply to the equipment for start-up or after shutdown, the operator must ensure that combustible dust cannot penetrate the enclosure or associated ducts in a concentration that poses a risk of ignition.

Although provided for in the standard, Ex pD in environments with combustible dust presents an additional risk when there is an air leak in some pipe connection. As the environment is subjected to combustible dust in layers near the air pipe, there is a possibility of a dust cloud forming with a concentration above its LEL, and measures must be taken to avoid ignition sources in the surroundings. It can be seen, therefore, that the Ex pD type, despite its simple design, requires adequate expertise from the designer and also specific training for the operators.

5.4.8.5 Differences between IEC 60079–2 and NFPA 496

To meet NFPA 496 requirements for purging and pressurization:

- Four complete air exchanges must be done prior to pressurization, while IEC 60079–2 requires five air exchanges prior to pressurization. The volume of air needed for replacement is based on the interior volume of the enclosure (Height × Width × Depth), the type of protective gas used, and the ambient conditions.
- An exterior label is provided, for the system specifying the exact time needed for purging prior to pressurization and powering equipment.
- Any penetrations, or leak areas, will have protective gas exiting the enclosure rather than hazardous gas, or dust, migrating into it. This segregates any external explosive or hazardous material from the energized internal equipment.
- NFPA496 requires a minimum pressure of 0.1 inches of water column (25 Pa) for Class I applications and 0.5 inches of water column (125 Pa), while IEC 60079-2 requires a minimum pressure of 50 Pa (0.2 inches of water column) for types px and py, and 25 Pa (0.1 inches of water column) for type pz.

IEC 60079–2 expects the purge air to be between –20°C and +40°C and offer limited guidance on how to handle lower extremes. Most purge systems are certified for use between –20°C and +55°C.

5.4.8.6 Differences between IEC 60079–2 and UL 60079–2

The IEC 60079–2 does not meet the requirements outlined in the UL 60079–2, in the areas of:

- Reference standards
- Marking pressurization systems

IEC 60079–2 exceeds the requirements outlined in the UL 60079–2, regarding the safety devices to detect overpressure.

5.4.8.7 Applications

Examples of equipment that use Ex p protection include electric panels that in normal operation have internal apparatus with sparks, electric arcs or hot surfaces, and large electric machines. In addition to being applied to electric panels, pressurization can be also applied to control rooms, process analyser containers, and other bulky constructions.

Ex p is also used to pressurize control and analyser rooms. IEC 60079–13 defines requirements for the design, construction, assessment, verification, and marking of pressurized rooms. In this method, the interior of the room is kept pressurized at a higher pressure than the external atmosphere, preventing the entry of flammable gases. This is achieved by continuously supplying air into the environment.

One advantage that this type of protection offers is lower manufacturing costs. However, on the other hand, it is necessary to check the costs of supplying pressurized air, which may include the construction of a considerable length of piping from the air compressor, which is generally installed outside the classified area.

5.4.9 Ex q

Powder filling protection, as established in IEC 60079–5, is a technique used to mitigate the risk of ignition of explosive atmospheres. This method consists of encapsulating the desired equipment in a casing filled with sand, or inert granular material, creating a physical barrier between the equipment and the surrounding explosive atmosphere, preventing the equipment from being able to ignite it.

The sand (quartz) used is non-conductive and heat-resistant, acting as an effective protection against sparks, heat, or hot surfaces that may arise inside the equipment. In addition, the sand also offers mechanical resistance, protecting the equipment against physical damage and abrasion. The sand filler must meet specific requirements with respect to grain size, purity, moisture content, and fracture resistance. A filler material other than quartz is permitted, if it meets the requirements.

IEC 60079–5 applies to electrical equipment, parts of electrical equipment, and Ex components with:

- A rated supply current less than or equal to 16 A
- A rated supply voltage less than or equal to 1,000 V
- A rated power consumption less than, or equal to, 1,000 W.

Electrical equipment with type of protection powder filling Ex qb can be used in hazardous areas Zones 1 and 2.

5.4.9.1   Differences between IEC 60079–5 and UL 60079–5

IEC 60079–5 meets the requirements outlined in the UL 60079–5 in the following subject areas: Degree of protection of the container, fuse (marking), power supply prospective short-circuit current (marking), and marking procedures.

IEC 60079–5 does not meet the requirements outlined in the UL 60079–5 in the following subjects:

- Reference standards
- Fuse (overload test requirements)
- UL 60079–5 requires that a flameproof "d" cable gland that complies with UL 2225 be provided for the powder-filled electrical equipment, and it also states that an increased safety "e" cable gland may not provide adequate pressure sealing to the powder-filled "q" enclosure.

5.4.9.2   Applications

Equipment such as ballasts and drivers for lighting fixtures are common examples of devices that can be designed with Ex q to be installed in areas where there is a risk of flammable atmospheres.

5.4.10   Ex т

This is a type of protection for explosive dust atmospheres, where electrical equipment is given with an enclosure that provides dust ingress protection and a means to limit surface temperatures. The entry of dust inside enclosures, where electrical energy may be released, is prevented by specially designed enclosures.

IEC 60079–31 sets out the requirements for this type of protection. However, it does not apply to Ex equipment or Ex components intended for use in the underground sections of mines or in the surface sections of such mines that are at risk of firedamp and/or combustible dust.

The inside of the enclosure is typically considered a non-classified region, but the configuration of equipment inside is carefully done to ensure that the external surface temperature does not exceed the Temperature Class.

Ex equipment with the Ex t type of protection has the following EPL:

- Ex ta (EPL "Da"), for use in Zones 20, 21, and 22
- Ex tb (EPL "Db"), for use in Zones 21 and 22
- Ex tc (EPL "Dc"), for use in Zone 22.

Ex ta equipment shall be rated for connection to a circuit with a prospective short-circuit current of not greater than 1.5 kA. Ex tb or Ex tc equipment which is intended for mains connection, and intended to interrupt fault current above 10 kA, shall have a rated maximum short circuit withstand current defined in IEC 60079–31 and be adequately marked. For Ex ta equipment which contains a cell or battery, only a sealed cell or battery shall be used. For Ex tb and Ex tc equipment, where there are sparking contacts or hot surfaces, and also which contain a cell or battery, only a sealed cell or battery shall be used.

5.4.10.1   Differences between IEC 60079–31 and UL 1203

There are three Dust Groups (E, F, and G) in North America. Dust Group E consists of combustible metal dusts; Dust Group F consists of combustible dust other than combustible metal dust; and Dust Group G consists of solid particles, including fibres.

The most common protection technique used in Class II, Division 1 hazardous locations is "dust ignition proof", based on UL 1203 or FM 3616. Please note that the requirements for dust ignition, dust-tight protection, and use in an underground atmosphere (e.g. mines), i.e. Classes II and III and Groups I and II, are similar to those in IEC 60079–31.

5.4.10.2   Applications

Ex t can be used for inverter-controlled motors, provided that they have been tested together for the requested duty and with a protective device or using direct temperature protection with embedded temperature sensors, giving sufficient margin to protect bearings and the rotor. They can be certified for use in locations under the presence of combustible dust either in suspension or in layers up to 5 mm thick.

The action of the protective device shall cause the motor to be disconnected. Only external surface temperature needs to be considered for Temperature Class, but no hot surfaces outside the enclosure are allowed to appear in both rated and fault conditions.

5.4.11   Ex s

Special protection does not mean an Ex type of protection but a design concept that allowed a given equipment to be considered safe when used in a given hazardous atmosphere, but does not conform to any of the other recognized techniques of protection. The requirements for equipment to be recognized with this type of protection are in IEC 60079–33, that is applied to:

- Electrical equipment employing a method of protection not covered by any existing standard in the IEC 60079 series
- Electrical equipment employing one or more recognized types of protection where the design and construction are not fully compliant with the standard for the type of protection
- Electrical equipment where the intended use is outside the parameters of the scope of the standard for the type of protection.

To know where the equipment can be installed, and if special conditions apply, it is imperative to read the Ex certificate. Although some equipment were certified as Ex s, nowadays, it is more common to find equipment certified with combined types of protection.

5.4.11.1 Differences between IEC 60079–33 and UL 60079–33

In UL 60079–33, the references to ISO and IEC standards were changed to refer American standards.

5.4.11.2 Applications

New developments like flameproof enclosures with explosion vents and pressure reduction grid plates in the side wall.

Other examples of equipment that were labelled as Ex s in the market are:

- A factory-sealed fluorescent hand lamp with flexible cable
- A potted solenoid for valve operation, complete with cable.

5.4.12 Ex v

Artificially ventilated environment is also a safety technique for classified areas. In this case, the environment is kept safe through constant artificial ventilation to dilute a release of flammable substance, that the required EPL inside is reduced from either Gb or Gc to non-hazardous or from Gb to Gc. This technique is often used in locations where pressurization is not feasible or practical.

Rooms with artificial ventilation "v" and located in a hazardous area shall have:

- Ex vc (EPL Gc): The ventilated room maintains artificial ventilation to dilute a release of flammable substance to reduce the internal classified area.

Relevant standard is IEC 60079–13, and this technique is not allowable for ventilation safety devices.

5.4.12.1 Differences between IEC 60079–13 and UL 60079–13

The UL 60079–13 does not permit the use of type of protection "v" for a room containing an internal source of release, if this same room has already been classified using mechanical or artificial ventilation in accordance with API RP 505, NFPA 497, or ISA 60079–10–1 as applicable. This is because such a use would rely upon the same concept for two independent safety functions.

However, for a larger room containing an internal source of release that has been classified using mechanical or artificial ventilation in accordance with API RP-505, NFPA 497, or ISA 60079–10–1 as applicable, another smaller room containing its own internal source of release may be included within the larger room that does utilize Type of Protection "v" (referred to as "a room within a room").

5.4.12.2 Applications

Examples of Ex v environments are battery rooms, process units' power substations, and extraction hoods where adequate ventilation is required to prevent the build-up of flammable atmospheres.

5.4.13 COMBINED EX TYPES

Combined types of protection seek the best cost-benefit ratio for installation in classified areas. For example, oxygen analysers are usually built by placing an intrinsically safe circuit inside an explosion-proof enclosure.

Although such a construction may, at first glance, be considered oversized, on certain occasions, the customer's specification requires such a level of safety; on others, it is necessary to take into account resistance to corrosive atmospheres, or it may even be the case that the manufacturer has used an auxiliary power supply that is not certified for isolated installation in a Zone 1 area.

5.4.13.1 Applications

There are many examples of equipment with combined Ex types of protection such as:

- Emergency luminaire certified as: Ex tb op is IIIC T135°C Db IP67
- Scale with optical data output capability as: Ex ia op is IIB T3 Ga.

5.5 MECHANICAL EQUIPMENT

Under ATEX 2014/34/EU, the majority of non-electrical equipment is covered by self-declaration of conformity by the equipment manufacturer. As a legal requirement, the manufacturer will typically need to lodge a sealed technical file with a notified body, which is retained for a period of 10 years.

Considering that there are mechanical devices that could become an ignition source, even when functioning as designed, due to hot surfaces, mechanical sparks (e.g. grinding and finishing, arc welding), and static electricity (e.g. fabric rubbing against plastics), the standards ISO 80079–36 and ISO 80079–37 were issued to specify basic methods and requirements for design, construction, testing, and marking of non-electrical Ex equipment and Ex components, under EPL Gb and Db. In the certification process, the equipment is subjected to a formal, documented ignition hazard assessment in order to identify all potential ignition sources that can occur.

Based on the aforementioned standards, Ex conformity certificates for mechanical equipment can be issued, generally on a voluntary basis, by an appropriate Accredited Certifying Body using the symbol "h". The protection techniques applied to Ex h are:

- Liquid immersion "k" is a protection technique where potential ignition sources are made ineffective, or separated, from the explosive atmosphere by either totally immersing them in a protective liquid, or by partially immersing and continuously coating their active surfaces with a protective liquid in such a way that an explosive atmosphere, which may be above the liquid or outside the equipment enclosure, cannot be ignited. Application example: Heat exchangers

- *Constructional safety* "c" is a protection technique where constructional measures are applied so as to protect against the possibility of ignition from hot surfaces, sparks, and adiabatic compression generated by moving parts. Application example: Robust and well-adjusted mechanical parts
- *Control of ignition source* "b" is a protection technique where mechanical or electrical devices are used in conjunction with non-electrical equipment to manually or automatically reduce the likelihood of a potential ignition source from becoming an effective ignition source. Application example: Pneumatic tools

Types of protection "c", "b", and "k" are not being marked on the product, because of possible user misinterpretation with EPL "c" and "b". Manufacturers indicate "Ex h" only and define the specific used method(s) in the user installation instructions.

5.5.1 APPLICATIONS

Examples of non-electrical equipment that may be used in hazardous areas could include mechanical disc brakes, mechanical couplings, mechanical pumps, fluid couplings, oil-filled gearboxes, oil-filled disc brakes, hydraulic and pneumatic motors, and any combination of them, which can be used as a machine, like fans and compressor assemblies.

5.6 OVERVIEW OF EX TYPES

The selection of the most appropriate type of protection for an installation must be the result of a careful analysis by the engineering team, combining their experience with the latest technology available to provide the best solution in terms of functionality, economy, and safety to the end user.

Table 5.15 gives the comprehensive scenario for Ex equipment related to Zones.

5.7 THE SELECTION PROCESS

The steps for the Ex equipment selection and purchasing process are:

1) Consult the area classification documents.
2) Obtain the Group and Temperature Class of the region.
3) Choose the type of Ex protection.
4) Select the appropriate EPL.
5) Check that the equipment chosen is certified; if it is not, choose another that is.
6) Demand that the installation and maintenance manuals be delivered together with the equipment purchased.

TABLE 5.15
Application of the Types of Protection Related to Zones

Type of ProtectiLon	Zones					
	0	1	2	20	21	22
Ex d—flameproof	Da	da db	da db dc			
Ex e—increased safety		eb	eb ec			
Ex i—intrinsic safety	Ia	ia ib	ia ib ic	ia	ia ib	ia ib ic
Ex m—encapsulation	ma	ma mb	ma mb mc	ma	ma mb	ma mb mc
Ex nC, Ex nR—non-incendive			nC nR			
Ex o—liquid immersion		ob	oc			
Ex op—optical protection	op_{is}	$op_{is}\, op_{pr}\, op_{sh}$	$op_{is}\, op_{pr}\, op_{sh}$	op_{is}	$op_{is}\, op_{pr}\, op_{sh}$	$op_{is}\, op_{pr}\, op_{sh}$
Ex p—pressurized		pxb pyb	pxb pyb pzc		pxb	pxb pzc
Ex q—powder filling		q	q			
Ex s—special	sa	sb	sc	sa	sb	sc
Ex t—protection by enclosure				ta	tb	tc

5.8 PURCHASING EX EQUIPMENT

The purchase of Ex equipment is not limited to the assessment of its technical characteristics, as there are legal requirements also that must be met. Each country defines its own legal requirements, which can range from a self-declaration of compliance with health and safety regulations to the issuance of a certificate of conformity by its accredited bodies.

5.8.1 IN THE EU

For example, countries in the European Union follow the ATEX Directive, which requires that electrical and electronic equipment meet basic health and safety requirements and that assessments in the certification process be carried out according to CENELEC standards, with the certificate issued by a notified body. In turn, the IECEx certification scheme carries out product assessments only according to IEC standards, without considering health and safety items, and their certificates are issued only by their accredited entities members, which may or may not be ATEX-notified bodies.

5.8.2 IN THE UNITED STATES

Regarding electrical installations in hazardous locations, in the United States, the government agency for safety and health also issues safety requirements for industrial installations, referencing the NEC. For example, for an expansion project of an industrial unit originally built according to the American standard, new electrical Ex equipment with IEC certification, although with similar functionality, may be prevented from being used if such expansion has an area classification study according to the concept of Classes and Divisions.

This is because the NEC establishes that Ex equipment with IEC marking can only be installed in facilities whose area classification study has been carried out according to the concept of Zones, as defined in its Article 505. The NEC provides that equipment marked in accordance with Article 505 for use in Class I, Zone 0, 1, or 2 is permitted in Class I, Division 2 locations for the same Gas Group, and with an appropriate Temperature Class. Equipment marked in accordance with Article 505 for use in Class I, Zone 0 locations is permitted in Class I, Division 1 and Division 2 for the same Gas Subdivision and with an appropriate Temperature Class.

Another detail that may go unnoticed is that in the United States, Ex equipment must also be evaluated for non-Ex electrical safety requirements. A simple manufacturer's declaration stating that the Ex equipment meets the general safety requirements (as permitted by ATEX 114 in the EU) is not accepted. And, as Ex equipment certified by IECEx are evaluated only by the requirements in IEC standards, excluding legal or other standards' requirements, they are not directly accepted in the United States.

In the United States, in order to get a North American Ex certificate, electrical equipment must comply with:

- General safety requirements for that type of equipment, also known as "ordinary location requirements", that encompass risk of electrical shock, fire hazards, and risk of injuries.
- The unique standards that apply to Ex products depending on class/division or Zone markings or both.

Different types of electrical equipment have also to follow the relevant standards for non-Ex applications. In some cases, testing for compliance with the non-Ex requirements, and also with the hazardous area requirements, may be conducted at the same time. So, buyers of Ex equipment need adequate training, as there is a risk of purchasing Ex equipment that, despite having similar functional characteristics, may be rejected by the end customer's inspection or even by inspectors from government agencies.

5.9   CLOSING REMARKS

The designer of the electrical installations for hazardous areas has many types of protection to choose, and knowing the characteristics of the industrial plant and the Ex equipment, he is able to make the selection that best suits the plant needs. The acquisition of the Ex equipment is not limited to their functional features, as there are legal requirements to be met, especially regarding the certification of conformity. In the case of imported equipment, greater care must be taken, as each country may establish special requirements, depending on whether trade reciprocity agreements have been established with the country where the equipment was manufactured.

It should be noted that the IECEx scheme is privately owned and does not have the capacity to replace trade agreements between countries that are, therefore, hierarchically superior.

On the other side, solely declaring a product as being in conformity with the ATEX Equipment Directive 2014/34/EU raises a question if the relevant requirements were fulfilled. Directive 2014/34/EU put forward obligations to be fulfilled according to the product (equipment, protective system, or device), Group, and Category, and the specific compliance must appear on the Ex-marking and also in the EU Declaration of Conformity.

In Brasil, there is a conformity certification system for Ex equipment, whose tests follow those stipulated by IEC standards. However, the certificate of conformity must be issued by an organization accredited by INMETRO, even if the product already has an ATEX or IECEx certificate. Therefore, we highlight that it is important to know the requirements of the country where the project is being developed in order to avoid delays in the delivery time or complications with local authorities.

BIBLIOGRAPHY

ANSI/ISA 60079-2—*Explosive atmospheres—Part 2: Equipment protection by pressurized enclosure "p"*. International Society of Automation, 2021.

ANSI/ISA 60079-7—*Explosive atmospheres—Part 7: Equipment protection by increased safety"e"*. International Society of Automation, 2017.

ANSI/ISA 60079-31—*Explosive atmospheres—Part 31: Equipment dust ignition protection by enclosure*. International Society of Automation, 2024.

ATEX 2014/34/EU—Directive of the European Parliament and of the Council on the harmonisation of the laws of the Member States relating to equipment and protective systems intended for use in potentially explosive atmospheres. 2014.

EN 61558-2- 6—*Safety of transformers, reactors, power supply units and similar products for supply voltages up to 1 100 V—Part 2–6: Particular requirements and tests for safety isolating transformers and power supply units incorporating safety isolating transformers*. CENELEC, 2010.

Explosionproof vs flameproof: Understanding the differences between ANSI/UL 1203 and IEC 60079-1. Intertek, White paper. 2020.

FM 3600—*Electrical equipment for use in hazardous (Classified) locations—General requirements*. FM Approvals, 2022.

FM 3600—*Examination standard for electrical equipment for use in hazardous (classified) locations—General requirements*. FM Approvals, 2022.

IEC 60079-0 – *Explosive atmospheres—Part 0: Equipment—General requirements*. International Electrotechnical Commission, 2017.

IEC 60079-1 – *Explosive atmospheres—Part 1: Equipment protection by flameproof enclosures "d"*. International Electrotechnical Commission, 2014.

IEC 60079-5 – *Explosive atmospheres—Part 5: Equipment protection by powder filling "q"*. International Electrotechnical Commission, 2015.

IEC 60079-6 – *Explosive atmospheres—Part 6: Equipment protection by liquid immersion "o"*. International Electrotechnical Commission, 2015.

IEC 60079-7 – *Explosive atmospheres—Part 7: Equipment protection by increased safety "e"*. International Electrotechnical Commission, 2015.

IEC 60079-11 – *Explosive atmospheres—Part 11: Equipment protection by intrinsic safety "i"*. International Electrotechnical Commission, 2023.

IEC 60079-13 – *Explosive atmospheres—Part 13: Equipment protection by pressurized room "p" and artificially ventilated room "v"*. International Electrotechnical Commission, 2017.

IEC 60079-14 – *Explosive atmospheres—Part 14: Electrical installation design, selection and installation of equipment, including initial inspection*. International Electrotechnical Commission, 2024.

IEC 60079-15 – *Explosive atmospheres—Part 15: Equipment protection by type of protection "n"*. International Electrotechnical Commission, 2017.

IEC 60079-18—*Explosive atmospheres—Part 18: Equipment protection by encapsulation "m"*. International Electrotechnical Commission, 2014.

IEC 60079-25—*Explosive atmospheres—Part 25: Intrinsically safe electrical systems*. International Electrotechnical Commission, 2020.

IEC 60079-28 – *Explosive atmospheres—Part 28: Protection of equipment and transmission systems using optical radiation*. International Electrotechnical Commission, 2015.

IEC 60079-31 – *Explosive atmospheres—Part 31: Equipment dust ignition protection by enclosure "t"*. International Electrotechnical Commission, 2022.

IEC 60079-33 – *Explosive atmospheres—Part 33: Equipment protection by special protection "s"*. International Electrotechnical Commission, 2012.

IEC 60364-1—*Low-voltage electrical installations: Fundamental principles, assessment of general characteristics, definitions*. International Electrotechnical Commission.

IEC 60529—*Degrees of protection provided by enclosures (IP code)*. International Electrotechnical Commission, 1989, COR1: 2019.

IEC 61140—*Protection against electric shock—Common aspects for installation and equipment*. International Electrotechnical Commission, 2016.

IEC 61508 – *SER – Functional safety of electrical/electronic/programmable electronic safety-related systems—All parts*. International Electrotechnical Commission, 2010.

IEC 61511—*SER – Functional safety—Safety instrumented systems for the process industry sector—All parts*. International Electrotechnical Commission, 2024.

IEC 63365—*Industrial process measurement, control and automation—Digital nameplate*. Ed. 1.0. International Electrotechnical Commission, 2022.

ISO 20653—*Road vehicles—Degrees of protection (IP code)—Protection of electrical equipment against foreign objects, water and access*. International Organization for Standardization, 2023.

ISO 80079-36—*Explosive atmospheres—Part 36: Non-electrical equipment for explosive atmospheres—Basic method and requirements*. International Organization for Standardization, 2016.

ISO 80079-37 – *Explosive atmospheres—Part 37: Non-electrical equipment for explosive atmospheres—Non electrical type of protection constructional safety "c", control of ignition source "b", liquid immersion "k"*. International Organization for Standardization, 2016.

ISO/IEC 80079-20-1—*Explosive atmospheres—Part 20–1: Material characteristics for gas and vapour classification—Test methods and data*. International Organization for Standardization, 2017.

NEMA 250—*Enclosures for electrical equipment (1000 Volts maximum)*. National Electrical Manufacturers Association, 2018.

NFPA 70—*National Electrical Code (NEC)*. National Fire Protection Association, 2023.

NFPA 496—*Standard for purged and pressurized enclosures for electrical equipment*. National Fire Protection Association, 2024.

Portaria INMETRO/ME 115—*Aprova os Requisitos de Avaliação da Conformidade para Equipamentos Elétricos para Atmosferas Explosivas—Consolidado*. Brasil. Mar. 2022. Available at: https://lnkd.in/gcgW9biM

Rangel Jr., Estellito—Brasil moves from Divisions to Zones. In: *49th Petroleum and Chemical Industry Committee Technical Conference*. New Orleans, US, 2002. Available at: https://ieeexplore.ieee.org/document/1044981

Rangel Jr., Estellito—Comprando equipamento Ex. *Section EMEx, Eletricidade Moderna Magazine*, São Paulo, ed. 476, Nov. 2013, p. 132.

Rangel Jr., Estellito—Ex ic. *Section EMEx, Eletricidade Moderna Magazine*, São Paulo, ed. 524, Nov. 2017, p. 58.

Rangel Jr., Estellito—Marcação de equipamento Ex. *Section EMEx, Eletricidade Moderna Magazine,* São Paulo, ed. 439, Oct. 2010, p. 192.

Rangel Jr., Estellito—ISO 80079-36 e ISO 80079-37. *Section EMEx, Eletricidade Moderna Magazine,* São Paulo, ed. 552, ano 48, Mar/Apr. 2020, p. 58.

Rangel Jr., Estellito—Nova IEC 60079-0. *Section EMEx, Eletricidade Moderna Magazine*, São Paulo, ed. 527, Feb. 2018, p. 49.

Rangel Jr., Estellito—Nova IEC 60079-0—Parte II. *Section EMEx, Eletricidade Moderna Magazine*, São Paulo, ed. 533, Aug. 2018, p. 59.

Rangel Jr., Estellito—O tipo Ex d. *Section EMEx, Eletricidade Moderna Magazine*, São Paulo, ed. 500, Nov. 2015, p. 76.

Rangel Jr., Estellito—Pressurização Ex pD. *Section EMEx, Eletricidade Moderna Magazine*, São Paulo, ed. 499, Oct. 2015, p. 86.

Rangel Jr., Estellito—Segurança aumentada. *Section EMEx, Eletricidade Moderna Magazine*, São Paulo, ed. 545, ano 47, Aug. 2019, p. 62.

Rangel Jr., Estellito—Seleção de tipos Ex. *Section EMEx, Eletricidade Moderna Magazine*, São Paulo, ed. 378, Sep. 2005, p. 156.

UL 1203—*Explosionproof and dust-ignition proof electrical equipment for use in hazardous (Classified) locations,* Underwriters Laboratories. 2023.

UL 2225—*Cables and cable fittings for use in hazardous (classified) locations.* Underwriters Laboratories, 2024.

UL 60079-13—*Explosive atmospheres—Part 13: Equipment protection by pressurized room.* Underwriters Laboratories, 2022.

UL 60079-33—*Explosive atmospheres—Part 33: Equipment protection by special protection "s".* Underwriters Laboratories, 2021.

6 Dangerous Sparking

6.1 INTRODUCTION

Industries that handle flammable products, liquids or powders, are at risk of fires and explosions. The presence of explosive atmospheres in these industries can result in devastating accidents if an ignition source is present.

Ignition sources are varied and can arise from various industrial operations. So, it is essential to identify the ignition sources that are present—or that may be present—on site.

The most easily identifiable ignition sources are possible hot spots and sparking electrical equipment. However, there is a considerable list of devices that can ignite an explosive atmosphere, some of which often escape the analysis of the installation designer. Let us examine them.

6.2 SOURCES OF IGNITION

The IEC 60079–0 mentions 13 types of ignition sources, and it is important to know them and where they may be present in order to ensure that the safety procedures mention the required attention on them.

6.2.1 HOT SURFACES

These ignition sources arise from heat losses in the production system from machines belonging to industrial systems, even in normal operation. Typical examples are process heaters—gas or electric—and heat exchangers, but it can also happen with motors and pumps, when overcharged and reaching temperatures sufficient to ignite flammable vapours.

These temperatures can generally be controlled, but it is important to analyse the consequences in the event of an abnormality in operation such as an overload on the bearings of large machines. The maximum temperatures that the substances involved in the installation will be subjected to, will be analysed and compared with their respective minimum auto-ignition temperatures.

The safety recommendation for carrying out the analysis will be that, even in the event of rare malfunctions, the temperature of all surfaces of the equipment, protection systems, and components that may be in contact with explosive atmospheres must not exceed 80% of the Minimum Ignition Temperature (MIT) of the flammable gas or liquid. As has been done for flammable gases' atmospheres, a reduction factor is determined for the minimum temperature at which the substances involved contribute to auto ignition, which in the case of dusty atmospheres will be TMI_c and TMI_L, as shown in Table 6.1.

DOI: 10.1201/9781003500001-6

TABLE 6.1
Safe Surface Temperatures in Combustible
Dust Explosive Atmospheres

Situation	Max. Surface Temperature
Cloud	$T \leq TMI_{cloud} \cdot 2/3$
Layer	$T \leq TMI_{layer} - 75°C$

Where:

  T   is the maximum operating temperature of the equipment or installation, which has a hot surface as a potential ignition source. In the hot spot check, this temperature must satisfy the conditions set out in Table 6.1, considering all surfaces that may come into contact with the dust cloud or layer.

  TMI_{cloud} is the Minimum Ignition Temperature from which the dust cloud begins to burn, with no ignition source, other than the temperature itself.

  TMI_{layer} is the Minimum Ignition Temperature from which the dust layer begins to burn, with no ignition source, other than the temperature itself. The dust layer can burn without the apparent presence of a flame; in this case, the temperature of the dust layer will rise above the temperature of the hot surface.

6.2.1.1   Fire Case

On June 1, 1974, occurred an explosion at Flixborough Works of Nypro Ltd., North Lincolnshire, England. A pipeline containing hexane, a highly flammable organic compound, ruptured at the nylon plant. The sudden release of large quantities of hexane created a flammable vapour cloud that spread rapidly throughout the area.

The investigation revealed that the pipeline rupture was caused by a number of factors, including unauthorized design changes, as the pipeline had been modified without proper authorization, which compromised its integrity. The source of ignition was probably a natural-gas-reforming furnace some distance away. It was estimated that 30–50 tonnes of cyclohexane escaped in the 50 seconds that elapsed before ignition occurred.

The explosion killed 28 people, injured 36 others, and destroyed much of the plant. In addition, the incident had a significant impact on the local community and the chemical industry in general.

6.2.2   Mechanically Generated Sparks

In order to assess the risk of ignition of explosive atmospheres that may be generated by the mechanical equipment involved in the operation of the installation, a checklist of critical points, including the characteristic of each type of equipment, is to be created.

Cutting equipment can generate sparks in certain situations expected under normal operating conditions, and therefore their installation in classified areas should be avoided. Mechanical failures, such as the breakage of parts due to excessive friction (e.g. due to a lubrication failure), should also be analysed.

A spark protection device is necessary for the mechanical equipment located in classified areas. It should be noted that, although most of the classified areas are Zone 2, any mechanical equipment containing Zone 0 or Zone 1 inside also needs to be analysed. This is because the configuration of the equipment itself may produce sparks of mechanical origin; agitators, for example, are a prime example of this.

6.2.2.1 Fire Case

On June 8, 1998, occurred an explosion at the DeBruce Grain elevator, described in the Guinness Book of Records as the world's largest grain silo, located 6 km southwest of Wichita, Kansas, USA, with a storage capacity of 729,468 m³ of grain (equivalent to 12 million of 60 kg bags of wheat). The explosion resulted in 7 dead and 10 injured.

The most likely source of ignition was created when a concentrator bearing seized due to lack of lubrication and caused it to lock, but the conveyor belt continued to move. This "blade sharpening" friction effect on the bearing raised its temperature to 260°C, well beyond the temperature of 220°C required to ignite the layer of dust, which was in abundance inside the bearing.

The investigation found a safety meeting minutes held on January 29, 1998, that reported a fire, but it was controlled, caused by the heating of a bearing that caused the fire and required replacement. This confirmation of heating of bearings was important because it confirmed the company's recognition of the link between the heating of bearings and the ignition of dust.

6.2.3 LIGHTNING

Lightning has enough energy to ignite an explosive atmosphere, and for this reason, the design of the atmospheric discharge protection system needs to follow special recommendations, especially in petrochemical and oil and gas plants. More details on this source of ignition will be discussed in Chapter 9.

6.2.3.1 Fire Case

A fire at a naphtha tank in Calcasieu Refining Co. in Lake Charles, Louisiana, on June 3, 2023, occurred. A lightning strike ignited the tank, leading to significant fire at the refinery. As a precaution, an evacuation was ordered for people within a 2 km radius of the facility. Firefighters managed to put out the fire by 4 am the next day, and the evacuation order was lifted shortly after.

6.2.4 STATIC ELECTRICITY

Regardless of the existence of voltage in the electrical system, sparks can occur due to static electricity. The accumulated energy can be released in the form of a spark and thus ignite an explosive atmosphere.

The accumulation of electrostatic charges can occur during the pumping, mixing, and transfer of flammable liquids. When a built-up charge is released, it can create a spark sufficient to ignite flammable vapours present.

In addition, when dealing with oil, the danger increases. Irregular actions such as cleaning the inside of the tank of a crude oil tanker may cause an electrostatic discharge, which may cause an explosion.

In the medical field, treatment using an oxygen tank or oxygen tent can lead to ignition due to electrostatic discharge caused by charging of clothes. Under normal operating conditions, discharges can occur in the following ways:

- Spark discharges: These can occur due to the charging of conductive and insulated parts (not connected to earth). To prevent this, all plant equipment need an electrical continuity to earth.
- Triboelectric discharges: They will be a source of ignition for almost all explosive atmospheres of gases and vapours, as well as for highly combustible dusts, and can occur in charged parts of non-conductive material, including most plastic materials. They can also occur in rapid separation processes (moving films on rollers, transmission belts, etc.).

It is important to note that such energy may be accumulated in non-electrical equipment on site as well, which requires an accurate assessment.

6.2.4.1   Fire Case

On November 20, 2017, occurred an explosion of the vapour of flammable liquids caused by static electricity at the Verla International, a plant in New Windsor, New York, the United States, which manufactured nail varnishes, creams, body lotions, and perfumes. The source of ignition was electrostatic discharge which originated at the moment a staff member wiped the IBC container with the flammable liquid. The container soon burst into flames, which became a reason for a serious fire.

The tragic event was registered by CCTV cameras. A man was killed, 125 were injured, and the company faced US$ 281,220 in proposed fines. The list of violations included failing to ensure proper electrical earthing and bonding to prevent flammable vapours from igniting, failure to develop and implement an emergency response plan, and failure to provide employees with first responder awareness level training. More details on this source of ignition will be discussed in Chapter 8.

6.2.5   Cathodic Protection

Cathodic protection systems promote the injection of low-intensity electric currents, which, depending on the configuration and maintenance of the electrical system, may result in potential differences between different earthing points. Cathodic corrosion protection is widely used in various industries to prevent metal structures from corroding, especially in the following industries that have hazardous areas:

- Oil and Gas Industry: Used to protect fuel pipelines, steel storage tanks, offshore platforms, and oil well casings
- Marine Industry: Applied to steel piles, piers, jetties, ship hulls, and offshore platforms.

To avoid cathodic protection systems becoming a source of ignition in explosive atmospheres, consider the following precautions:

- Ensure all components are properly earthed and bonded to prevent static electricity build up.
- Conduct regular maintenance and inspections to detect and mitigate any potential faults or failures.
- Continuously monitor the environment for explosive gases or vapours and take appropriate action if detected.

6.2.5.1 Fire Case

On January 2018, a 22-inch pipeline of Gasoducto Norte, Buenos Aires Province, Argentina, experienced an in-service rupture due to hydrogen-assisted cracking induced by the cathodic protection system. The malfunction resulted in an axial crack, which eventually caused a fire. Root cause analysis revealed that operating the cathodic protection system near the pipe-to-soil potential limit led to hydrogen generation and subsequent cracking.

6.2.6 RADIANT ENERGIES

This category includes ultrasonic systems and ionizing radiation. Equipment or systems that use these energies must be certified so that they can operate safely in hazardous areas.

Ionizing radiation can be produced, for example, by X-rays and radioactive substances. The possible presence of this type of ignition source needs to be assessed, because such systems are only used under controlled conditions (special work permits) and without the presence of products that may form explosive atmospheres.

6.2.6.1 Fire Case

There have been no documented cases of industrial plant fires directly caused by ionizing radiation. Ionizing radiation can cause damage to materials and potentially lead to fires in specific scenarios such as nuclear power plants.

However, it is important to note that ionizing radiation can indirectly contribute to fire hazards in industrial settings. For instance radiation exposure can degrade materials, making them more susceptible to ignition. Additionally, radiation accidents, while being rare, can lead to fires if they involve combustible materials or energy sources.

6.2.7 ULTRASOUND

The use of ultrasound waves means that a large amount of energy emitted by the electroacoustic emitter is absorbed by the solid and liquid substances that encounter the waves. In this way, the substance heats up and can represent an effective ignition source. Ultrasonic systems are commonly used in industrial settings for various applications such as cleaning and welding.

6.2.7.1   Fire Case

There have been no documented cases of industrial plant fires caused by ultrasound waves. While research has shown that under specific conditions, ultrasound can potentially ignite dust–air mixtures, this requires very specific circumstances and high-intensity ultrasound.

6.2.8   Adiabatic compression

Depending on the Pressure Ratio (PR), adiabatic or near-adiabatic compression can produce temperatures high enough to cause ignition. Shock waves are produced during the sudden discharge of high-pressure gases in pipelines; these propagate at speeds greater than that of sound, and very high temperatures can be reached in pipe bends, constrictions, connection flanges, closed valves, etc.

The recommendation to prevent this is to verify the design that the discharges are carried out without sudden changes in pressure, and, in addition, the installations through which gases circulate have been designed in such a way as to avoid the effect of bends, constrictions, etc.

Adiabatic compression is also a known risk in oxygen plants and other facilities handling compressed gases. Rapid compression of gases can generate significant heat, potentially leading to fires if not properly managed.

6.2.8.1   Fire Case

One example of an industrial plant fire caused by the sudden discharge of high-pressure gases is the explosion at the Chorzow Chemical Plant Azoty in Poland in the 1980s. This incident occurred when a synthesis gas mixture of hydrogen and nitrogen was released from a high-pressure vessel, leading to an explosion and subsequent fire. The accident resulted in fatalities and significant damage to the plant.

6.2.9   Chemical Reaction

Exothermic reactions can act as a source of ignition when the rate of heat release is greater than the rate of heat evacuation to the outside. As prevention, it is important to know the nature of the substances, the control and operating conditions, and the technical measures applied to the operations in order to identify if a chemical reaction can be expected that could represent an effective source of ignition.

6.2.9.1   Fire Case

The fire at the Bio-Lab Inc. facility in Conyers, Georgia, on September 29, 2024, was caused by a chemical reaction involving trichloroisocyanuric acid (TCCA) and sodium dichloroisocyanurate (DCCA), along with bromochloro-5,5-dimethylimidazolidine-2,4-dione (BCDMH). There were no casualties reported during the fire at the Bio-Lab, Inc. facility in Conyers; however, the incident did lead to significant offsite impacts. Approximately 17,000 people in the surrounding community were evacuated, and nearly 90,000 people in metropolitan Atlanta were advised to shelter-in-place. The fire caused a massive plume of potentially toxic smoke, which prompted these safety measures.

6.2.10 Electromagnetic Radiation

Radiation in this spectrum range (waves of 3 × 1011 to 3 × 1015 Hz) can constitute a source of ignition through absorption by explosive atmospheres or solid surfaces.

It is necessary to check if the equipment and installations of the unit analysed do not have a magnetic field strong enough to act as an ignition source. Despite this, the following sources of electromagnetic waves in the range of 3×10^{11} to 3×10^{15} Hz should be verified:

1) **Solar radiation**

    It is prevented from becoming an effective ignition source by eliminating the objects that allow the convergence of radiation: Windows with curved surfaces, light reflectors, etc.

2) **Intense light sources**

    Facilities lit using fluorescent luminaires, and where the necessary means of light diffusion are provided, are usually out of this risk. In industries with combustible dusts, and using LED luminaires, it is necessary to verify if the certificate of conformity confirms that they were evaluated under the requirements of IEC 60079–28, which were written in order to avoid the concentration of radiation so that the energy emitted by them does not represent an effective source of emission.

3) **Laser radiation**

    Specific activities that require the use of this equipment (e.g. the use of remote measuring devices) will need to be conducted under an appropriate work permit.

6.2.10.1 Fire Case

One example of an industrial plant fire caused by solar radiation is the fire at the Ivanpah Solar Electric Generating System, the largest solar plant, in California, which occurred on May 19, 2016. The fire was caused by a fault in the photovoltaic (PV) system, which led to the burning of electrical cables. This incident highlights the potential risks associated with solar radiation in solar power plants.

6.2.11 Electromagnetic Fields

Electromagnetic waves within the spectrum between 10^4 and 3×10^{12} Hz are used in high frequency systems, also called radio frequency (RF) systems. These systems emit waves, and all conductive parts located in the electromagnetic field behave as receiving antennas, so that they can constitute an effective ignition source if they meet any of the following conditions:

- The field is sufficiently powerful.
- The distance between the receiver and the transmitter is sufficiently small.
- The size of the antenna is sufficiently large.

In industrial plants, the presence of this type of radiation will occur with the use of radio transmitters or RF generators for heating, drying, hardening, welding, or oxycutting, etc. In case of using these industrial RF generating systems, adequate work permits will be necessary.

Usual radio systems for internal communication, with maximum apparent radiated power less than 100 W are not considered dangerous, provided that the antenna does not make contact with metal parts.

As a general safety measure, a safe distance must be maintained between the nearest radiating parts and the antenna receiving radio frequency waves (pager-receiving stations and radio communication switchboards).

6.2.11.1 Fire Case

Induction furnaces use RF generators to create high-frequency electromagnetic fields. These fields induce electrical currents in the material being heated, which generates heat due to the material's resistance to the flow of the currents.

On October 1, 2024, an explosion occurred at the Delta Centrifugal manufacturing plant in Temple, Texas. The explosion involved an induction furnace containing 4,599 kg of molten metal at 1,593°C. Five workers were severely burned in the incident.

6.2.12 Electrical Equipment

Electrical equipment can produce electric sparks and hot surfaces, which constitute sources of ignition. As a preventive and protective measure, it must be ensured that the electrical system design in hazardous areas is elaborated following the relevant technical standards and that only certified Ex electrical and electronic equipment have been specified, installed, and maintained in accordance with the certificates' observations and the corresponding manufacturers' manuals.

6.2.12.1 Fire Case

A fault in electrical circuits can also lead to environmental impacts, as that occurred on October 25, 2004, when a fire due to a short circuit in an electrical service room in the Arkema plant in Lannemezan, France, specialized in the production of hydrazine hydrate and its derivatives from hydrogen peroxide, resulted in a release of ammonia, a toxic gas. The plant was shut down for a week: Operating losses were in the thousands of euros.

Electrical systems present other risks such as arc flash and overvoltages.

A report from the Institution of Engineering and Technology (IET), the UK, has revealed that a particular type of electrical fault that results in the potential risk of fire is on the rise in the UK. The fault in question is from broken Protective Earth and Neutral (PEN) conductors, which are typically used to make a circuit via exposed metalwork such as gas, water, and oil pipes. These conductors are used on the Protective Multiple Earthing (PME), (in IEC, TN-C-S) network, which was introduced in the 1970s as a self-monitoring system to improve safety.

In hazardous areas, the correct selection of electrical and electronic equipment is also very important to safety, as described in Chapter 5.

6.2.13 Open Flames, Hot Gases, and Hot Particles

Flames and combustion products (hot gases and, in the case of flames containing solid particles or soot, also incandescent solid particles) can ignite an explosive atmosphere and are among the most effective ignition sources. For this reason, the presence of naked flames and hot gases must be carefully analysed in hazardous areas. Flame-bearing devices (e.g. heating systems) are only permissible if they meet the following requirements:

1) The flame is safely contained, and the outside temperatures are not likely to generate a source of ignition by a hot surface.
2) The equipment ensures that flame propagation will not occur in the hazardous area and that the enclosure is sufficiently resistant to the effect of flames.

In Zones 1, 2, 21, and 22, open flames may only be tolerated, provided that appropriate protective measures have been applied (implementation of isolation or "disconnection": Devices for gases, vapours and mists, dusts, or hybrid mixtures, as appropriate).

Hot gases may be introduced if it is guaranteed that the MIT of the explosive atmosphere will not be reached. As example, for natural gas burners that allow the heating of the air used in atomizer dryers, if located near a Zone 22 area, it will be necessary to check if their proximity to a Zone 22 does not entail a special risk of ignition source.

With regard to incandescent solid particles, if there is a possibility of their presence in Zones 1, 2, 21, and 22, appropriate preventive measures are necessary to be applied (especially, the use of flame arresters). In a similar way, flames caused by special work (e.g. welding, oxycutting, use of radial saws, etc.) are controlled by applying organizational measures (hot work permits). Flames caused by cigarettes are generally controlled by a smoking prohibition throughout the plant.

6.2.13.1 Fire Case

In August 2, 2014, an explosion at Kunshan Zhongrong Metal Production Company in Kunshan, Jiangsu, China, was attributed to an open flame used improperly during a maintenance operation in a dust-filled room, resulting in 146 fatalities, 114 injuries, and extensive property damage. Kunshan Zhongrong, which polishes aluminium wheels for car manufacturers including General Motors, failed to properly store dangerous goods, did not have appropriate ventilation or dust removal systems, and had a substandard electrical system, according to the government investigation cited by Xinhua.

The main recommendations were:

- Implementation of a safety training programme on combustible dusts risks
- Acquisition of appropriate fire safety system
- Updating the electrical system.

6.3 PREVENTIVE MEASURES

6.3.1 FOR PLANTS WITH FLAMMABLE LIQUIDS

Preventing fires and explosions in flammable liquid industries presents challenges due to the volatility of flammable vapours, which can disperse and accumulate in unexpected areas. In brief, effective mitigation requires:

- Strict control of the sources of ignition and elimination of all ignition sources
- Continuous monitoring, in order to detect flammable vapours and changes in environmental conditions
- Ongoing and updated training of workers on safe handling practices and emergency response.

6.3.2 FOR PLANTS WITH COMBUSTIBLE DUSTS

Prevention in combustible dust industries is particularly challenging due to the insidious nature of combustible dust, which can accumulate on surfaces and form explosive clouds when disturbed. In brief, effective mitigation includes:

- Implementing ventilation and dust collection systems to minimize dust accumulation
- Rigorous cleaning programmes to remove accumulated dust from all surfaces
- Selecting dust-explosion-protected equipment and special tools for all operations.

6.4 CLOSING REMARKS

Both flammable liquid and combustible dust industries face significant challenges on preventing fires and explosions. However, combustible dust industries may face a slightly greater challenge due to the difficulty of controlling and completely eliminating combustible dust, which can accumulate in hard-to-reach places and form explosive atmospheres with even minor disturbances. Strict implementation of safety measures, regular maintenance, and ongoing worker training are essential to preventing such disasters in any type of industry.

Examples of incidents in flammable liquid and combustible dust industries demonstrate the severity of the consequences when preventive measures are not properly implemented. Industrial safety must be a constant priority to minimize risks and protect lives and property.

BIBLIOGRAPHY

CSB incident update 2024-04-1-GA – Chemical Decomposition, Fire, and Toxic Gas Release at Bio-Lab, Inc. Conyers, Georgia. Incident date: Sep. 29, 2024. Available at: www.ichemsafe.com/sites/www/uploads/userfiles/files/Investigation%20Report/20240929%20Investigation_Update_Bio-Lab_Final_Nov2024_%28005%29.pdf

CSB report: Didion Milling Company explosion and fire. 2023. Available at: www.csb.gov/didion-milling-company-explosion-and-fire/.

CSB report: Philadelphia Energy Solutions (PES) refinery fire and explosions. 2022. Available at: www.csb.gov/file.aspx?DocumentId=6202.

CSB report: West pharmaceutical services dust explosion and fire. 2004. Available at: www.csb.gov/west-pharmaceutical-services-dust-explosion-and-fire/.

Eckhoff, R. K.—*Dust explosions in the process industries.* Gulf Professional Publishing, 2003.

Fire breaks out at Ivanpah solar plant. Engineering News-Record, May 19, 2016. Available at: www.enr.com/articles/39541-fire-breaks-out-at-ivanpah-solar-plant-in-calif.

Gyenes, Zsuzsanna—Sharing information on lightning strikes causing tank fires. *Loss Prevention Bulletin 292,* 2023. Institution of Chemical Engineers. Available at: www.icheme.org/media/20706/lpb292_pg19.pdf.

Hattwig, M. and Steen, H. *Handbook of explosion prevention and protection.* Wiley-VCH, 2004.

IEC 60079-0—*Explosive atmospheres—Part 0: Equipment—General requirements.* International Electrotechnical Commission, 2017.

IEC 60079-28—*Explosive atmospheres—Part 28: Protection of equipment and transmission systems using optical radiation.* International Electrotechnical Commission, 2015.

Kunshan tragedy reveals long-term as well as immediate dangers of working in a dust-filled factory. *China Labour Bulletin,* 2014. Available at: https://clb.org.hk/en/content/kunshan-tragedy-reveals-long-term-well-immediate-dangers-working-dust-filled-factory.

Oleszczak, P. and Wolansk, P.—Ignition during hydrogen release from high pressure into the atmosphere. *Schock Waves,* 2010. https://doi.org/10.1007/s00193-010-0291-x.

Pipeline rupture and fire. Westcoast Energy Inc. Prince George, British Columbia. Transportation Safety Board of Canada, Oct. 9, 2018. Available at: www.tsb.gc.ca/eng/enquetes-investigations/pipeline/2018/p18h0088/p18h0088.html

Rangel Jr., Estellito—Equívocos que comprometem a segurança das empresas de petróleo e gás. In: *Rio Oil& Gas Technical Conference.* Rio de Janeiro, RJ—virtual. IBP, 2020. Available at: http://bit.ly/3uINGYD

Rangel Jr., Estellito—Causas das explosões. *EMEx Section, Eletricidade Moderna Magazine,* São Paulo, ed. 573, Sep./Oct. 2023, p. 68.

Rangel Jr., Estellito—Eletricidade como fator de incêndio. In: *V ESW Brasil—Seminário Internacional da Engenharia Elétrica na Segurança do Trabalho.* IEEE, São Paulo, 2011.

Rangel Jr., Estellito—Fontes de ignição. *EMEx Section, Eletricidade Moderna Magazine,* São Paulo, ed. 367 Oct. 2004, pp. 206–207.

Rangel Jr., Estellito—Safety in electrical installations in petroleum industry. In: *1st Petroleum and Chemical Industry Committee Mexico—PCIC Mexico.* Mexico City, 2013. Available at: https://bit.ly/3Pe0mCi

Rangel Jr., Estellito, Naegeli, Guilherme S. T. and Esposte, Jorge L. D.—Conscientização para atmosferas explosivas. In: *I Encontro Petrobrás sobre Instalações Elétricas em Atmosferas Explosivas.* EPIAEx, Rio de Janeiro, 1994.

Umar, Abubakar Saidu and Hamid, Yunus Khalid—Summaries of causes, effects and prevention of solar electric fire incidents. *International Journal of Engineering and Applied Physics (IJEAP),* vol. 3, no. 1, Jan. 2023, pp. 663–672.

Zhang, N., Shen, S. L., Zhou, A. N. and Chen, J.—A brief report on the March 21, 2019 explosions at a chemical factory in Xiangshui China. *Process Safety Progress,* vol. 38, no. 2, 2019, p. e12060. Available at: https://doi.org/10.1002/prs.12060.

7 Earthing

7.1 INTRODUCTION

The term "earthing" refers to the earth itself or to a large mass that is used in its place. When it is said that something is "earthed", it means that at least one of its elements is purposefully earthed. In general, electrical systems do not need to be earthed to function, and in fact, not all electrical systems are earthed. However, in electrical systems, when voltages are referred, they are usually referred to earth. In this way, earth represents a reference point (or a point of zero potential) to which all other voltages are referenced.

In fact, as computerized equipment communicates with other equipment, a "zero" reference voltage is critical to its proper operation. The earth, therefore, is a good choice as a zero reference point, as it surrounds everything, everywhere.

When someone is standing in contact with the earth, his body is approximately at earth potential. If the metal structure of a building is earthed, then all of its metal components are approximately at earth potential.

So, electrical earthing is the process of connecting an electrical or electronic system to the earth through a conductor in order to provide a low-resistance path for unwanted electrical currents. In a typical earthing system, components such as earth rods, copper cables, and connections are used to ensure an effective connection to the earth, which is considered the zero reference point.

Equipment earthing means the connection of conductive materials that are not intended to carry electrical current (such as conduits, cable trays, junction boxes, cabinets, and motor housings) to the earth. Earthing an electrical circuit means connecting the "neutral" terminal (such as that of a transformer, rotating machine, or electrical system) to earth, either directly or through a current-limiting device. Figure 7.1 shows the equipment earthing, the system earthing, and the bonding jumper.

7.2 DEFINITIONS

In order to establish a common perspective, some definitions and short explanations of terms are presented.

a) Earthing

It is the process in which the instantaneous discharge of the electrical energy takes place by transferring charges directly to the earth through low resistance conductor. The main types of earthing systems used in industrial and commercial power systems are: Solid earthing, low resistance earthing, high resistance earthing, and isolated.

DOI: 10.1201/9781003500001-7

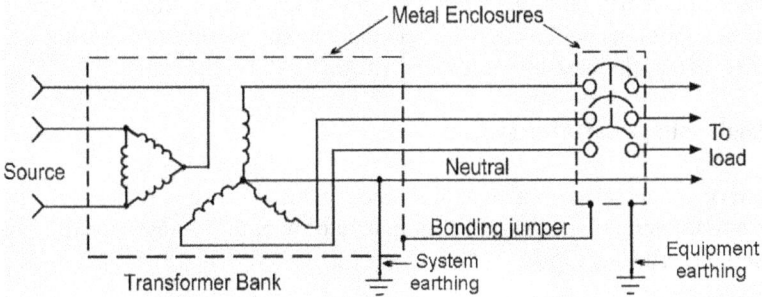

FIGURE 7.1   Earthing system diagram.

In some countries, like the United States, "earthing" is called "grounding".

b) Low voltage earthing system

An earthing system connects specific parts of an electric power system with the earth, typically the equipment conductive surface, for safety and functional purposes. Identification of the types of earthing systems is defined in IEC 60364 by two letters:

- The first letter for connection of the transformer neutral:
  T for "earthed" (from the French: Terre)
  I for "isolated"
- and the second letter for the type of connection of the exposed conductive parts of the installation:
  T for "directly" earthed
  N for "connected to earthed neutral" at the origin of the installation.

The combination of these two letters gives the possible configurations for earthing systems: TT, TN, and IT.

Note:

Each earthing system can be applied to an entire low voltage electrical installation. However, several schemes can jointly exist in the same installation.

c) Isolated system

A system, circuit, or apparatus, without an intentional connection to earth, except through potential indicating or measuring devices, or other very high-impedance devices. Note that although called "isolated", this type of system is in reality coupled to earth through the distributed capacitance of its phase windings and conductors.

In absence of an earth fault, the neutral of an isolated system, under reasonably balanced load conditions, will usually be held there by the balanced electrostatic capacitance between each phase conductor and earth.

d) High voltage earthing system

A system of conductors in which at least one conductor or point (usually the middle wire or neutral point of a transformer or generator winding) is intentionally earthed, either solidly or through impedance.

d.1) Direct (solidly) earthed
   Connected directly through an adequate earth connection in which no impedance has been intentionally inserted.
d.2) Indirect earthed
   When established via additional resistive, inductive, or capacitive resistances.
d.2.1) Resistance-earthed
   Earthed through impedance, the main element of which is resistance.
d.2.2) Inductance earthed
   Earthed through impedance, the main element of which is inductance.
e) Effectively earthed

Earthed through low impedance such that $X_0/X_1 < 3.0$ and $R_0/X_0 < 1.0$, where R_0 and X_0 are zero-sequence resistance and reactance; X_1 is the positive-sequence reactance of the power system.

f) Earth-fault current

It is the current passing to earth, or earthed parts, when an earth fault exists at only one point at the site of the fault (earth-fault location). This is:

- The capacitive earth-fault current in networks with isolated neutral
- The earth-fault residual current in networks with earth-fault compensation
- The zero-sequence current with low-resistance neutral earthing.

It also includes networks with isolated neutral point or earth-fault compensators in which the neutral point is briefly earthed at the start of the fault.

g) Earthing current

It is the total current flowing to earth via the earthing impedance. The earthing current is the component of the earth-fault current, which causes the rise in potential of an earthing system.

7.3 IMPORTANCE OF EARTHING SYSTEMS

In industrial facilities with flammable product processes, earthing, although usually not required for the functionality of electrical circuits, is essential for the safety of people and equipment. Among their benefits, we highlight some points as given in the next subsections.

7.3.1 PROTECTION AGAINST ELECTRIC SHOCKS

An electric shock is the pathophysiological effect of an electric current through the human body. Its passage impacts essentially the muscular, circulatory, and respiratory functions and sometimes results in serious burns. The degree of danger for the victim is a function of the magnitude of the current, the parts of the body through which the current passes, and the duration of current flow.

In the event of a failure in the insulation of electrical equipment, earthing directs the current to the earth, reducing the risk of electric shock to people and animals that may touch the equipment when it is faulty. All the system earthing arrangements guarantee equal protection against electrical shocks, provided that they are implemented and used according to standards, as shown in Figure 7.2.

Electrical systems that are not properly earthed can present electrical shock hazards, putting workers' lives at risk. In addition, electrical faults can interrupt production, resulting in lost productivity and profits. An effective earthing system ensures the continued and safe operation of the facility.

Figure 7.2 shows the current path during an isolation failure in a three-phase equipment with an earthed PE conductor.

7.3.2 PROTECTION AGAINST LIGHTNING

Earthing protects the electrical equipment against damage caused by voltage surges, acting as a "shock absorber" for oscillations caused by lightning. A lightning strike

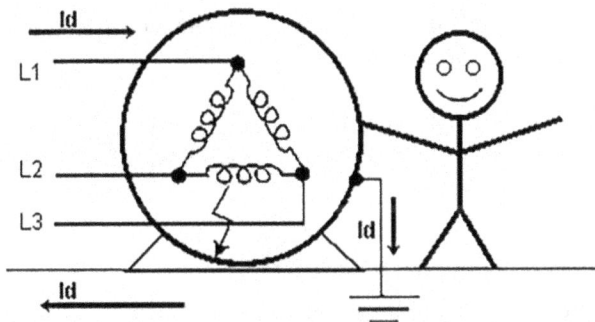

FIGURE 7.2 Earthing protects people from electrical shocks.

can not only cause fires but also damage sensitive electronic equipment and control systems that are essential to industrial operations. Lightning protection minimizes these disruptions, ensuring that industrial processes can continue with minimal disruption. Equipment integrity is preserved, resulting in less downtime and lower repair or replacement costs.

Lightning can cause catastrophic damage to structures and equipment and, most importantly, pose a significant risk to the safety of workers. The earthing system for lightning protection is designed to provide a low-resistance path for the lightning current, diverting it to the earth and thus minimizing damage.

7.3.2.1 Storage Tanks

Related with the hazardous locations, many instances of lightning-induced hydro-carbon fires have been recorded, especially at atmospheric storage tanks. As most storage tanks release flammable vapours at seals and vents, they are susceptible to lightning-induced fires. Direct lightning strikes can ignite flammable contents of cone roof tanks, unless the roof is provided with bonding for the structural parts.

Floating-roof storage tanks are safer than fixed-roof tanks because they prevent the formation of a vapour space above the hydrocarbon liquid. They also float on the stored product, rising and falling as inventory levels change. They limit the area of vapour release to the circumferential seal.

Low flash point liquids should always be stored in tanks which will not allow the creation of vapours in sizable quantities.

Floating-Roof Tanks (FRT) can cost twice of fixed-roof tanks; however, by reducing emissions, the increased cost can be justified on the basis of reduced product loss through evaporation and impact to the environment. FRT with seal hangers in the vapour space may be ignited indirectly when changes on the roof are released by a nearby lightning stroke. So, they are commonly protected against lightning ignition by:

- Bonding the floating roof to the seal shoes
- Use of insulating section in the hanging linkages
- Covering sharp points on hangers with insulating material
- Installation bonds across each pinned hanger joint.

Regular inspection and maintenance are crucial to ensuring that Lightning Protection Systems (LPSs) are functioning as designed, providing a layer of safety in hazardous areas.

7.3.3 Protection against Overvoltages

In addition to lightning protection, proper earthing is vital to the safety and efficiency of electrical systems in industrial facilities. In areas where flammable products are processed, the need for a safe and stable electrical system is even more critical.

Electrical systems are subject to various forms of overvoltages, which can be caused by fluctuations in the electrical network, equipment failures, or even by distant lightning strikes. Overvoltages can cause equipment failures, short circuits, and fires. The earthing system works as a "shock absorber", which seeks to stabilize the voltage during transients in the electrical system, caused by earth faults, switching, etc., in such a way that dangerous overvoltages are lowered during these periods that could cause the insulation of electrical equipment to break down.

An efficient earthing system provides a stable reference point for electrical system voltages and also, for the operation of electronic equipment, "dissipating" these overvoltages, protecting equipment and preventing damage. These overvoltages, when exceeding the limits that the equipment can withstand, can damage them. This means the loss of sensitive electronic devices, their programmes, communication between systems, and, ultimately, significant direct and indirect losses.

The set of measures to protect the electrical installation against transient overvoltages consists of the existence of an efficient earthing electrode, the presence of local equipotential connections that ensure the lowest possible potential difference between the components involved (this includes the installation of line surge arresters and voltage-switching surge protective devices (SPD)), as well as the reduction of induced voltages that enter the installation, carried out by means of voltage-attenuating SPD.

It is necessary to determine the number and quality of necessary equipotential zones in order to implement the protection devices required (surge arresters, etc.) on the lines of the various incoming and outgoing electrical systems. It is important to consider that:

- The IT system earthing more often requires the use of surge arresters.
- No system earthing completely does away with these measures.
- In the IT system earthing, protection against overvoltages due to MV faults must be provided by a surge limiter.

All SPD have in common the fact that they "divert" excess voltage from the power supply circuit that could cause damage to the installation. This "deviation" always uses the earthing system as the preferred path. Here, once again, the earthing electrode used for this protection is the same (only) one used in the electrical installation and the LPS mentioned earlier. Typically, a surge protection device (SPD) is connected between the live conductors (L1, L2, L3) and the earthed conductor (neutral), or the equipment earthing conductor (PE), or between a live conductor and the earthing electrode conductor. Figure 7.3 shows the Neutral, the Protective Earth (PE) conductor, and the Protective Earth and Neutral (PEN), connected to the earthed Equipotential Bar (EB).

In some standards, the neutral must also be connected to a SPD, as shown in Figure 7.4.

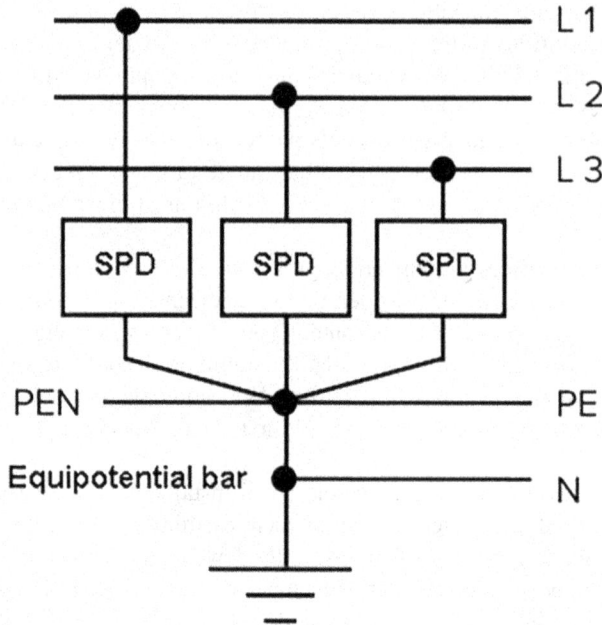

FIGURE 7.3 Neutral, PE, and PEN connected to the earthed Equipotential Bar.

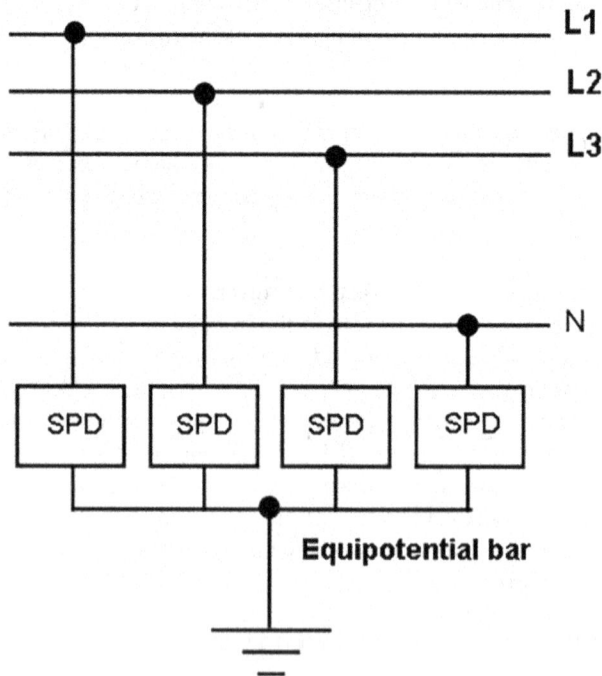

FIGURE 7.4 All three phases and the neutral protected by SPD connected to the Equipotential Bar.

7.3.4 Protection against Electrostatic Discharges

In industrial environments with flammable products, the presence of static charges poses a significant risk, because static charges are easily generated, and certain materials such as hydrocarbons usually have a low conductivity. For example, moving flammable liquids can generate static electricity. Without a proper earthing system, this charge can build up and eventually cause a spark, potentially leading to an explosion if an explosive atmosphere is present.

Explosions and fires in oil and gas tankers have occurred where static has been blamed as the cause. After a tank has been emptied, vapour may be present having been mixed with air that has been sucked into the tank as a result of emptying. This provides almost ideal conditions for static to cause ignition during the start of cleaning operations.

Earthing safely dissipates these static charges. All exposed metal parts must be properly earthed to prevent the accumulation of static electricity.

7.3.5 Protection against Electromagnetic Disturbances

Any system earthing can be chosen:

- For all differential mode disturbances
- For all disturbances (common or differential mode), with a frequency greater than 1 MHz.

The TT, TN-S, and IT system earthing can thus satisfy all electromagnetic compatibility criteria. However, it should be noted that the TN-S system generates more disturbances during the insulation fault, as the fault current is higher.

On the other hand, the TN-C and TN-C-S system earthing are not recommended, as in these systems, a permanent current due to load unbalance flows through the PEN conductor, the exposed conductive parts, and the cable shielding. This permanent current creates disturbing voltage drops between the exposed conductive parts of the sensitive equipment connected to the PEN. The presence of third-order multiple harmonics has considerably amplified this current in modern installations.

7.4 EARTH FAULTS

An earth fault is an unwanted connection between the conductors of the electrical system and the earth. Earth faults, if the system does not have adequate protection, can go unnoticed and cause damage to the production processes of the industrial plant.

Undetected earth faults can have the following consequences:

- Potential risks to the health and safety of workers
- Safety risks such as fire and electric shock
- Damage to equipment and production processes, seriously impacting financial results.

7.5 EARTHING SYSTEMS

The three earthing systems defined in IEC 60364 are:

- TT, where the electrical supply and customer loads are separately earthed
- TN, where the electrical supply is earthed and the customer loads are earthed via neutral
- IT, where only the customer loads are earthed.

Some countries use a modified Protective Multiple Earthing (PME) system called "Multiple Earthed Neutral (MEN)", where the neutral is earthed at each consumer service point, thus effectively bringing the neutral potential difference to zero along the entire length of the LV lines. In IEC 60364 terminology, this is called TN-C-S. In North America, the term "Multi-grounded Neutral system" (MGN) is also used.

7.5.1 TT SYSTEM

The source neutral is connected directly to earth, with the installation masses connected to an earthing electrode, independent of the source electrode.

Figure 7.5 shows a two-phase equipment, with the PE conductor connected to its metallic enclosure during a fault, where I_d is the earth-leakage current, and there is a resistance path of 10 Ω between the equipment and the source. In this case, the path of a phase-to-earth current includes the earth, which greatly limits the value of the current due to the high value of the earth resistance.

This current usually is insufficient to activate circuit breakers or fuses but sufficient to endanger a person. Therefore, it must be detected and eliminated by more sensitive devices called Residual Current Devices (RCD).

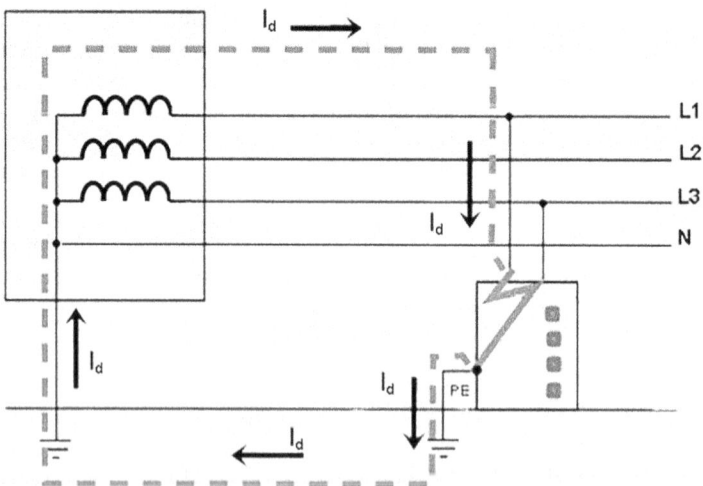

FIGURE 7.5 TT system during a fault.

7.5.1.1 During a Fault

In the presence of an insulation fault, the fault current I_d is limited for the most part by the earthing resistances (if the earthing connections for the exposed conductive parts, and for the neutral, are not combined). This fault current induces a fault voltage in the load earthing resistances. As the earthing resistances are normally low and of the same order of magnitude (@ 10 W), this voltage of around $U_o/2$ is dangerous.

Considering $U_n = 230$ V and R1 = R2 = 5 Ω, $I_d = 230/5 + 5$ A = 23 A (earth-leakage current).

The part of the installation concerned by the fault must therefore be automatically disconnected by an RCD.

7.5.1.2 In Hazardous Areas

If the TT type of system earthing (separate earths for power system and exposed conductive parts) is used, then it shall be protected by a residual current device, but, where the earth resistivity is high, such a system may not be acceptable. In the TT system earthing, when the first insulation fault occurs, the current generated by this fault is low, and the risk of fire is slight.

All exposed and external conductive parts must be connected to the equipotential bonding system.

7.5.2 TN System

The source's neutral is connected directly to earth, with the installation's masses connected to this point by means of metal conductors (protection conductor). In this case, the path of a phase-to-earth current has very low impedance (copper), and the current can reach high values, sufficient to be detected and interrupted even by circuit breakers or fuses.

The TN system, according to IEC 60364, has sub-systems:

- TN-C: If the N (neutral) and PE (protection) conductors are combined (PEN).
- TN-S: If the N (neutral) and PE conductors are separate.
- TN-C-S: It is a mixed scheme; where there is a TN-S downstream of a TN-C (the opposite is forbidden).

In Brasil, the TN system is the most common, when dealing with installations directly supplied by the public low-voltage grid of the electricity utility. In this case, the neutral conductor coming from the utility is connected to the earthing electrode of the utility entrance box and continues to the interior of the building, where it is connected to the main Equipotential Bonding bar (EB), generally located in the installation's main switchboard or close to it. Therefore, this conductor has a dual function, protection and neutral (PEN conductor), and such section of the installation has the configuration of a TN-C system.

Figure 7.6 shows a TN-C system with two three-phase equipment and a PEN conductor, where its dual function is been used in the equipment on the left. In this arrangement, the PEN conductor also has the purpose of equipotentializing the

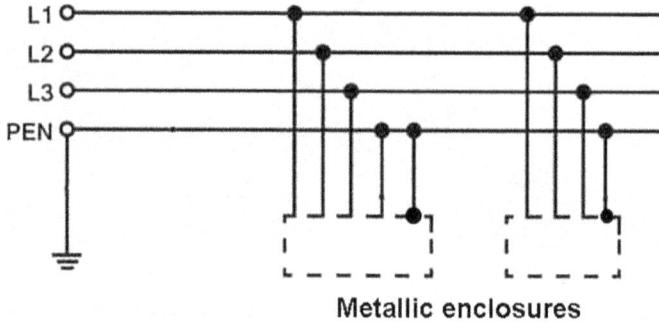

Metallic enclosures

FIGURE 7.6   TN-C system.

earthing electrode of the utility entrance box with the earthing electrode of the building's internal electrical installation.

From the EB, the neutral conductor starts to serve the phase-neutral terminal circuits, while the protection conductor is distributed to all circuits, thus forming a TN-S scheme.

Note that if the neutral conductor is lost before the utility entrance box (for example, if it breaks due to a vehicle accident on the street), or if this conductor is the PEN in the section after the entrance box, the system will become TT. This leads us to the conclusion that, even in TN systems, it is always convenient to use RCD devices to ensure the protection of people against electric shocks. Figure 7.7 shows a two-phase equipment with the PEN conductor connected to its metallic enclosure, and with a low earth resistance, during a fault.

Note that some standards can require a minimum copper equivalent cross-section area of 10 mm² for the main Protective Bonding Conductor (PBC), when the copper equivalent cross-sectional area of the PEN conductor can be up to 35 mm².

Figure 7.8 shows a TN-C-S system, where there is a TN-S downstream of a TN-C system.

7.5.2.1   During a Fault

In the presence of an insulation fault in TN systems, the fault current I_d is only limited by the impedance of the fault loop cables. In the event of an impedance fault, the TN system earthing implemented without residual current devices does not provide sufficient protection, and the use of the TN-S system earthing combined with residual current devices is recommended.

For 230/400 V systems, this voltage of the order of $U_o/2$ (if $R_{PE} = R_{ph}$) is dangerous as it is greater than the limit safety voltage (UL = 50 V), even in dry environments.

Considering $U_n = 230$ V and $R_g = 0.1\ \Omega$, $I_d = 230/0.1 = 2,300$ A.

The installation or part of the installation must then be immediately and automatically de-energized by an RCD. As the insulation fault is similar to a phase-to-neutral short circuit, breaking is performed by the overcurrent protection devices.

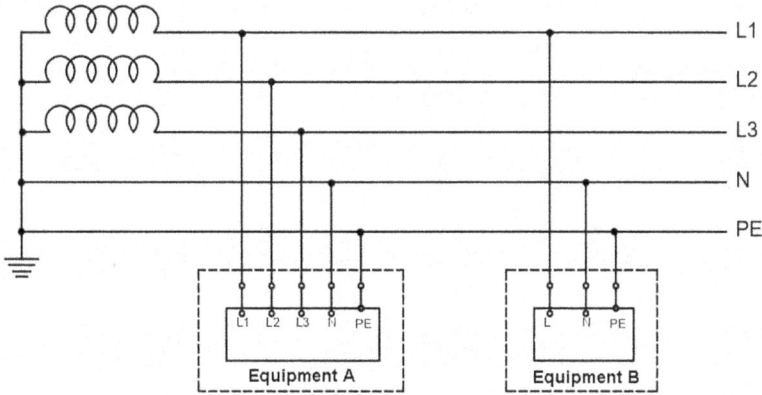

FIGURE 7.7   TN-S system.

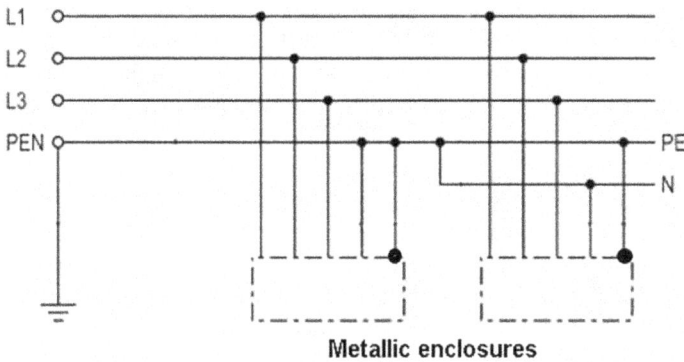

Metallic enclosures

FIGURE 7.8   TN-C-S system.

7.5.2.2 In Hazardous Areas

In normal operation, the TN-C system earthing arrangement presents a higher risk of fire than the others, because the load unbalance current permanently flows through not only the PEN conductor but also the devices connected to it: Metal frameworks, exposed conductive parts, shielding, etc.

When a short circuit occurs, the energy lost in these stray trajectories increases considerably. So, if the type of earthing system TN is used in a hazardous area, it shall be type TN-S (with separate neutral N and protective conductor PE), i.e. the neutral and the protective conductor shall not be connected together, or combined in a single conductor, in the hazardous area, because there could be a potential difference between the main earthing terminal of the installation and the general mass of earth, and incendive sparking could then occur between the earth of the electrical installation and any extraneous metalwork which is in contact with the general mass of earth. For this reason, the TN-C and TN-C-S (PME) systems' earthing arrangements are forbidden in hazardous areas.

At any point of transition from TN-C to TN-S, the protective conductor shall be connected to the equipotential bonding system in the non-hazardous area.

7.5.2.2.1   Safety Note on Solidly Earthed Systems

Solidly earthed systems have a power source in which the neutral, or X_0 point of the transformer or generator, is connected to earth through a solid bonding jumper. This jumper has minimal resistance or impedance to earth—it does not appreciably limit earth-fault current. Common examples are three-phase 208/120 V wye or 480/277 V wye configurations, as well as single-phase transformers with the secondary neutral connected to earth (in rare instances, corner-earthed delta).

Many countries use solidly earthed systems for household power, typically a split-phase 240/120 V system with its neutral bonded solidly to earth. One disadvantage of this earthing method is the large prospective earth-fault current. Fire, electrical-component damage, or personnel injury can occur. Nevertheless, a tripped overcurrent device (circuit breaker or fuse) enables the electrician to quickly locate the faulted circuit. Corrective action will often take place after the fault has occurred, and the damage is done. Preventive maintenance is not necessarily associated with the solidly earthed system.

In the case of earthed systems, i.e. TN/TT systems, the insulation resistance is determined indirectly via the magnitude of the fault current. However, when a shutdown is a problem for operations, and the availability has priority, then residual current monitors (RCM) are usually used.

The function of a RCM is to monitor an electrical installation, or circuit, for the occurrence of a residual current and to indicate by an alarm if this exceeds a specified value. The RCM should be configured so that an alarm message (e.g. acoustic, visual, or via e-mail) occurs before the standard-compliant automatic switch-off.

7.5.3   IT System

It is a system similar to the TT; however, the source is earthed by inserting a high-value impedance (resistance or inductance), as shown in Figure 7.9. This limits the fault current to a desired value, so as not to allow a first fault to shut down the system. Generally, this current is not dangerous to people, but as the installation will be operating under a fault condition, devices must be used to monitor the insulation of the conductors, preventing an excessive degradation of the installation components.

The use of IT systems, as with unearthed systems, is restricted to cases where a first fault cannot immediately shut down the power supply, interrupting important processes (such as in surgical rooms, certain metallurgical processes). Industrial plants where the continuity of service is critical often use High Resistance-Earthed (HRE) systems. Other facilities, such as in mining, use HRE systems that trip on an earth fault, for maximum safety. Typically, these are in three-phase, three-wire (no neutral)-wye-configured power systems but sometimes are connected to an "artificial neutral" of a zigzag transformer.

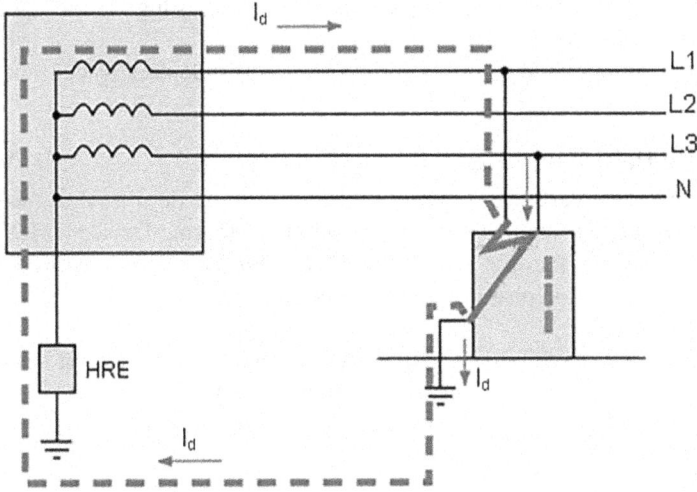

FIGURE 7.9 IT system.

A main advantage of the resistance-earthed system is that the resistor limits the amount of current available to an earth fault, dramatically reducing the point-of-fault damage and the probability of an arc flash.

7.5.3.1 During a Fault

Earth-fault current in a resistance-earthed system is limited in magnitude. This is the main difference when compared to the solidly earthed systems. The earth-fault magnitude depends on the Neutral-Earthing Resistor (NER) rating and the return-path and earth-fault impedances. Typical HRE and NER ratings are 5, 10, or 25 A.

For non-tripping systems, the resistor must be rated for continuous duty use. For tripping systems, it is usually 10 s rated.

The resistance-earthed system also allows for selective earth-fault tripping which is achieved by monitoring all feeders and with adjustable time delays on cascaded relays.

- **Behaviour on the first fault**

As the neutral is unearthed, there is no flow of a fault current I_d. Voltage is not dangerous, and the installation can therefore be kept in operation. As the Insulation Monitoring Device (IMD) has detected this first fault, it must be located and eliminated before a second fault occurs.

- **Behaviour on the second fault**

The fault concerns the same live conductor: Nothing happens and operation can continue. The fault concerns two different live conductors: The double fault is a short circuit, and the current generated by the insulation fault is high in the TN type system, resulting in serious damage.

Breaking is performed by the overcurrent protection devices, as considering $U_n = 230$ V and $R_g = 0.1\ \Omega$, $I_{d2} = 230/0.1 = 2{,}300$ A.

7.5.3.2 In Hazardous Areas

If a type of system earthing IT (neutral isolated from earth or earthed through sufficiently high impedance) is used, an Insulation Monitoring Device (IMD) shall be provided to indicate the first earth fault. If the first fault is not removed, a subsequent fault on the same phase will not be detected, possibly leading to a dangerous situation.

In hazardous areas, all exposed and external conductive parts must be connected to the equipotential bonding system as described in IEC 60364-4-41.

Safety notes:

1) In the case of unearthed systems, the insulation resistance between the active conductors and earth is continuously monitored using an Insulation Monitoring Device (IMD). If the value measured is below a specific resistance, then an alarm is triggered.

In accordance with IEC 60364-4-41, a shutdown is not necessary on the occurrence of a first fault, such that the operation can continue uninterrupted. This aspect is of crucial importance in safety-related areas, e.g. industrial plants or electric mobility. As the IT system supplied is in operation, the IMD measures the total insulation resistance of the system, including all loads switched on that are electrically connected to the IT system.

An RCM is distinguished from an IMD in that it is passive in its monitoring function, while an IMD is active in its monitoring and measuring functions.

2) Another important consideration on resistance-earthed systems is that an open-NER failure leaves the power system unearthed and, therefore, without earth-fault detection. Similarly, the failure of the bonding connection to earth at a feeder, or piece of equipment, leaves a portion of the system without protection. As a result, a hazardous touch voltage can exist. This problem is especially important for mobile or movable loads, which rely on the cable earth conductor as the earth return path. Earth-check monitors protect against the open-earth-path condition by monitoring the continuity of the circuit, consisting of an earth-check (pilot) conductor and earth conductor in the cable, terminated at the load by a specialized component that is recognized by the monitor.

The earth-check fault condition is directly related to an earth-fault hazard, and there are in the market, some relays with both earth fault and earth check functions. If earth-fault current exceeds the pickup value, the alarm relays change state after the selected delay time.

The earth-check circuit monitors the resistance of the earth-check loop (earth-check conductor, earth conductor, and termination device). The earth-check circuit detects and trips when the earth-check loop is shorted or opened or if the maximum resistance is exceeded.

7.6 EQUIPOTENTIAL BONDING

Equipotential bonding ensures that all metal parts of the installation have the same electrical potential, avoiding potential differences that could cause electric shocks and insulation failures. Consistent equipotential bonding must be established in all potentially explosive atmospheres to prevent potential differences between different and extraneous conductive parts. Building columns and structural parts, pipes, containers, etc. must be integrated in the equipotential bonding system so that a voltage difference does not occur even under fault conditions.

According to IEC 60079–14, supplementary equipotential bonding is required and must be properly established, installed, and tested in line with the IEC 60364-4-41 and IEC 60364–5-54. Equipotential bonding connections should only be made through designed connection points (not rely on fortuitous contact) and should be secure against self-loosening. This requires the use of materials that are designed for that particular application and are fit for purpose.

It is worth noting that equipotential bonding conductors would not be required where insulation ensures that circulating currents cannot flow. However, provision shall be made for adequate earthing of isolated exposed-conductive-parts. The insulation of such parts shall be capable of withstanding a test of 100 V_{rms} for 1 min.

When using surge protective devices, it is necessary to check the relevant standard, as it can require that the cross-section of the copper earthing conductor for equipotential bonding be, at least, 4 mm^2. Drum filling is a good example to illustrate the principle. Where the liquid enters the drum leaving the supply pipe nozzle, the flow profile deposits a charge on the nozzle, and the opposite charge accumulates in the drum, depositing itself on the metal of the drum. The charge appears between the nozzle and the filling cap rim.

To equalize the charge, it is necessary to bond the drum to the nozzle such that the charge is dissipated. As the nozzle is moved towards the drum at the start of the filling cycle, and away at the end, this is where any initial potential discharge may cause a spark. Bonding must be in place before and after the filling event to ensure safety, as shown in Figure 7.10.

7.6.1 EARTHING ALREADY CHARGED OBJECTS

It is generally understood that if a conductive object, such as a drum, is earthed with a suitable clamp and cable, before it is used in an operation, then "it can never become charged with static electricity". However, if the earth connection was forgotten, and the object was allowed to become electrostatically charged, then approaching it with an earthed item, such as a earthing clamp, will cause a spark to be drawn from it, just before the earth contact is made. This spark discharge may cause the ignition of

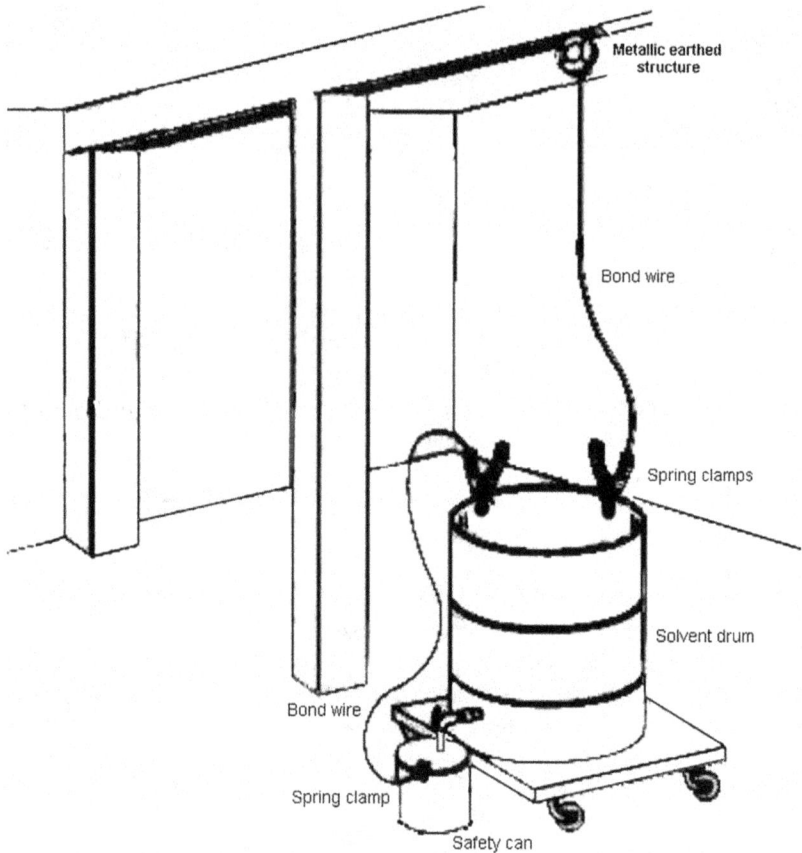

FIGURE 7.10 Bonding in drum operations.

flammable gases, vapours, or even some dusts, if they are present in optimum concentrations in the surrounding atmosphere.

A number of devices, or methods, are in use which rely on the fact that the earthing clamp is not directly in contact with earth, in order to prevent discharges. The two most common types are explained in the next subsections.

7.6.1.1 The Open Earth Cable

Here, the earth connection is broken by an isolation switch, usually in a flameproof enclosure built into the clamp or situated on the wall close by. The concept is that the clamp will not draw a spark from an object that may have become charged as the clamp is not earthed until after it has been applied, and the isolation switch has been turned to make the earth connection. It is said that "any sparking will occur in the Ex d enclosure".

But, this is a myth! A spark will occur not only between a charged object and an earthed clamp but also between a charged object and anything with enough electrical capacitance and at a lower electrical potential! Hence, when the clamp comes close to the charged object, the capacitance of the clamp and cable will allow a spark to jump from the object, regardless of whether the isolation switch has broken the earth connection or not.

7.6.1.2   Earth Cable with High Resistance

Here, the earthing clamp contains a resistor, typically of around 1 MΩ. The belief is that the clamp will not draw a spark from an object that may have become charged, as the resistance will impede the connection to earth and prevent it.

This is another false assumption! Experimentation has shown that a 1 MΩ resistance is not able to prevent incendive sparks. Even raising the resistance to 100 MΩ, or above, will not do this, and as the resistance is increased, the clamp and cable's ability to carry out their intended function of providing a good earth connection is diminished. Moreover, an incendive spark is just as likely to jump across to the conductive body of a clamp, owing to the fact that it has its own electrical capacitance, when the clamp is brought into close proximity with the charged plant item.

7.6.1.3   Safety Note

Both of these methods lead the user into a false sense of safety. In truth, it is essential to ensure that objects are earthed prior to being used in a static-generating operation. However, if an object has inadvertently been allowed to become charged in a potentially flammable area, the only safe ways to proceed are as follows:

- Test the area with a gas detector (without getting too close to the drum), and if no flammable atmosphere is present, attach the clamp
- Leave the object for some considerable length of time, so that the charge can relax naturally. In practice, this may take a long time, depending on the situation.

Considering the various cases of fuel tanker trucks that caught fire, or exploded, during loading and unloading operations, even in properly equipped terminals, with trained workers and following safety procedures, it is important to consider that it might have happened because a person approached a "charged" tanker, drawing a considerable spark, as detailed in Section 12.9. So, it is not a matter of the approach of an earthed clamp alone, as anything with sufficient capacitance and at a lower electrical potential can cause a spark to occur in these situations.

7.7   CLOSING REMARKS

It is important to note that different earthing systems can exist side by side in the same electrical installation. This highlights that the best solution for safety and availability needs will be found for every case. But, it is necessary to check that the choice of earthing system is not imposed by standards or legislation. This choice must be based on the installation characteristics, the operating conditions, and the legal requirements.

It is pointless trying to operate an unearthed neutral system in part of an installation which, by its very nature, has a low insulation level (a few thousand ohms) like old and extended installations and installations with external lines. Likewise, it would be a contradiction in the industry where the continuity of supply is essential, and fire risks high, to choose a multiple earthed-neutral system.

Table 7.1 shows, in brief, the general performance of the earthing systems.

There is no simple recipe for selecting an earthing scheme that can be summarized in a single table. Other considerations, depending on the characteristics and resources provided for the industrial unit, can decisively influence the definition of the earthing system during the design of the electrical system.

Taking into account the continuity of supply characteristics, Table 7.2 gives orientation, and Table 7.3 summarizes the characteristics of the various earthing methods.

TABLE 7.1
Earthing Systems' Performance × Design Objectives

Objective	Earthing System Performance			
	TT	TN-C	TN-S	IT
Protection of life	Good	Good	Good	Good
Protection against fire	Good	Poor	Average	Good
Protection of equipment (insulation fault)	Good	Poor	Poor	Good
Continuity of supply	Average	Average	Average	Excellent
Electromagnetic disturbances (external and internal of the electrical system)	Average	Poor	Average	Average

TABLE 7.2
Selection of Earthing Schemes Considering the Facility's Profile

Facility's Profile	TN	TN-S	TT	IT
Continuity of supply, with good maintenance resources				X
Continuity of supply, with weak maintenance resources [*1]			X	
Continuity of supply is not essential, with good maintenance resources [*2]		X		
Continuity of supply is not essential, without maintenance resources			X	
Risk of fire, with good maintenance resources [*3]				
Very extensive system, or with a high leakage current		X		
Use of standby power supplies			X	
Loads with sensitive-to-high fault currents (motors) [*4]			X	X
Loads with low natural insulation (furnaces) or with a large HF filter (large computers)	X			
Supply of control and monitoring systems [*5]			X	X

Notes:

[*1] There is no completely satisfactory solution. Prefer a TT system for which discrimination on tripping is easier to implement, and which minimizes damage compared with a TN system;

[*2] Prefer a TN-S system, which requires rapid repairs and extension according to standards;

[*3] IT and use of 0.5 A RCD;

[*4] TT or IT systems are good options.

[*5] Prefer an IT (continuity of supply) or TT system (enhanced equipotentiality of communicating devices).

TABLE 7.3

Characteristics of the Various Earthing Methods

Situation	Essentially Solid Earthing			Reactance Earthing	Resistance Earthing	
	Unearthed	Solid	Low-Value Reactor	High-Value Reactor	Low Resistance	High Resistance
The current for a phase-to-earth fault expressed as a percentage of the current for a three-phase fault.	Less than 1%	Varies, may be 100% or greater	Usually designed to produce 60 to 100%	5 to 25%	5 to 20%	Less than 1%
Transient overvoltages	Very high	Not excessive	Not excessive	Very high	Not excessive	Not excessive
Automatic segregation of faulted Zone	No	Yes	Yes	Yes	Yes	No
Lightning arresters	Unearthed neutral type	Earthed-neutral type	Earthed-neutral type if current is 60% or greater	Unearthed neutral type	Unearthed neutral type	Unearthed neutral type
Remarks	Not recommended due to overvoltages and nonsegregation of fault	Used on systems 600 V and below	Used on systems over 15 kV	Not used due to excessive overvoltages	Generally used on industrial systems of 2.4 to 15 kV	Generally used on systems 5 kV and below

It is worth to say that measurements in earthing systems are generally performed at low frequency, but in case of atmospheric discharges, high-frequency components will appear (up to 4 MHz), and, therefore, the impulsive impedance of the earthing electrode will may be higher or lower than the values measured at low frequency. The behaviour of the earthing electrode basically depends on the geometry of the electrode, the resistivity of the soil, and the injection point of the atmospheric discharge current.

The relationship between earthing impedance and resistance at low frequency, up to 50 kHz, remains unitary, that is the values of earthing resistance and impedance are equal. However, for higher frequencies, the relationship may vary higher or lower, depending on the behaviour of the electrical discharge, whether it tends to be inductive or capacitive. Therefore, measuring 10 Ω of earthing resistance at low frequency once does not guarantee that the earthing resistance will always be "good".

Implementing an efficient earthing system is crucial for electrical protection in industries with hazardous areas. With regular maintenance and careful inspection practices, the earthing system continues to effectively protect against short circuits and overloads, contributing to the safety of workers, the preservation of assets, and the continuity of industrial operations.

BIBLIOGRAPHY

Aterramento eletrico. Moreno, Hilton and Fernandes, Paulo C. – Instituto Brasileiro do Cobre, Sao Paulo, Brasil, 1999.

Britton, L. G.—*Avoiding static ignition hazards in chemical operations*. CCPS, American Institute of Chemical Engineers, 1999.

Britton, L. G.—Using material data in static hazard assessment. *Plant/Operations Progress*, vol. 11, no. 2, 1992.

Dunki-Jacobs, J. R., Shields, F. J. and St. Pierre, C. *Industrial power system grounding design handbook*. Thomson-Shore, 2007.

IEC 60079-14—*Explosive atmospheres—Part 14: Electrical installation design, selection and installation of equipment, including initial inspection*. International Electrotechnical Commission, 2024.

IEC 60364-1—*Low-voltage electrical installations—Part 1: Fundamental principles, assessment of general characteristics, definitions*. International Electrotechnical Commission, 2005.

IEC 60364-4-41—*Low-voltage electrical installations—Part 4-41: Protection for safety—Protection against electric shock*. International Electrotechnical Commission, 2005.

IEC 60364-5-54—*Low-voltage electrical installations—Part 5-54: Selection and erection of electrical equipment—Earthing arrangements and protective conductors*. International Electrotechnical Commission, 2011.

Lightning Protection Guide—3rd edition. Dehn International, 2014. Available at: www.dehn-international.com/sites/default/files/media/files/lpg-2015-e-complete.pdf

Pratt, T.—*Electrostatic ignitions of fires and explosions*. American Institute of Chemical Engineers, 2000.

Rando, Ricardo—*Aterramento em atmosferas explosivas: Práticas recomendadas*. Editora Blucher, 2021.

Rangel Jr., Estellito—A atualização das normas aplicáveis às instalações Ex brasileiras. *Lumière electric Magazine*, Editora Lumière, ed. 196, Aug. 2014, pp. 44–48.

Rangel Jr., Estellito—Instalações elétricas para áreas com risco de explosão. *Incêndio Magazine*, Editora Cipa, Apr. 2012, pp. 20–31.

Rangel Jr., Estellito—Instalações elétricas: ponto-chave da segurança na indústria do petróleo. *Tn Petróleo Magazine*, Benício Biz Editores Associados, no. 29, Apr. 2003, pp. 48–50.

Rangel Jr., Estellito—Segurança nas instalações de petróleo e gás: é necessário ir além das normas! *Petro & Química Magazine*, Editora Valete, ed. 373, Dec. 2017, pp. 24–27.

Rangel Jr., Estellito and Rando, Ricardo – Considerações sobre aterramento de sistemas BT. In: *V Encontro de Engenharia Elétrica Petrobras*, 1999.

Rangel Jr., E., Costa, P. F. and Borel, J. E. B.—Aterramento do neutro em áreas classificadas: A tecnologia do resistor de alto valor. In: *III Encontro Petrobras sobre Instalações Elétricas em Atmosferas Explosivas*, 2002.

Rangel Jr., E. and Silva, M. V.—O perigo invisível na indústria do petróleo: A eletricidade estática. In: *VIII Encontro de Engenharia Elétrica Petrobras*, Brasil, 2006. Available at: https://bit.ly/2LhPV2X

Zipse, D. W.—Grounding: The good, the bad and the stupid. In: *60th Petroleum and Chemical Industry Conference*, Chicago, 2013.

Zipse, D. W.—The hazardous multigrounded neutral distribution system and dangerous stray currents. In: *50th Petroleum and Chemical Industry Conference*, Houston, 2003.

8 Risks of Static Electricity

8.1 INTRODUCTION

One of the sources of ignition of explosive atmospheres—which usually goes unnoticed—is static electricity. To avoid damage due to possible discharges or the simple presence of static electricity, it is necessary that the voltages on the surfaces of materials and devices (both those that conduct electricity and those that insulate) remain low.

Controlling surface voltage is not easy, particularly in industrial environments, such as in the clean rooms of pharmaceutical industries. This is because many materials need to be made of plastic to meet the established requirements, in order to avoid contamination with the chemical. In these places, special care is taken to control the wear and tear of the special uniforms worn by workers.

Among the measures that can be implemented to dissipate static charges are:

- Use of conductive materials
- Proper earthing of equipment (and even, workers)
- Use of antistatic devices.

8.2 CHARGE GENERATION

In industrial plants, the movement of powders and the flow of liquids can generate large amounts of static electricity.

Electrical machines and equipment, in certain situations, can generate electrostatic charges on the products processed in the unit.

Figures 8.1, 8.2, and 8.3 illustrate the charge generation during the operation of conveyors, tank filling, and when mixing products.

Bonding and earthing are needed when dispensing flammable liquids from storage drums to smaller electrically conductive containers. Similarly, whenever these liquids are transferred between conductive containers in any work area, for example, when filling or draining dip tanks, mixers, rinse tanks, or other equipment, bond both containers together and earth one of them. Earthing connections should be checked regularly to ensure they are in good condition.

People can also generate static electricity when working, moving equipment, or simply walking. A large part of the static electricity in the industrial plant is generated in the so-called "triboelectrification" process, which is the frequent formation and interruption of surface contact between materials. The electrostatic charge can be transferred between materials that come into contact or rub against each other. The influence of the charge stored on a surface and on nearby bodies is directly related to the tension on the surface and local electric fields.

DOI: 10.1201/9781003500001-8

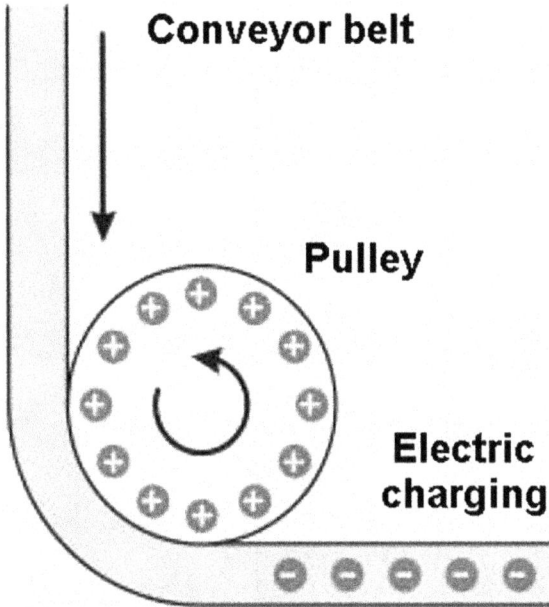

FIGURE 8.1 Charge generation in conveyors belts.

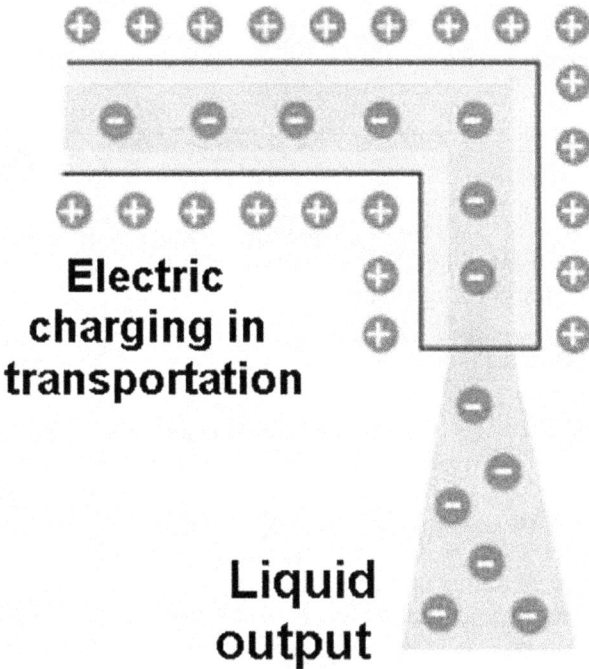

FIGURE 8.2 Charge generation in tank filling.

Charge generation

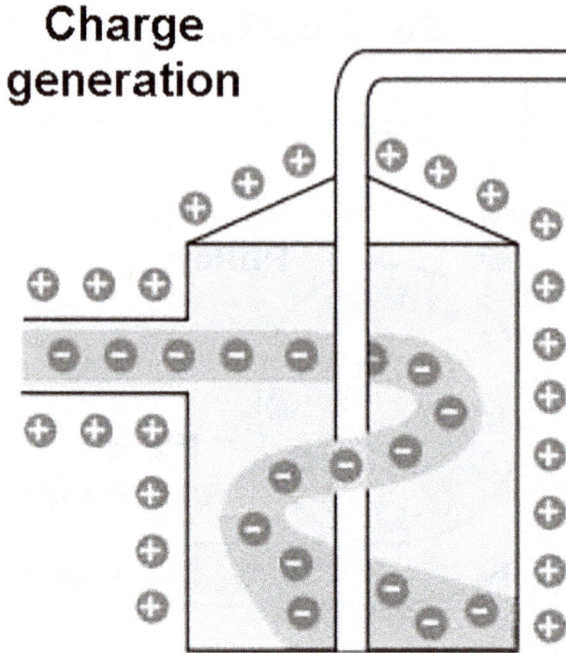

FIGURE 8.3   Charge generation during mixing operations.

The principal mechanisms are:

a. Contact and separation of solids
b. Fragmentation of solids having no uniform surface charge densities
c. Shear at liquid–solid, liquid–gas, and two-phase liquid interfaces
d. Separation by gravity of suspended material with unequal size and charge
e. Induction charging
f. Ionic charging.

Mechanisms (e) and (f) require the presence of large electric fields and represent processes whereby charge can be transferred between systems that are electrically isolated from each other.

Charge generation may be a discrete process, in which case it is associated with a transferred charge resulting in a net charge density, or can continue indefinitely, in which case the rate of production of charge is equivalent to a charging current. An illustration of the contact and separation mechanism is shown in Figure 8.4.

Considering that at least one of the materials features low conductivity, and the separation process being faster than charge compensation, Figure 8.4 shows at the

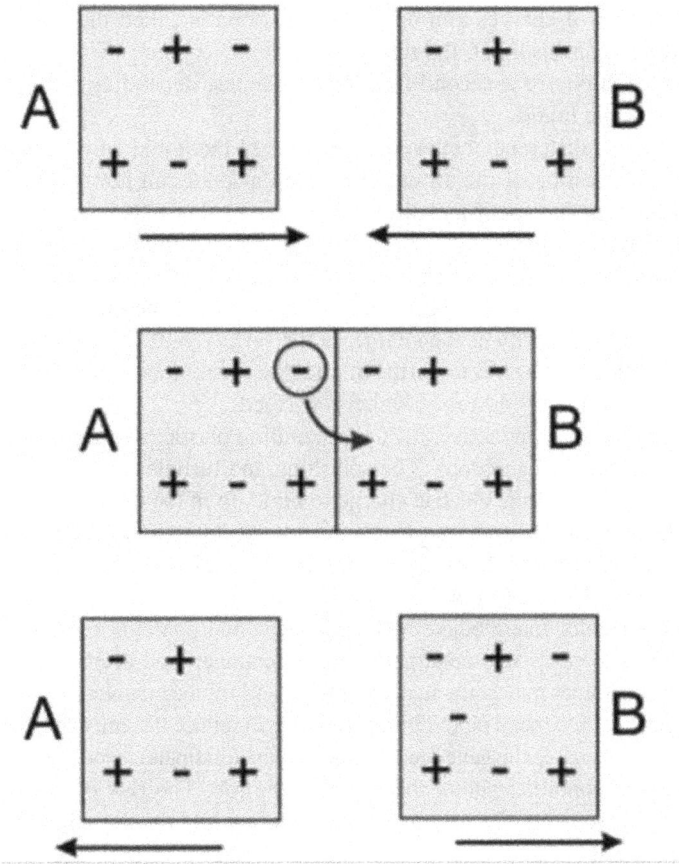

FIGURE 8.4 Charging mechanism during contact and separation of solids.

top, materials A and B in neutral state before contact. In the middle, due to the contact and/or friction, charge (electron) moves from A to B. At the bottom, after separation, material A becomes positive charged, and material B becomes negative charged. Whenever charged surfaces are separated quickly, electrons can be prevented from moving to neutralize opposite charges if the conductivity of the material through which the electrons must pass is low.

The resistance to electron movement through a metal is extremely low, and neutralization easily occurs. Static electricity is rarely a problem when both surfaces are metallic. With other materials, such as insulators, electron movement will be prevented, and after separation, the excess electrons remain on the surface of the insulator.

When a metal tank is earthed, it is considered electrically neutral. The charged liquid in the tank will have a charge on its surface. This surface charge will attract an opposite charge in the tank that is earthed. Eventually, the opposite charges will come together, and the charged liquid becomes neutral.

This attraction of charges aiming the neutral state is called the relaxation time. When a liquid is an insulator, the relaxation time is very long. The relaxation time varies from fractions of a second to several minutes, depending on the electrical conductivity of the liquid.

If the potential difference between the surface of the liquid and the metal tank is too high, the ionization of the air can occur and a spark can jump to the tank. If a mixture of flammable vapour and air is present, an explosion can occur.

In a tank insulated from earth, when the tank is being filled with liquid, a charge appears on the surface. This charge on the surface will attract a charge of opposite polarity inside the wall of the metal tank. The outside of the tank will have a free charge of the same polarity as the charge on the surface of the liquid. This charge is capable of producing a spark to earth. In a tanker truck, this spark can be between the open tank and the filling valve, which is earthed.

When the liquid is conductive, filling or handling plastic or other non-conducting containers can still be hazardous. The splashing and turbulence of the liquid in the container can cause a static electric charge to build up in the liquid or on conductive parts on the container that are not earthed. A spark with enough energy to ignite a vapour–air mixture in its flammable range (an incendive discharge) can originate from the liquid or from the container.

For portable tanks, intermediate bulk containers, and non-bulk containers, the recommendation is to earth any metal parts on the container (and nearby conductive surfaces that the container may come in contact with) and fill the container from the bottom through a long, earthed metal pipe. This procedure will reduce the amount of static charge produced and will enable the generated charge to relax (dissipate) through the metal pipe.

The smallest container should be used for the job. The risk of fire from static electricity increases with the volume of the container and the volatility of the liquid being used.

When filling non-conducting portable containers, the recommendation is that an earthed dip pipe, or earthed wire, be in the liquid in the container while it is being filled. The filling rate should be minimized, especially if there is a filter in the line.

Any metal parts of the container and metal funnel, if one is used, should also be earthed. When filling containers with low conductivity liquids (i.e., ones with a conductivity less than 50 picoSiemens, pS), one should keep the earthed dip rod in the liquid for around 30 seconds after the filling is completed.

Similarly, filling an unearthed portable fuel tank on a plastic-lined truck bed can cause spark-induced gasoline fires. For that reason, portable fuel tanks should be removed at a safe distance from the vehicle (which, of course, is turned off) and be filled on the earth. The nozzle should be held in contact with the container while it is being filled.

8.3   CHARGE ACCUMULATION

After being generated, the accumulation of electrostatic charge is the process that raises the potential to dangerous values. The charge can accumulate in the process materials, in practically any component that does not have a path for the charge to flow to the earth.

This occurs whenever the rate of charge generation exceeds that of charge dissipation. One of two results may occur for a leaky capacitor system with constant charging current: either a steady state is attained in which the rate of current dissipation balances the charging current, or a static discharge occurs before such a condition can be reached.

Figure 8.5 illustrates the situation, when a high-resistivity liquid, gas, or powder, being electrostatically charged during process operations, will charge an electrically isolated conductive plant equipment and materials in direct contact with, or in close proximity to, it.

The risk associated with static electricity in explosive atmospheres is closely linked to the energy released by the discharge between two surfaces.

Each substance has a specific value for its Minimum Ignition Energy (MIE). A suitable mixture of hydrogen and air, for example, requires only 0.02 mJ to ignite. Therefore, a careful evaluation of the characteristics of industrial processes, and of the materials used, needs to be done.

In Figure 8.5, I_c is the electrostatically charged powder or liquid that is making contact with the object C1 (that is earthed), which could be a road tanker or any process equipment that is being charged. R represents the electrical resistance between the charged object and the earth. The object that is being charged, C1, if for some reason is isolated from the earth, this isolation is caused by something that places a high resistance, R, between the object and the earth.

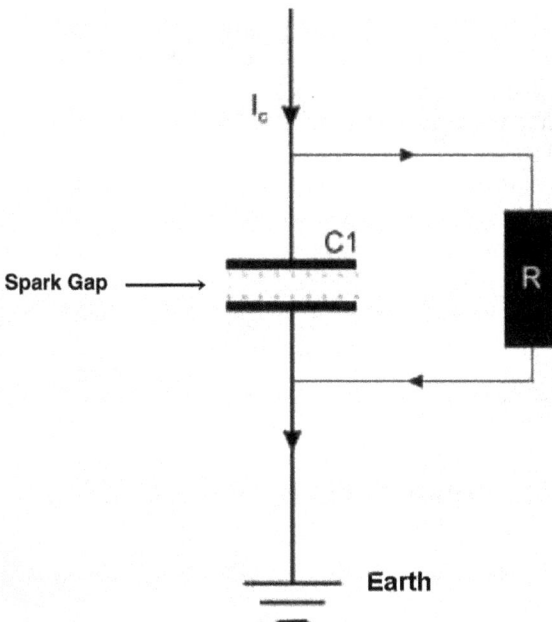

FIGURE 8.5   Representation of charging in a process equipment.

Where:
   - The capacitor represents the object becoming increasingly charged due to con-
     tact with the charged powder or liquid.
   - The resistor represents the resistance between the charged object and a connec-
     tion to the true earth.

If C1 had a low-resistance connection to earth, the charge would flow directly
to earth, because the general mass of the earth has an infinite capacity to balance
electrical charges, which would result in no voltage being present on the object C1.
If the resistance to earth is high, it will impede the flow of charge from the object to
earth. The charge will, instead, rapidly accumulate on C1, the object. As more charge
is deposited onto C1, its voltage will increase rapidly.

Although the magnitude of the charging current, I_c, can be very small (typically
no more than 100 µA), the voltage on the object can be very high, easily entering the
kV range. The relationship between voltage, charge, and capacitance can be summed
up in Equation 8.1.

$$V = \frac{Q}{C} \tag{8.1}$$

Where:
   V—voltage of charged object [Volt]
   Q—total quantity of charge on the object [Coulomb].
   C—capacitance of charged object [Farad].

So, it can be seen that if the object being charged is, as an example, a metal drum
with a capacitance C of 100 pF and has 1.25 µC of charge Q deposited onto it by an
electrostatically charged liquid, its voltage V will be 12.5 kV. If more charge is deposited onto the drum, its voltage will continue to rise.

The potential for an Electrostatic Discharge (ESD) to ignite flammable liquids
and combustible dusts is real. As shown in Table 8.1, the amount of energy required
to ignite common flammable materials is rather low.

It is important to note that, in practice, charging currents may not be constant,
and the system considered may be significantly changed during the charging process.
Another charging process is a powder loading to a hopper, where the effect of rising
level inside the container must be included in any analysis.

TABLE 8.1
Electrostatic Sensitivity Levels of Some
Flammable Materials

Material	Ignition Energy [J]
Methane	2.9×10^{-1}
Cyclopropane	1.8×10^{-4}
Ethylene	8.0×10^{-5}

TABLE 8.2
Powder Classification on the Basis of Its Volume Resistivity

Volume Resistivity [Ωm]	Powder Classification
Up to about 10^6	Conductive
10^6 to 10^9	Medium resistivity
Above 10^9	High resistivity

Generally, powders are divided into three groups depending on their ability to retain static charge, even if the powder is in contact with an electrically earthed conductive object. This ability is known as Volume Resistivity, as shown in Table 8.2.

The hazard of charge accumulation usually increases with increased container volume. For example, if a tall, cylindrical container is filled with uniformly charged, non-conductive powder, the maximum electric field appears at the wall, and it is proportional to the radius of the container, while the maximum potential appears on the axis and it is proportional to the square of the radius.

It follows that for any assumed powder charge density, the electric field eventually exceeds the breakdown field of air as container radius increases. In cases where charge dissipation occurs at a significant rate, such as liquid tank filling operations, it becomes impractical to fill very large tanks fast enough to offset the rate of charge dissipation. Therefore, hazardous charge accumulation is typically not seen in very large storage tanks, but, instead, the hazard maximizes at an intermediate tank size, depending primarily on flow rate, inlet pipe diameter, and liquid conductivity.

Owing to the competitive processes of charge generation and dissipation, the rate at which a container is filled is an important factor whenever significant current leakage can occur. Effects can be additive. For example, during drum filling with a non-conductive liquid, higher charge densities result from higher velocities and rates of shear at the pipe wall, while there is less time for charge inside the drum to dissipate to the walls.

If the drum is poorly earthed, the combination of a high flow rate and charging current is likely to result in hazardous charge accumulation on the drum. If the charging current is very high, charge dissipation is limited by conduction through the liquid itself and static discharges may occur inside a properly earthed drum.

In hazardous areas, in order to avoid explosions, the objective is to ensure that the equipment potential difference does not rise during operation.

As the charge accumulation can only take place if there is a resistance present between the equipment and general mass of earth, the connections to earth should be provided by high-integrity connectors. These high-integrity earth connectors should be providing protection against lightning strikes and electrical faults with plant equipment, providing a satisfactory path for static electricity.

For safety reasons, it is necessary to ensure that any plant equipment, whether it is mobile or fixed, never becomes isolated from the designated earthing points. The common causes for equipment to become isolated are shown in Table 8.3.

TABLE 8.3

Electrical Isolation (Capacitance) of Some Objects

Object	Probable Points of Isolation
Portable drums	Protective coatings, product deposits, rust
Piping	Rubber and plastic seals, anti-vibration pads, and gaskets
Flexible intermediate bulk container (FIBC)	Non-conductive fabric

8.3.1 Consequences of Charge Accumulation

Among the most common effects of the accumulation of electrostatic charge, we have sparks. The accumulation of charge can reach values higher than the dielectric strength of air, causing "sparks", which usually "jump" a few millimetres. They can occur, for example, in metal drums during filling or emptying. Liquids with low conductivity, when agitated in a reactor, can experience sparks, which can even damage the visual monitoring windows. Sparks can be observed as a result of:

a) **Brush discharges**

They occur in insulating materials that have been charged, such as plastic drums. Brush discharges are limited in energy but can easily ignite the vapours of most solvents. They are not easy to detect, and therefore careful assessment should be made in the industrial environment.

Important differences exist between brush discharges and sparks. The first is that as only a part of a charged insulator contributes to the brush, it is possible to obtain many discharges from a single large charged surface. Second, because of the low mobility of charge carriers, in an insulator, it is impossible to remove the charge by simply connecting to earth. The discharges themselves also vary significantly in the way the energy is released with brushes producing lower currents and consequently lower temperatures.

b) **Propagating brush discharges (PBD)**

They occur when there is a charged insulator near a conductor, and also when powder flows through a plastic pipe or even when the powder accumulates in a plastic-lined container. Its energy is sufficient to ignite any explosive atmosphere. Such discharges can emit bright light and noise and cause damage to the surfaces of insulating materials.

In the industry, PBD may occur where large areas of insulating sheets are subjected to high levels of electrostatic charge generation (note that both of these conditions are necessary). More often than not, these insulating sheets are in the form of a lining or layer, with an earthed metal backing. An example of this situation is a Teflon or glass-lined reactor vessel.

c) Conical discharges

They are a special case of brush discharge that occurs on the surface of a powder that accumulates from a conductive funnel. These discharges tend to have a radial direction and have enough energy to ignite clouds of combustible powders and flammable vapours. Conditions for cone discharges are a charged powder of resistivity exceeding 10^{10} Ω.

In earthed metal silos, the discharges travel radially towards the silo wall across the surface of the powder cone during filling and may have an effective energy of up to 20 mJ. Particle size and charge density will both impact the resultant ESD energy.

Although they are rarely observed with the naked eye, the appearance of some disturbances along the ducts allows people to perceive the action of the accumulation of electrostatic charges and the probable occurrence of conical discharges.

8.4   CONDITIONS FOR IGNITION

The electrostatic discharge, depending on its liberated energy, can be a source of ignition in hazardous areas. In order for static electricity to be a source of ignition, four conditions must be fulfilled:

- The rate of charge generation must exceed the rate of dissipation, so charge can accumulate
- A static discharge must coincide in time and space with a flammable atmosphere
- The effective energy of the static discharge must exceed the ignition energy of the local mixture
- A locally ignited flame must propagate into a surrounding flammable atmosphere.

Due to their higher energy content, propagating brush discharges and cone discharges can ignite dust clouds. Sparks from unearthed conductors, however, remain the most common source of electrostatic ignition in industry.

It is possible for a sensitive flammable atmosphere and electrostatic discharges to coexist without ignition. If the discharge energy is increased, threshold energy at which the discharge becomes incendiary is eventually obtained either partially or wholly through the volume of the flammable medium.

This threshold energy is known as the MIE and is usually expressed in millijoules (mJ). So, charge accumulation and static discharges are commonplace in some systems, such as pneumatic silo filling with powders, where ignition may be delayed for many years by failure to meet all the four conditions. For example, a non-uniform dust cloud's ignition energy varies by orders of magnitude with time and position, while a large range of effective energies may be manifested by static discharges, also

TABLE 8.4
Capacitance of Selected Objects

Charged Items	Capacitance [pF]	Potential [kV]	Energy [mJ]
Flange/bucket	10	10	0.5
Small tank (< 50 l)	500–100	8	2–3
Human being	100–200	12	7–15
Large tank	100–1,000	15	11–2

TABLE 8.5
Minimum Ignition Energy of Selected Products

Product	Minimum Ignition Energy [mJ]
Acetone	0.55
Ethanol	0.28
Methanol	0.20
Hydrogen	0.016

varying with time and position. For flame propagation to result, these two random events must coincide in time and place such that a local ignition occurs, then the small flame kernel so formed must propagate into a surrounding flammable mixture without being quenched.

Table 8.4 shows the accumulated energy due to the capacitance of items that became charged to the given potential, and Table 8.5 shows the MIE of some flammable gases. These data prove the risk of explosions caused by electrostatic discharges in hazardous areas.

8.5 FIRES AND EXPLOSIONS DUE TO ELECTROSTATIC DISCHARGES

There are many accidents involving electrostatic discharges as sources of ignitions that led to significant damage. Some examples are given next.

8.5.1 OIL TANKERS

Explosions and fires in oil and gas tankers have occurred where static electricity has been blamed as the cause. After a tank has been emptied, vapour may be present having been mixed with air that has been sucked into the tank as a result of emptying. This provides almost ideal conditions for static electricity to cause ignition during the beginning of cleaning operations.

Drum filling is a good example to illustrate the principle (see Figure 8.2). Where the liquid enters the drum, leaving the supply pipe nozzle, the flow profile deposits a charge on the nozzle and the opposite charge accumulates in the drum, depositing itself on the metal of the drum.

The charge appears between the nozzle and the filling cap rim. To equalize the charge, it is necessary to bond the drum to the nozzle, such that the charge is dissipated.

As the nozzle is moved towards the drum at the start of the filling cycle and away at the end, this is where any initial potential difference may cause a spark. Bonding must be in place before and after the filling event to ensure safety.

Preventive measures in order to avoid explosions in oil tankers due to electrostatic discharges are included in IEC TS 60079–32–1 as:

a) The bonding resistance between the chassis, the tank, and the associated pipes and fittings on the truck is recommended to be less than 1 MΩ. For wholly metallic systems, the recommended resistance should be 10 Ω or less, and if a higher value is found, further investigations should be made to check for possible problems of e.g. corrosion or loose connection.

b) An earthing cable should be connected to the truck before any operation (e.g. opening man lids, connecting pipes) is carried out. A recommendation is to provide a resistance of less than 10 Ω between the truck and the gantry's designated earthing point, and this should not be removed until all operations have been completed.

c) It is recommended that the earth cable required in b) be part of a static earth monitoring system that continuously monitors the resistance between the truck and a designated earthing point on the gantry and activates interlocks to prevent loading when this resistance exceeds 10 Ω.

It is further recommended that the static earth monitoring system should be capable of differentiating between connection to the truck's tank (or earth connection point) and other metal objects. This type of system will prevent operators from connecting the earthing system to objects (e.g. the mudguards) that may be electrically isolated from the truck's container.

8.5.2 OIL AND GAS INSTALLATIONS

From 1983 to 2013, 99 electrostatic accidents occurred during the process of oil–gas storage and transportation. It was shown that about 85% of the accidents occurred in tank farms, gas stations, or petroleum refineries, and 96% of the accidents included fire or explosion.

The results show that three major reasons were responsible for accidents, including:

- Improper operation during loading and unloading oil
- Poor grounding
- Static electricity on human bodies.

Those accounted for 29%, 24%, and 13% of the accidents, respectively.

Safety actions suggested to help operating engineers to handle similar situations in the future are:

1. Ensure that all equipment and tanks are properly earthed to prevent the build-up of electrostatic charges.
2. Implement continuous monitoring systems for electrostatic charges during flammable liquid transfer operations.
3. Train workers on the dangers of electrostatic discharges and safe handling practices.

8.5.3 CHEMICAL INDUSTRIES

An explosion occurred inside the polypropylene silo at the Yeosu Industrial Complex, Korea, causing a fire in the polypropylene stored inside the silo, resulting in the burning of one aluminium silo on July 10, 2017. Fortunately, there was no human injury due to the shift time.

It is possible that the unreacted components, such as propylene and ethylene impregnated in the polypropylene powders, were desorbed and accumulated in the silo during the synthesis process. The flammable gas was measured to be 3,764 ppm after the fire suppression. Although the gas concentration meter for measuring propylene concentration was present in the process, the calibration of the gas sensor was set without multiplying the gas correction value for methane, and it was mistakenly assumed that it did not exceed the Lower Explosive Limit (LEL) at that time.

As the pellet product was conveyed by air pressure, friction between the pipe and the pellet occurred. As a result, friction static electricity was generated, and polypropylene dust floating in the silo was charged. Under normal conditions, it was transported to the silo, and, even if charged, the residual fine dust was discharged through an air purification device. However, it was presumed that the dust with static electricity was introduced into the storage silo, the lower part of which was open at the time of the accident, and the discharged pellet product acted as an ignition source for the flammable gas stagnation space.

Safety actions in order to avoid such events:

• Silos used in the production and storage of various resin products must be evaluated and monitored for the occurrence of gas concentrations at a normal time, due to process characteristics
• Oxygen concentrations must be lower than the explosion minimum concentration through the inclusion of an inert gas such as nitrogen. If a change in oxygen concentration is observed, it is considered to be a dangerous situation, and the operation should be immediately stopped.

8.6 CONTROLLING THE IGNITION RISK

As sparks due to static electricity cause several industrial fires and explosions every year, it is a priority to identify weak points and establishing control methods in order to prevent explosions and fires in industrial facilities. Electrostatic events in the

industrial unit must always be reported, and investigations by specialized professionals must be carried out to ensure that they do not become precursors to a fire or
explosion, which can be devastating.

Some efficient preventive measures are described in the next subsections.

8.6.1 EARTHING AND BONDING

Earthing and bonding are very effective ways to control static electricity and are
widely used as a first line of defence. The principle is to provide a conductive path
for static charge to flow between two objects or an object and earth.

Bonding consists of effectively connecting two conductive objects together to
minimize the potential difference between them. This eliminates the gap between
the two objects and provides a conductive path through which static charges can
flow to recombine and neutralize. Therefore, the objects will always be at the same
potential and no spark can occur.

Earthing consists of effectively connecting a conductive object to earth. The conductive path between an object and earth allows any static charge generated on the
object to flow to earth, thus eliminating the accumulation of a charge. If no charge is
allowed to accumulate, no spark can occur.

8.6.1.1 Resistance

Effectively earthing (or bonding) an object in the realm of static electricity is different than that of power distribution. To prevent the accumulation of static charge, the
resistance between two objects, or an object and the earth, can be as much as 1 MΩ.

This is based on the following rationale. Using Equation 8.2, the maximum
acceptable resistance of a connection to earth can be calculated as:

$$E = 1/2\ CV^2 \tag{8.2}$$

Where:
  E is the energy in joules;
  C is the capacitance in farads between a charged body and earth, and
  V is the voltage to which the capacitor is charged.

$$V = IR \tag{8.3},$$

therefore:

$$E = 1/2\ C\ (IR)^2 \tag{8.4}$$

Assuming that the lowest level of energy necessary to ignite most flammable vapours
is 0.1 mJ, the resistance to earth must not exceed

$$R = \sqrt{\frac{2 \cdot 10^{-4}}{C \cdot I^2}} \Omega \tag{8.5}$$

Using a tank truck as an example, with a charging current of 1.00×10^{-6} A and a capacitance of 1.00×10^{-12} F, it can be assumed to put things in perspective. According to this equation, the resistance to earth could be as high as $4.47 \times 10^6 \; \Omega$.

Considering that earthing and bonding are the first preventive measures, it is important to know the requirements for earthing clamps, which provide connection via cables to earthing points and are important for preventing electrostatic discharge in mobile or fixed equipment in industrial plants with explosive atmospheres. Although they are visually simple construction products, we must consider that in certain installations, they may be used hundreds of times a day, making it essential that they ensure a good electrical connection at all times. The effectiveness, reliability, and durability of electrostatic earthing, including the clamp and associated cables, are essential for process operations to be carried out safely, avoiding fires due to possible sparks.

In process plants, it is common to find containers, drums, and trucks with rust or even completely painted. Such situations can cause the formation of an insulating layer that can render certain clamps normally used for earthing, ineffective.

In some countries, the design of the earthing clamp must be evaluated by certification bodies to allow its use in hazardous areas. In Europe, earthing clamps must meet a specific criterion to be certified as suitable for use in hazardous areas. For example, earthing clamps made of aluminium must be anodized for use in Zone 0 or Zone 20 areas, and there is also a limitation on the amount of plastic that can be used on the clamp body, all analysed to avoid the accumulation of static charges.

The European guidelines also require an assessment of possible sources of stored energy and their ability to cause a spark in the hazardous area. Therefore, the clamps must contain any stored energy within their structure to prevent it from causing a spark in the surrounding environment.

In the United States, other requirements for earthing clamps are applied, such as the maximum resistance at the point of contact of the clamp with the earthing point, to declare the product as suitable for use in hazardous areas.

Among other tests that the clamps are subjected to, in order to obtain approval from the certifying bodies, are:

- Separation force—to verify that the clamps are not easily removed during operation
- Contact pressure—to verify that the clamps can break through foreign films, such as rust and coatings, and make an effective connection of the equipment to the earth
- Vibration—to verify that contact will be guaranteed even when the clamp is subjected to external interference.

8.6.2 ANTISTATIC MATERIALS

Another interesting solution, suitable for industrial processes, is the use of antistatic materials in process equipment. Most antistatic additives use an electrically conductive material, such as carbon fibre, to allow static discharge to take the path of least resistivity. A properly manufactured and installed conductive device can provide the path of least resistivity in the same way that a lightning rod can protect a property by diverting atmospheric discharges to a safe, isolated path.

There are two types of antistatic products that act directly to vary the speed of static discharge that may be present in a given process.

a) **Static-conductive**

It acts quickly because it drains the charge almost instantly. However, it presents a considerable risk in explosive atmospheres, as it can generate arcs and sparks. The surface resistivity in this case is less than 10^5 Ω/square.

b) **Static-dissipative**

The speed is lower than in the conductive alternative; however, the formation of arcs and sparks is avoided. The surface resistivity must be at least 10^5 Ω/square but must not exceed 10^{12} Ω/square.

The unit ohms/square is very convenient when working with a thin, flat piece of conductive material, like a sheet of graphite. If the piece is divided into small, identical squares, no matter which square is picked, the resistance for a current flowing from one side of a square to the other (side-to-side, not diagonally) remains the same.

That is "ohms per square". So, if a material has a surface resistivity of 10 Ω/square, and a current flows across one square, the resistance it encounters is 10 Ω.

And, if the current flows across ten squares in a row (still in a straight line), the resistance would be 10 Ω/square multiplied by the number of squares (10), totalling 100 Ω. And, the resistance per square does not change with the square's size; it is consistent for any square configuration. This unit helps in assessing and comparing the resistive properties of materials, without the need for complex calculations considering their dimensions.

It is also important to note that resistance and resistivity are related but distinct concepts in electrical engineering. Resistance measures how difficult it is for an electric current to flow through a material. It is dependent on the material's dimensions (length and cross-sectional area) and its resistivity. The Equation 8.6 gives resistance (R) as

$$R = \rho \times L/A \ \Omega \qquad (8.6)$$

Where:
ρ—resistivity,
L—length, and
A—the cross-sectional area.

Resistivity is an intrinsic property of the material, indicating how strongly it resists the current flow. It is independent of the material's shape or size, and its unit is Ωm. High resistivity means the material resists current well; low resistivity means it conducts well. In brief, resistivity is like an inherent "oppositional nature" of the material, while resistance is the result of combining that nature with the shape and size of the material.

To avoid dangerous electrostatic effects, the surface resistance of equipment enclosures and components should not exceed 10^9 Ω (earthing in an electrostatic sense), or the surface areas of electric isolating parts should be limited (depending on the MIE of the combustible substance, e.g. to a maximum area of 100 cm^2).

8.7 CLOSING REMARKS

All facilities that contain hazardous (classified) areas should be carefully reviewed to determine if the possibility exists for the generation of static electricity, and if a subsequent static discharge could result in the ignition of a flammable atmosphere. If such a situation exists, a static electricity earthing policy should be developed and implemented.

A static earthing policy should include all of the requirements and procedures necessary to ensure safe operating conditions and to reduce the probability of static electricity ignition of flammable atmospheres. All terms in such a policy should be appropriately defined, including the requirements concerning the wearing of shoes and clothing, as well as the proper earthing of portable equipment, the electrical continuity of solvent and exhaust hoses, and the use of plastics. In addition, the specific procedures relating to the use, the testing, and the maintenance of the elements required for the mitigation of static electricity should be included.

BIBLIOGRAPHY

API RP 2003—*Recommended practice for protection against ignitions arising out of static, lightning, and stray currents.* American Petroleum Institute, 2015.

CSB Investigation Report No 2008-05-I-GA – *Sugar dust explosion and fire: Imperial Sugar Company.* Chemical Safety and Hazard Investigation Board, Sep. 2009. Available at: www.csb.gov/userfiles/file/imperial%20sugar%20report%20final%20updated.pdf

Hattwig, M. and Steen, H. – *Handbook of explosion prevention and protection.* Wiley-Vch, 2004.

Hu, Yuqin et al. – A case study of electrostatic accidents in the process of oil-gas storage and transportation. *Journal of Physics: Conference Series,* vol. 418, 2013, p. 012037.

IEC TS 60079-32-1 – *Explosive atmospheres – Part 32-1: Electrostatic hazards, guidance.* International Electrotechnical Commission, 2013.

Luttgens, G. and Wilson, N.—*Electrostatic hazards.* Oxford: Butterworth-Heinemann, 1997.

NFPA 77 – *Recommended practice on static electricity.* National Fire Protection Association, 2019.

Pak, Seonggyu, Jung, Seongho, Roh, Changhyun and Kang, Chankyu – Case studies for dangerous dust explosions in South Korea during recent years. *Sustainability,* vol. 11, no. 18, 2019. Available at: https://doi.org/10.3390/su11184888

Pratt, Thomas - *Electrostatic ignitions of fires and explosions.* American Institute of Chemical Engineers, 2000.

Rangel Jr., Estellito – Eletricidade estática. *EMEx Section, Eletricidade Moderna Magazine,* São Paulo, ed. 526, Jan. 2018, p. 58.

Rangel Jr., Estellito—Eletricidade estática. *EMEx Section, Eletricidade Moderna Magazine,* São Paulo, ed. 374, May 2005, p. 178.

Rangel Jr. Estellito—Perigos da estática. *EMEx Section, Eletricidade Moderna Magazine,* São Paulo, ed. 561, Sep/Oct. 2021, p. 66.

Rangel Jr., Estellito—Explosões na indústria farmacêutica: dados e fatos. In: *IV Seminário Internacional de Prevenção e Proteção contra Explosão na Indústria*, São Paulo, 2008.

Rangel Jr., Estellito—Garras de ateramento. *EMEx Section, Eletricidade Moderna Magazine*, São Paulo, ed. 497, Aug. 2015, p. 92.

Rangel Jr., Estellito and Silva, M. Valério.—Risks due to static electricity at fuel loading terminals. In: *I Petroleum and Chemical Industry Conference*, IEEE, 2006.

Rangel Jr., Estellito—Riscos da eletricidade estática. *EMEx Section, Eletricidade Moderna Magazine*, São Paulo, ed. 443, Feb. 2011, p. 160.

9 Lightning Protection

9.1 INTRODUCTION

Thunderstorms come into existence when warm air masses containing sufficient moisture are transported to great altitudes. This transport can occur in a number of ways. In the case of heat thunderstorms, the earth is heated up locally by intense insolation. The layers of air near the earth heat up and rise.

Lightning is a natural phenomenon which develops when the upper atmosphere becomes unstable due to the convergence of a warm, solar heated, vertical air column on the cooler upper air mass. These rising air currents carry water vapour which, on meeting the cooler air, usually condenses, giving rise to convective storm activity.

Lightning can also be produced by frontal storms, where a front of cold air moves towards a mass of moist warm air. The warm air is lifted, thus generating cumulonimbus clouds and lightning. One major differentiation of this type of event is that the cold front can continue its movement and result in cumulonimbus clouds spread over several kilometres of width.

The surface of the earth is negatively charged, and the lower atmosphere takes on an opposing positive space charge. As rain droplets carry charge away from the cloud to the earth, the storm cloud takes on the characteristics of a dipole with the bottom of the cloud getting negatively charged and the top of the cloud positively charged.

Electrostatic charge separation processes, e.g. friction and sputtering, are responsible for charging water droplets and particles of ice in the cloud. Positively charged particles accumulate in the upper part, and negatively charged particles in the lower part of the thundercloud. This separation of positive and negative charges within the clouds, due to the collision of ice particles and water droplets in different stages of freezing, causes electrification.

When the difference in electrical potential between two regions of the cloud, or between the cloud and the earth, is sufficiently high, the air ionizes and forms a conductive path for the electrical discharge. The ionized path allows the charges to move rapidly, generating heat and light visible as lightning and an acoustic shock wave that is heard as thunder.

So, lightning is a natural phenomenon resulting from electrical discharges in the atmosphere, which occurs when there is a significant accumulation of electrical charges in cumulonimbus clouds and can be formed between different parts of the same cloud (intracloud), between different clouds (intercloud), or between the cloud and the earth (discharge to the earth), as shown in Figure 9.1.

DOI: 10.1201/9781003500001-9

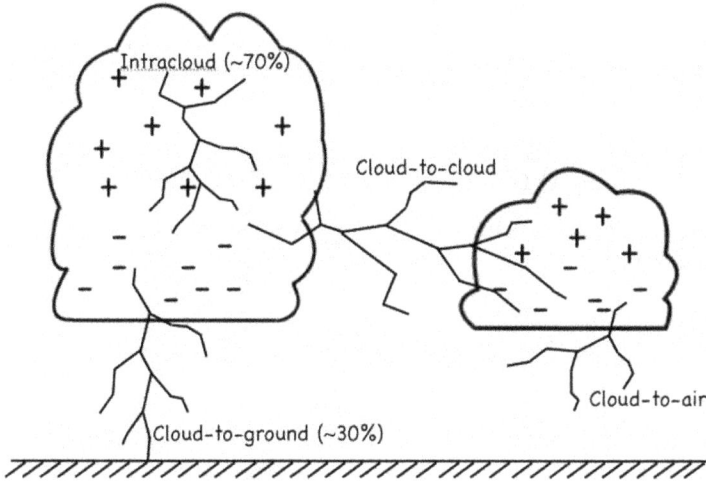

FIGURE 9.1 Atmospheric discharges.

9.2 IMPORTANCE OF LIGHTNING PROTECTION

Lightning poses considerable risks to industrial installations, with potential to cause heavy losses, especially in hazardous areas. Among them, we highlight the following:

- Fire and/or explosion triggered by heat of lightning flash or even melted metals
- Punctures of structure roofing due to plasma heat at lightning point of strike
- Failure of internal electrical and electronic systems.

Industries with hazardous areas are risk locations due to the presence of flammable gases or vapours. Among the benefits of a Lightning Protection System (LPS) in these industries, are:

a) *Fire and explosion prevention:*
    A lightning strike in a hazardous area can cause catastrophic fires and explosions. An LPS minimizes this risk by intercepting and conducting the lightning energy safely to the earth.
b) *Protection of critical infrastructure:*
    Lightning can cause direct and indirect damage to equipment and facilities essential to the operation of, as an example, petrochemical industries with their storage tanks, processing units, and control systems.

c) *Operational continuity:*

Production interruptions due to lightning damage can result in significant financial losses. An LPS contributes to operational continuity by avoiding unplanned shutdowns.

9.3   FIRES AND EXPLOSIONS DUE TO LIGHTNING

Some recent examples of damages caused by lightning in industries with hazardous areas are shown below.

a) **El Palito Refinery**

Location:	Carabobo, Venezuela
Year of occurrence:	2012
Approximate damage:	US$ 10 million
Event description:	

On September 20, 2012, lightning struck two of the oil storage tanks at the Venezuelan refinery. About 120 firefighters worked to put out the blaze. The refinery processes 135,000 barrels/day.

b) **Calcasieu Refinery**

Location:	Lake Charles, Louisiana, US
Year of occurrence:	2023
Approximate damage:	US$ 9 million
Event description:	

A tank that contained naphtha, an extremely flammable hydrocarbon product, caught fire on June 3, 2023, when lightning struck it at Calcasieu Refining, Louisiana. When the fire started, the tank contained around 46,000 barrels (there are 42 gallons in a barrel).

c) **Balongan Refinery**

Location:	Balongan, Indonesia
Year of occurrence:	2021
Approximate damage:	US$ 500 million
Event description:	

On March 29, 2021, lightning struck the Balongan refinery, operated by PT Pertamina, causing an explosion and a massive fire. The fire spread rapidly, impacting several storage tanks and causing the evacuation of hundreds of nearby residents.

Indonesia's Balongan refinery, one of six operated by the state-owned oil firm Pertamina, contributes about 12 per cent of Indonesia's petroleum refining capacity or 125,000 barrels a day. The net profit margin of Balongan was some US$ 225 million in 2017, equivalent to about 4.3 per cent of Pertamina's total profits.

d) **Matanzas Supertanker Base**

Location:	Matanzas, Cuba
Year of occurrence:	2022

Approximate damage: US$ 15 million
Event description:

A massive, lightning-sparked fire feeding on two oil storage tanks in the Cuban city of Matanzas raged out of control on August 6, 2022, and injured at least 121. Roughly 1,300 people living near the fire at the Matanzas Supertanker Base have been evacuated, with an additional 600 workers told to leave.

e) **Gas line**
   Location: Collierville, Tennessee, US
   Year of Occurrence: 2024
   Approximate damage: US$ 10 million
   Event description:

On December 16, 2024, lightning struck the gas line in Collierville. The company MLGW said that the suspected cause was lightning striking an 8-inch gas line causing a rupture.

9.3.1 FIRES IN TANKS

a) **Fixed-roof and horizontal tanks**

Most tank explosions that have occurred as a result of lightning strikes have been attributed to the following:

- Roof openings, such as gauge hatches, that have been left open
- Vents that have not been protected by flashback devices such as pressure-vacuum vent valves
- Corrosion holes or thinned areas of tank roofs.

Gas and oil storage are not the only products stored in above ground tanks that are at risk of lightning strikes. Tanks with chemicals, such as benzene, toluene, and xylene are also at risk.

Storage tanks with fixed cone roofs are available in designs that prevent explosions. A weak seam built into the top allows the roof to lift out of the way during overpressure. This design allows fire departments to access the fire at the fuel surface. With easy access, fireman can suppress the flames with a foam spray application.

Maintenance errors are the primary cause of fires in cone and fixed-roof tanks. Poor grounding and liquid leaks provide the vapour/oxygen mix that a lightning strike will ignite. Routine tank inspections are recommended to verify proper operation of venting equipment, overfill prevention, and leak detection systems.

b) **Open Floating-Roof Tanks**

Fires have occurred in the sealed space of open FRT as a result of lightning-caused discharges. Ignition can happen from a direct strike or from the sudden discharge of an induced (bound) charge on the floating roof. The induced charge is released when a charged cloud discharges to the earth somewhere in the vicinity of the tank.

The most effective defence against ignition by lightning is a tight seal and properly designed shunts. Submerged shunts are metallic straps placed at intervals of not more than 3 metres on the circumference of the roof, which bond the floating roof to the shell and permit any lightning-related current to propagate to earth, without generating a spark in an area likely to ignite vapour.

c) Internal Floating-Roof Tanks

Internal FRT with conductive steel roofs are inherently protected against internal ignitions from lightning-induced sparking by the Faraday-cage effect. As a result, the internal floating roof does not require bonding to the shell or tank wall for lightning protection. However, the floating roof or cover still requires bonding to the shell for protection against electrostatic charges due to product flow.

Limited experience to date indicates that aluminium geodesic domes do not impose an increased risk.

Because internal FRT have open vents, it is important that the potential for a flammable atmosphere between the fixed roof and floating roof be minimized by:

- Providing a tight seal around the floating roof
- Providing adequate venting in accordance with API 650
- Avoiding landing the roof during normal operations (refloating the roof displaces the vapours into the vapour space)
- Avoiding initial filling of the tank during an electrical storm
- Testing the vapour space for combustibles on a routine basis (in order to check the condition of the seal).

FRT are especially vulnerable to lightning strikes. Fires in FRT are typically started because of arcing between the floating roof and the tank shell.

API RP 2003 recommends three measures for FRT:

- Install submerged shunts between the roof and shell every 3 metres around the roof perimeter and remove any existing above-seal shunts
- Electrically insulate all seal assembly components (including springs, scissor assemblies, seal membranes, etc.) and all gauge and guide poles from the tank roof
- Install bypass conductors between the roof and shell no more than every 30 metres around the tank circumference. These bypass conductors should be as short as possible and evenly spaced around the roof perimeter.

9.3.2 Fires in Silos

A lightning bolt hit the primary grain elevator, leaving a hole in it and starting a fire on November 14, 2015, in Watonwan Farm Service, Minnesota, US. No one was injured.

A concern in agricultural installations is the position of the cell phone antennas, usually at the highest point, which increases the risk of lightning for the host structure. The electromagnetic emissions of the antennas can disrupt nearby installations in or on the host structure or, in the worst case, cause an explosion in the case of silos.

For structures as the grain storage facilities, where there are hazardous areas, the most important thing is to perform not only a lightning protection study but also a safety survey to validate that electromagnetic disturbances generated by the antennas will not create any damage (Electromagnetic Compatibility).

Silos must be effectively protected against the risks generated by electrostatic discharges and lightning effects. They must not have collective transmission or reception antennas on roofs, unless a technical study justifies that devices installed are not a source of sparkover, fire, or risk of dust ignition, considering the lightning study conclusions also.

Electromagnetic radiation may disturb the good operation of safety devices installed on or in the structure. The level of the radiated field needs to be kept below the threshold susceptibility of these devices (determined in the EU by the EMC Directive 2014/30/EU). As a matter of fact, these devices must operate normally in the presence of an electric field with a maximum value fixed at 10 V/m.

9.4   THE LIGHTNING PROTECTION SYSTEM

The concept of LPS was originally treated in IEC 61024, and it was ruled by a set of constructional directions according to the desired Lightning Protection Level (LPL) from I to IV. Years ago, the objective of the LPS consisted of the protection of the structure and its contents against direct lightning flashes. The method for determining the LPL of the LPS was based on an efficiency factor, E, first introduced by that standard. It was defined to show the percentage of possible flashes, which could be controlled by the LPS without resulting in damage.

IEC 62305 series replaced IEC 61024 and introduced two big upgrades:

- While IEC 61024 only assessed the structure's overall lightning risk, by installing an external LPS of an appropriate LPL, IEC 62305 also analyses the internal part of the structure, permitting to separate the LPS design in two different approaches: External and internal LPS, which are discussed in their parts 3 and 4. A new set of measures for lightning protection and the new concept of *Lightning Protection Zones* (LPZ) were also included.
- The risk management methodology used in IEC 61024, with weighting factors and probabilities that translate the characteristics of the structure and its contents, became standardized in IEC 62305, aiming to give a safer solution for lightning protection.

But, the acceptance of the scientific community of the new methodologies introduced by IEC 62305 series took time. As an example the NFPA 780, that used the weighting factors of IEC 61024, only adopted the detailed risk assessment implemented by IEC 62305 in 2011, five years after the issuing of IEC 62305-1 ed. 1.0 in 2006.

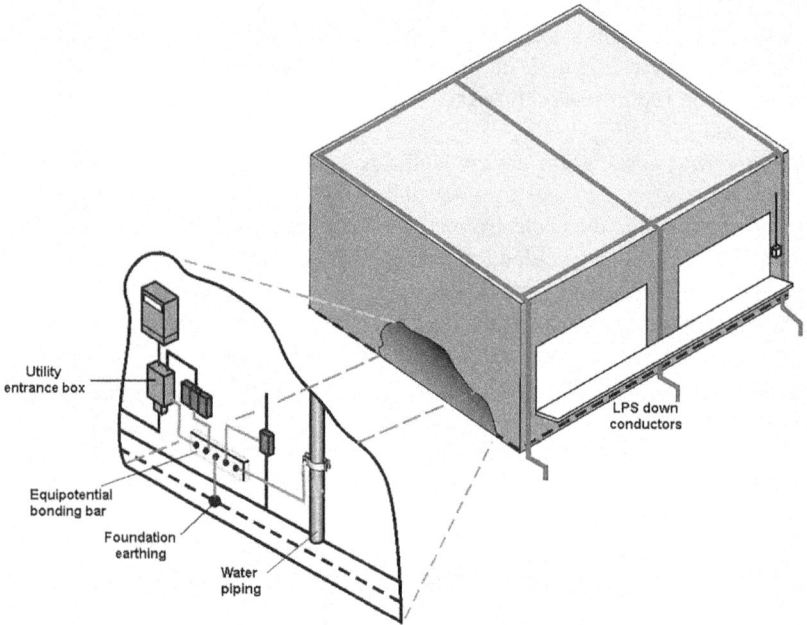

FIGURE 9.2 Exploded view showing the internal and external LPSs.

Where:

| | Surge protection device for AC circuits |
| | Surge protection device for telephone line |

The function of an LPS is to protect structures from fire or mechanical destruction and persons in the buildings from injury or even death. An LPS consists of an external and an internal LPS, as shown in Figure 9.2. The functions of the external LPS are:

- To intercept direct lightning strikes via an air-termination system
- To safely conduct the lightning current to the earth via a down-conductor system
- To distribute the lightning current in the earth via an earth-termination system.

The function of the internal LPS is

- To prevent dangerous sparking inside the structure. This is achieved by establishing equipotential bonding or by maintaining a separation distance between the components of the LPS and other electrically conductive elements inside the structure.

The internal LPS will protect the electrical installation from voltage disturbances such as lightning strikes, power switching, and faults, which can lead to sudden increases in voltage, known as spikes and surges. These phenomena can travel through power lines, inducing extremely high-voltage surges and potentially damaging connected electronic and electric devices. Surge protective devices (SPD) also filter out the disturbances caused by Electromagnetic Interference (EMI) and Radio Frequency Interference (RFI), ensuring a stable power supply and protecting sensitive electronics as measuring and control systems.

In industrial systems, quick switching of electrical loads, such as turning on/off machinery, causes voltage spikes, and SPD suppress them, preventing damage to connected equipment. In summary, the SPD play a crucial role in mitigating the impact of various factors that can lead to voltage spikes and surges. They act as a defence mechanism by filtering out disturbances, suppressing spikes, and safeguarding sensitive electronic equipment from potential damage within the electrical grid system.

Lightning equipotential bonding reduces the potential differences caused by lightning currents. This is achieved by connecting all isolated conductive parts of the installation directly by means of conductors or SPD.

The LPS consists of four components that work together to safely intercept, conduct, and dissipate the energy of the lightning discharge.

1) **Lightning arresters (air terminals):**

This constitutes the protection system against the lightning strikes, which in its absence would struck the building or the structure to be protected. Their function is to intercept the atmospheric discharge before it reaches the protected structure. These are metal devices, usually pointed, installed at high points of the structures. The arresters are connected to the rest of the protection system by means of metal conductors.

2) **Down conductors:**

These are conductive cables that connect the arresters to the earthing system. They conduct the lightning current to the earth, minimizing the risk of fire or explosions.

3) **Earthing grid:**

It is probably the most important part of the protection. A good earthing system is absolutely necessary. Two other parts of the protection (air terminals and surge protectors) would have no effect without the possibility of "evacuation" for the unwelcome extra voltages and extra currents. It consists of metal rods, or meshes, buried in the earth, designed to dissipate the lightning current safely into the earth. An efficient earthing system is crucial to the effectiveness of the LPS. Earthing plays a vital role in all electrical systems. The main reasons for earthing are:

- To protect people
- To protect equipment
- To allow the equipment to function correctly
- To ensure the reliability of electrical services.

A good earth connection should possess the following characteristics:

- Low electrical resistance between the electrode and the earth. The lower the earth electrode resistance, the more likely the lightning or fault current will choose to flow down that path in preference to any other, allowing the current to be conducted safely to and dissipated in the earth.
- Impedance value (inductance) should be the lowest possible to minimize the electromotive force. For this effect, it is strongly recommended to avoid earthing with only one underground horizontal wire or only one vertical electrode. It is not advisable to use a deep well to catch moisture, as this system shows high impedance at depths of over 10 metres. The most effective design involves multiplying the number of horizontal wires and vertical earth electrodes.
- Good corrosion resistance, because it will be buried in soil for many years.

4) **Surge protective devices:**

At last, but not least, the fourth component of the "LPS" has as function, to protect the interior installations and electric/electronic equipment, as well as the people, against voltage spikes induced by lightning. SPD are often incorrectly thought to offer protection against a wide range of power distribution system disturbances, but their function is to divert excess energy to the earth, thus protecting internal electric/electronic circuits.

The IEC 61643-11 relates to the tests the SPD must be able to meet. There are three classes of tests.

- Class I tests (Type 1 SPD) are intended to simulate partial conducted lightning current with a 10/350 μs current wave. This represents a standard lightning impulse.
- Class II tests (Type 2 SPD) involve impulses of shorter duration and are carried out with a 8/20 μs current wave.
- Class III, tests (Type 3 SPD) involve also impulses of shorter duration, that are carried out with a combination of voltage waves (1.2/50 μs) and current waves (8/20 μs).

A SPD may be classified according to more than one test class. Where this is the case, the tests required for all declared test classes shall be applied to device.

Among the technologies used in surge protectors, we will highlight three:

- **Zener diode**
  ○ Characteristics are close to the ideal curve
  ○ Response time is fast (10–12 picoseconds) for an accurate voltage setting
  ○ Leaking current can be neglected
  ○ The main inconvenience is the low energy dissipation. This type of concept cannot be used at the head of an installation and only as last protector for individual devices' protection.

- **Discharge tubes**

  They can be only electrodes in ambient air or electrodes within a hermetic casting filled with gas. When overvoltage happens, air or gas is ionized and provokes a strike between electrodes. The air electrodes are subject to external interferences, which can cause unwelcome functioning. After a strike, a residual leaking current (Is) can last some tens of seconds and induce an increase of potential.

  These systems have low reliability, and they were used on HV overhead lines but progressively replaced by direct protection devices. The advantages of this technology include significant energy dissipation and a negligible residual current (after the initial residual current has been evacuated).

- **Varistor**

  This technology uses a voltage-dependent resistor (VDR), where the resistance value fluctuates according to the voltage across. With nominal voltage, resistance is high, and as soon as voltage increases, resistance value goes down quickly, having as advantages:

  - Low construction cost
  - Low response time (nanoseconds)
  - High energy dissipation.

TABLE 9.1
Technologies' Comparison of Surge Protectors

Characteristic Curve	Component	Leaking Current I_f	Leaking Current Is	Residual Voltage	Conducted Energy	Conduction Delay
	Ideal surge protector	0	0	Low	High	Low
	Discharge tube	0	High	Low but U high	High	High
	Varistor	Low	0	Low	High	Average
	Zener	Low	0	Low	Low	Low

As disadvantages:
- After several low-energy overvoltages, heating occurs and lifetime shortens
- A too high overvoltage destroys the component up to short circuit
- Explosion can even occur with high-high voltage.

Note: Leaking current, initially low, will increase after each peak of voltage. After several strikes, the component has to switch off by itself. An indicator of "end-of-life" should equip those devices for proper replacement in due time.

Table 9.1 compares these technologies. In the United States, the UL 1449 relates the tests to SPD, with differences when compared with IEC 61643.
Where:

If = Initial leaking current
Is = Leaking after "actuation".

9.4.1 SURGE RATING

As a result of a lightning flash to the structure, the lightning current will be conducted and injected into the earth. Some of this current will flow from the structure on metallic services to more remote earth points. In the cases of electrical/electronic services, this current flow will occur "up" through the SPD and out of the services to remote earth points, such as to a remote distribution transformer earth.

The IEC 62305 series takes a worst-case approach to the sizing of the SPD, where 50% of the lightning current is assumed to be injected into the local earth, and 50% flows out of the services.

Where the services are conductive, the current is assumed to divide equally between them. Within an electrical service, the current is assumed to divide equally between the conductors. For SPD selection, the injected current is determined from the maximum current relating to the class of LPS, as shown in Table 9.2.

9.4.2 RISK ANALYSIS

Lightning protection is an issue of statistical probabilities and risk management. A system designed in compliance with the standard should statistically reduce the

TABLE 9.2
Total SPD Surge Rating Requirements, Assuming No Other Services

Class of LPS	Total SPD Surge Current (Wave Shape = 10 / 350 µs) [kA]
I	100
II	75
III	50
IV	50

risk to below a pre-determined threshold. The IEC 62305–2 risk management process provides a framework for this analysis.

The risk of a lightning strike and the necessity of an LPS for an object to be protected are determined according to IEC 62305–1 and IEC 62305–2. Technically and economically optimal protection measures can be selected and implemented under the orientations from IEC 62305–3 and IEC 62305–4, depending on the risk.

9.5 LIGHTNING PARAMETERS

The type and amount of lightning damage that an object suffers depends on both the characteristics of the lightning discharge and the properties of the object. The physical characteristics of lightning that are of most interest are the current waveform and the radio frequency electromagnetic fields.

Four distinct properties of the lightning current waveform are considered important in producing damage, which are explained next.

1) **Peak current**

The current associated with the return stroke has large values (~kA) that potentially can be very dangerous to various systems, both outside and inside of the structure. The name of this current is *peak return stroke current* or simply *peak current*.

For objects or systems that present an essentially resistive impedance, the voltage V on the object or system with respect to remote earth will be proportional to the current I, via Ohm's law: $V = RI$, where R is the effective resistance at the strike point. Examples of those objects can be a long power line, a tree, earth rods driven into earth, and so on. In a case when a lightning return stroke with a peak current of 30 kA strikes a power line with a surge impedance of 400 Ω, it can produce a prospective overvoltage of 6,000 kV, assuming the division of current. The large voltage produced can lead to an electric discharge from the struck phase conductor to adjacent phase or neutral conductors, or to earth across insulating materials, leading to possible loss of service of these lines and ultimately of services in the structure.

2) **Maximum current changing rate (maximum current steepness)**

The occurrence of damage of internal systems due to the indirect effect of lightning discharges has assumed a lot of importance in recent times, essentially due to the increasingly sensitive electric and electronic equipment which are highly vulnerable to such electromagnetic effects. These electronic devices are normally connected to different electrical services such as the mains supply and the data link, which, depending on the line routing and earthing inside the structures, can built up large open loop networks.

The maximum current steepness is responsible for the maximum of the magnetically induced voltages in such open loops.

Objects that exhibit a primarily inductive impedance, like wires in an electronic system, will experience a peak voltage that is directly proportional to the maximum rate of change of the current flowing through them, especially under circumstances like a lightning strike. ($V = L \, dI/dt$, where L is the inductance of the length of wire, and V is the voltage difference between the two ends of the wire).

Even a very small fraction of the lightning current circulating in earthing and bonding wires can cause damage to solid-state electronic circuits that have communication, power, and other inputs earthed at different locations.

3) Integral of the current over time

The charge in Equation 9.1:

$$Q = \int i.\, dt \qquad (9.1)$$

is responsible for melting effects at the attachment points of the lightning channel.

Charge is the integral of the current over time, so most of the charge-induced damages in structures, or lines, are due to the so-called "long" continuing current that follows some of return strokes. Generally, large charge transfers are due to long-duration (tens to hundreds of ms) lightning currents, such as long continuing currents, whose magnitude is in the 100 to 1000 A range, rather than return strokes having larger currents, but relatively of short duration and hence producing relatively small charge transfers.

The severity of heating, or burn-through of metal sheets, such as metal roofs, is to a first approximation proportional to the lightning charge transferred, which is in turn proportional to the energy delivered to the surface. This is because the input power to the conductor surface is the product of the current and the almost current-independent voltage drop at the arc-metal interface, which usually ranges from 5 to 10 V.

The voltage drop at the interface is close to being a contact potential between the two materials, or the difference in the work functions of the two materials, although the situation is considerably more complex than the case of two contiguous solid conductors in that the metal surface is partially melted, and the arc produces an atmosphere with metal vapour. Additionally, even those impulse currents which do have relatively large charge transfers cause only relatively minor surface damage on metal sheets, apparently because the current duration is too short to allow the penetration of heat into the metal.

4) Specific energy (integral of the current squared over time)

The specific energy is represented in Equation 9.2 as the time integral of the square of the current, and it is a measure of the heat generated by a lightning strike in an object of resistance R. The heating and melting of resistive materials, which may or may not be relatively good conductors, and the explosion of poorly conducting materials are, to a first approximation, related to the value of the specific energy, that is, the time integral of the Joule heating power, when R = 1.

$$W/R = \int i^2.\, dt \qquad (9.2)$$

Experiments have shown that about 5% of negative first strokes in earth flashes have specific energy exceeding 5.5×10^5 A²s, and about 5% of positive strokes have specific energy exceeding 10^7 A²s. In the case of most poorly conducting materials, this heat vaporises the internal material, and the resultant gas pressure causes an explosive fracture.

9.6 THE ELECTROGEOMETRIC MODEL

The Electrogeometric Model (EGM) describes the attachment of the downward leader to the earthed object, and at its core is the concept of "striking distance". The striking distance can be defined as the distance from the tip of the downward leader to the object to be struck at the instant the upward connecting leader is initiated from this object.

As the distance between the earth and the downward stepped leader becomes smaller, the electric field at the top of the earthed structure increases steadily, and when it reaches a critical value, an upward connecting leader arises from the structure. With that, if a critical value of electric field is defined, the striking distance increases with the electric charge of the downward leader, which is related with the return peak current value. Therefore, a relation between the striking distance and the return peak current can be found.

Nowadays, the most popular striking distance expression, especially used in international lightning protection standards, is represented in Equation 9.3:

$$r_s = A \times I^b \qquad\qquad (9.3)$$

Where:
  I is the first stroke peak current in kA
  r_s is the striking distance in metres;

The constants A and b are empirical.

It is important to note that the EGM was first introduced for the protection of power transmission lines. With the need for protection of structures against lightning discharges becoming more and more a necessity, it was fundamental to create a response for this problem. So, as EGM was so successful for power transmission lines' protection, the obvious solution was to use its concept of striking distance and apply it to structures. The Rolling Sphere Method (RSM) was the resulting solution.

9.7 THE ROLLING SPHERE METHOD

The concept of the *RSM* is directly related to the EGM in that it is based on the assumption that a stepped leader has to approach a critical distance (i.e. striking distance), before it will be attracted to the structure. In other words, this concept assumes that there is a spherical region with radius equal to the striking distance, and located around the tip of the stepped leader, with the property that the first point of a earthed structure that enters into this spherical volume will be the point of attachment of the stepped leader.

So, the main application of the RSM is the positioning of air terminals on an ordinary structure, so that one of the terminals, rather than a roof edge or other part of the structure, initiates the upward leader that intercepts the downward leader and, hence, becomes the lightning attachment point. For that reason, the RSM is suggested in IEC 62305–3 to be used for the positioning of air-termination systems. This is a method suitable in any case regarding the structure to be protected.

IEC 62305 standards series gives orientation on designing the RSM, which is based on two assumptions:

- The point of strike of lightning is determined when the downward leader approaches the earth or a structure with a striking distance.
- Lightning strikes the nearest earth object from the orientation point, and so its worst position is the centre of a sphere which attaches several earth objects.

The idea is that no lightning will strike the structure to be protected, if its striking distance is greater than the radius of the sphere. While the efficiency of the air-termination system is not determined by the radius of the sphere, the RSM is the most used design procedure for air-termination systems.

9.7.1 Lightning Protection Level

IEC 62305 assumes that for different requirements for LPS, four LPLs are defined, resulting in four classes of LPS: *I, II, III,* and *IV,* which are determined using a set of construction rules including dimensioning requirements. Each set comprises class-dependent (e.g. radius of the rolling sphere, mesh size) and class-independent (e.g. cross-sections, materials) requirements.

For each LPL, a set of maximum and minimum lightning current parameters is fixed:

- The maximum values of lightning current parameters relevant to LPL I will not be exceeded with a probability of 99%; they are reduced to 75% for LPL II, and to 50% for LPL III and IV. With these values, a correct design of various lightning components can be made, like the cross-section of conductors, thickness of metal sheets, current capability of SPD, and separation distance for preventing dangerous sparking.
- The minimum values of lightning current amplitude for the different LPLs are used to derive the rolling sphere radius r_s, which is fixed between 20 m and 60 m.

Hence, planning with the rolling sphere leads to the possible point of strikes—where the air terminals have to be placed.

To offer a greater LPL, as classified in IEC 62305 series, a smaller rolling sphere radius would be used. This would result in a reduced spacing between air terminals, thus positioning the air terminals to capture smaller lightning flashes and increasing the total percentage of lightning flashes captured. The LPL determines the spacing of the mesh, radius of rolling sphere, protective angle, etc. It should be noted that while lightning protection is typically implemented as a bonded network of air terminals and down conductors, other methods are available.

The main challenge is the choice of radius R for the sphere. If a radius r1 is chosen, it corresponds to a protective current i1 as per the electro-geometrical model. When the striking current is smaller than i1, it can go across the protection.

The recommended practice in this case is to consider the smaller striking current possible, being 2 kA, and R is then equal to 15 m. Under this theory, lightning strike intensity—1 to 400 kA—is function of the sphere radius.

It is important to highlight that the trajectory of the lightning channel is not directly impacted by the height of the objects located on the soil. It is only at the very last stage of its descent that the strike "decides" its point of impact.

To ensure the permanent availability of complex information technology systems, even in case of a direct lightning strike, additional measures, which supplement the lightning protection measures, are required to protect electronic systems against surges.

9.7.2 ZONES OF PROTECTION

According to IEC 62305–1, there are four main zones of protection against atmospheric discharges:

ZPR 0_A
A zone where the elements are subjected to direct discharges. They must conduct the entire discharge current, and there is no attenuation of the electromagnetic field.

ZPR 0_B
A zone where the elements are not subject to direct impacts of the discharge, but there is no attenuation of the electromagnetic field.

ZPR 1
A zone where the elements are not subject to direct impacts of the discharge. The discharge currents must be reduced compared to ZPR 0_B and the electromagnetic field must be sufficiently attenuated by the shielding.

ZPR 2, ... n
Additional zones, if the complexity of the structure and the installations require a greater reduction of the lightning current.

It is important to note that a lightning air termination does not attract the lightning at the time of its formation; it collects the electric discharge only if it is already, spontaneously, very near.

When we consider the system cloud-earth as a big capacitance, it is known that a charge of millions of volts is ready to arc. Considering this huge potential at the centre of a sphere, where the radius of this sphere is the function of the charge, the more potential, the bigger the radius. This centre is searching in space its opposite potential towards which it will strike, somewhere at the earth level. If it is a mountain, a high building, or a striking rod, the discharge will go preferably through this easiest and closest way. That is why striking rods are pointing towards the clouds on the highest parts of the buildings (or on a mast): To be in the radius of the discharging range.

Figure 9.3 is referencing to the height $H = 60$ m (the highest value that the rolling sphere radius can assume) due to the fact that until this height, the standard predicts that the occurrence of side flashes is negligible; above 60 m, consideration should be given to install a lateral air-termination system on the upper part of tall structures (typically the top 20% of the height of the structure).

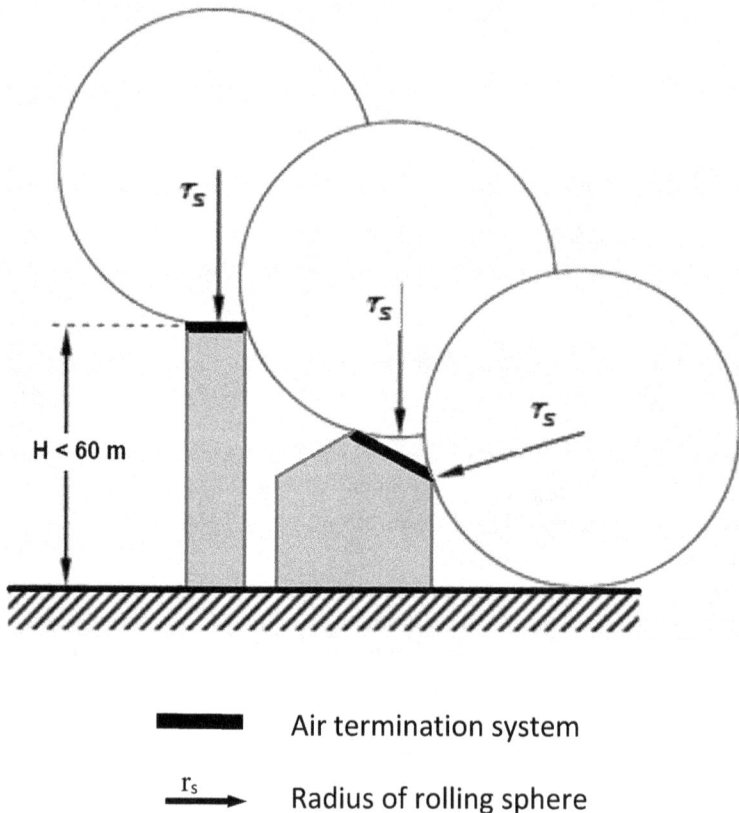

FIGURE 9.3 The Rolling Sphere Method.

Where:

▬▬▬	Air-termination system
➤	Radius of rolling sphere

TABLE 9.3

Relation between Lightning Protection Level, Interception Probability, Final Striking Distance h$_B$, and Minimum Peak Value of Current I.

LPL	Probabilities for the Limits of the Lightning Current Parameters		Radius of the Rolling Sphere (Final Striking Distance h$_B$) r [m]	Minimum Peak Value of Current I [kA]
	> Minimum Values	< Maximum Values		
IV	0.84	0.95	60	16
III	0.91	0.95	45	10
II	0.97	0.98	30	5
I	0.99	0.99	20	3

Source: IEC 62305–1.

The external protection (rods) catches about 80% of the strikes. No LPS is 100% effective. A system designed in compliance with the standard does not guarantee immunity from damage.

The relationships between LPL/class of LPS, interception effectiveness of the air-termination systems, final striking distance/radius of the rolling sphere, and current peak value are shown in Table 9.3.

9.8   ATTENTION ITEMS

The designer of the LPS must take care of the points given next, in order that people and properties are protected when a lightning strikes.

9.8.1   STEP POTENTIAL CONTROL

It is recommended to locate the down conductors and earthing electrodes in areas not accessible to people. If this is not possible, then additional measures should be used to limit step and touch potentials, as shown in Figure 9.4. This is important for locations where a large number of people may gather.

To reduce step voltages, IEC 62305–3 requires, for example potential control that can be achieved by various ways, as examples:

- Asphalt, 50 mm thick covering the area within 3 m of the electrode
- Gravel, 150 mm thick covering the area within 3 m of the electrode.

9.8.2   TRANSIENT OVERVOLTAGES

When a lightning flash occurs to a structure or service, the injected lightning current causes a rapid rise in voltage due to the impedance of the conductor. This voltage has the potential to surpass the insulation rating of the service, leading to a flashover, and/or it may exceed the tolerance of the connected equipment. Voltage transients are one type of power distribution system disturbance that can damage electrical and electronic equipment or impact its operation.

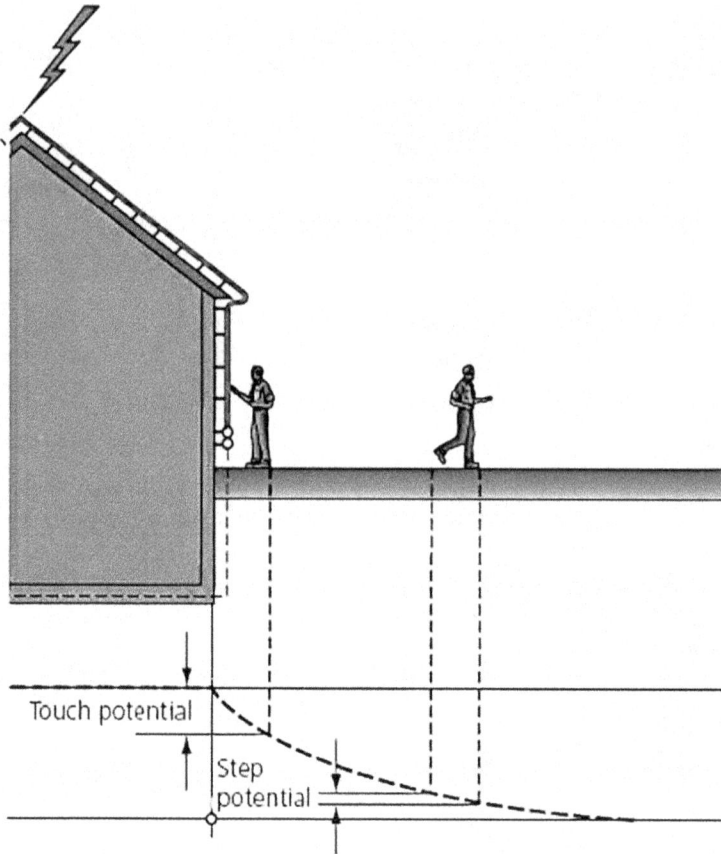

FIGURE 9.4 Step and touch potentials.

It is important to note that transient overvoltages are conducted into the sensitive circuitry of electronic equipment on power and data communication, signal, and telephone lines. So, protection must include:

- All cables that enter or leave the building (except fibre-optic ones)
- The local power supply to important equipment
- Electronic equipment outside the main building.

Note that data communication, signal, and telephone lines are not usually exposed to switching transients and, owing to the protection afforded by their cable screening, do not normally require additional local protection.

Transient overvoltages can be caused by lightning strikes between clouds or between cloud and earth (and objects upon it), coupling on to electrical cables and, hence, into the sensitive electronic equipment connected to them. To protect the electronic equipment inside a building, all cables that enter or leave the buildings must be protected against transient overvoltages, where they could have far-reaching effects. However, where the cable run to equipment exceeds 20 metres, transient overvoltages may appear on the mains after the protector at the main low voltage incomer, resulting from:

- Electrical switching of large inductive loads within the building
- Lightning strike to the building due to induction from the lightning currents through down conductors
- Transient overvoltages on to nearby power cables
- Inductance and capacitance of long cable runs, reflecting the voltage "let-through" the protector at the main low voltage incomer.

The larger the transient overvoltage reaching the electronic equipment, the greater the risk of interference, physical damage, and, hence, system downtime. Consequently, the transient overvoltage let through the protector should be as low as possible and, certainly, lower than the level at which interference or component degradation may occur.

Transient overvoltages can exist between any pair of conductors:

- Phase to neutral
- Phase to earth and neutral to earth on mains power supplies
- Line to line, and line(s) to earth, on data communication, signal, and telephone lines.

Thus, a good SPD must have a low let-through voltage between every pair of conductors.

9.8.3 FLAMMABLE LIQUIDS' STORAGE TANKS

Tanks exposed to lightning strikes, or related surges, can suffer significant damage, including total loss of the tank and its contents. Metallic tanks, equipment, and structures found in the petroleum industry that are in direct contact with the earth are commonly considered "self-protected", if they are built with materials of appropriate thickness (not less than 4.8 mm as per NFPA 780, not to be punctured by a direct strike), and the increase in temperature of the internal surface at the point of impact does not constitute a risk of explosion.

In the "self-protection" philosophy scheme, the lightning stepped leader is allowed to intercept with the tank roof, thus total lightning current will be injected into the metal structure of the tank. In this way, supplemental earthing by means of driven earth rods neither decreases nor increases the probability of being struck, nor does it reduce the possibility of ignition of the contents. Supplemental earthing is necessary, however, where direct earthing is not provided.

The first return stroke of negative lightning strikes, the commonest among cloud-to-earth flashes, has impulse currents with a peak value Ip of 30 kA, on an average. Thus, the interception of lightning leader, and consequent passage of impulse current, may generate highly localized hotspots and mechanical stresses in the material, due to the flow of fast varying currents.

The protection system to the tank against direct impacts consists of three elements:

- Air terminals to intercept direct discharges
- Down conductors to guide the currents to a safe place
- The earth grid SPT, where the discharge energy will be dissipated.

The RSM will be appropriate to locate the air terminals, with a radius of 30 m for the sphere (a typical value in tropical areas).

To avoid discharges that impact from the side as protection to the equipment attached to the tank, a separation distance of 20% of the highest part of the tank will be managed.

Among the preventive measures to avoid considerable damage due to lightning strikes, are:

- All metal parts of the tank must be rigidly connected and electrically interconnected. Particular attention must be paid to the pipe inlets on the tank roof and to the hatches, equipotentializing them with the roof plate.
- It is recommended to install a metal strap on the pipe for good contact, connecting it to the roof plate through a metal tape. The conductive materials used must be corrosion resistant.
- The metal hatch covers must be connected to the roof plates using flexible cables, ensuring an electrical connection even when open or not screwed in.
- All sensors and their power supply panels must be equipotentialized to the tank. Appropriate SPD must be installed to prevent potential differences that could generate surges inside the tank or in hazardous areas.

9.8.3.1 Earthing

According to API RP 2003, a tank is considered adequately earthed if the tank bottom is resting on the earth or foundation. This applies whether or not there is an elastomeric liner in or under the tank bottom.

Ignition cannot occur unless flammable vapours are present together with an oxygen concentration that places those vapours within the flammable range. Four areas of fire vulnerability for FRT include:

- Flammable vapours which are exposed by a failed floating roof
- Tank overfill situation or storing product with a vapour pressure and exceeding the tank's limit
- Failing seals, lack of buoyancy, and other damages to the floating deck
- Venting other than as required by API 650.

9.8.4 ALTERNATIVE PROTECTION METHODS

Insufficient scientific data is currently available to quantify the effectiveness of any of the systems as given here. Any claims made about the ability of an earth-based LPS to trigger or attract lightning strikes from clouds should be viewed with caution.

Selection and installation of any LPS should be based on a thorough analysis that considers the probability of lightning strikes at the site, the likelihood of success for the proposed protection system, and the potential consequences should the protection system fail to function as desired.

Alternative, non-conventional, methods for direct stroke lightning protection include the following:

- **Charge Transfer (or Streamer Delaying) System**

    A charge transfer system, or streamer delaying system, consists of installing a suitable array composed of a multitude of well-earthed sharp conductors around the area to be protected. According to one manufacturer's theory, the system diverts lightning away from the protected area by forming a space charge above the array. The space charge is a collection of charged particles (ions) in air that modifies the local electrical field.

    In the charge transfer system, a space charge is generated by point discharge from the sharp conductors. Once formed, the space charge reduces the electric field locally above the protected area and deflects any incoming lightning stroke. Should the space charge generated prove to be insufficient to deflect a lightning stroke of unusually high intensity, the array is also designed to function as a lightning receiver and provides a metallic path of adequate cross-section to propagate to earth the stroke with minimum damage. The number of sharp conductors needed in the array to provide adequate protection and the construction details of the array are proprietary to the companies that have developed these systems.

- **Early streamer emission air terminal**

    An early streamer emission air terminal LPS consists of a suitable number of lightning rods equipped with a device that is said to trigger the early initiation of the upward connecting leader that serves to complete the lightning stroke path to earth. The triggering devices that have been used are either radioactive sources (that emit weak alpha particle radiation) or electrical devices (that apply a fast high-voltage pulse directly to the lightning rod or to a spark gap electrode arrangement at the top of the lightning rod).

    According to the vendors, the early streamer emission air terminals are meant to attract the lightning stroke by enhancing the local electrical field enough to create a local discharge either continuously (radioactive triggering device) or at the most opportune time (electrical triggering device) to initiate the upward connecting leader. Early streamer emission air terminal LPSs do not protect against indirect lightning currents or induced voltages.

9.9 INSPECTIONS

Considering that the structure, the persons therein, and the electrical and electronic systems need to be protected all the time, it is necessary that the mechanical and electrical characteristics of an LPS remain completely intact for the whole life. To ensure this, a coordinated inspection and maintenance programme for the LPS should be laid down by the owner of the structure. If non-conformities are found during the inspection of an LPS, the owner of the structure is responsible for the immediate correction.

The inspection of the LPS must be carried out by a competent person, that is to say, a professional engineer able to design, install, and inspect LPS due to his technical training, knowledge, and familiarity with applicable standards, regulations, and safety directives. The criteria are usually fulfilled after several years of work experience and current occupation in the field of lightning protection. A report of the LPS containing the design criteria, design description, and technical drawings should therefore be available to the inspector.

The inspections to be carried out are distinguished as follows:

9.9.1 At the Design Stage

The inspection at the design stage should ensure that all aspects of the LPS with its components fulfil the standards' requirements.

9.9.2 At the Construction Stage

Parts of the LPS which will be no longer accessible when the construction work is completed must be inspected as long as this is possible. These include foundation earth electrodes, earth-termination systems, reinforcement connections, concrete reinforcements used as room shielding, as well as down conductors and their connections laid in concrete. The inspection comprises checking of technical documents, on-site inspection, and assessment of the work carried out.

9.9.3 Acceptance Term

The acceptance term is signed when the LPS has been completed. Compliance with the protection design and the work performed conforming to the standards must be thoroughly inspected.

9.9.4 Periodic Inspections

Regular inspections are required to ensure that the LPS remains effective. Inspection intervals may be specified in government regulatory requirements or national standards. Note that possible impacts due to aggressive environmental conditions can necessitate shorter inspection intervals.

In structures exposed to severe weather and harsh atmosphere (as petrochemicals and oil refineries), more frequent inspections and testing are recommended. IEC 62305–3 provides recommended procedures for inspection frequency and testing, as shown in Table 9.4.

TABLE 9.4

LPS Inspection Interval, after Inspection and Testing Done at the Installation Time

Protection Level	Visual Inspection (years)	Complete Inspection (years)	Critical Systems Complete Inspection (years)
I and II	1	2	1
III and IV	2	4	1

9.9.4.1 Measurements

Measurements are used to test the continuity of the connections and the condition of the earth-termination system. They must be made to check whether all connections of air-termination systems, down conductors, equipotential bonding conductors, shielding measures, etc., have low-impedance continuity. The recommended value is less than 1 Ω.

The contact resistance to the earth-termination system at all test joints must be measured to establish the continuity of the lines and connections (recommended value < 1 Ω). Furthermore, the continuity with respect to the metal installations (e.g. gas, water, ventilation, heating), the total earth resistance of the LPS, and the earth resistance of single earth electrodes and partial ring earth electrodes must be measured. The results of the measurements must be compared with the results of earlier measurements.

In case of old earth-termination systems, the condition and quality of the earthing conductor and its connections can only be visually inspected by exposing it at certain points. It is necessary to check for possible locations where rising currents may cross hazardous areas such as near the tank vents. Ends, screws, rods, guardrails, and other metal parts on the tank roof can act as points of occurrence for rising currents. A thorough inspection of the tanks must be carried out, identifying locations with potential for explosive atmospheres or gas leaks.

Another recommendation is to verify if floating seals are installed on tanks, especially those with an aluminium geodesic dome (which are more vulnerable to the effects of lightning strikes). Between the seal and the dome roof, these seals hinder the local atmosphere's ability to attain its LEL by reducing the air–fuel mixture.

9.9.4.2 Report

A report must be prepared for each inspection, which must be kept together with the technical documents and reports of previous inspections at the installation. The following technical documents must be available to the inspector when assessing the LPS:

- Design criteria
- Design descriptions
- Technical drawings of the external and internal LPS
- Reports of previous maintenance and inspections.

The recommended topics to be included in the report are as follows:

Inspection Information:

- Name of the inspector
- Inspector's company
- Name of person accompanying the inspector
- Date of inspection
- Signature of the inspector
- Signature of the company's representative.

Location information:

- Owner
- Address
- Year of construction
- Type of construction
- Type of roofing
- Lightning protection level (LPL).

Information on the LPS:

- Manufacturer
- Material and cross-section of the conductors and of the connecting lines between the single earth electrodes
- Number of down conductors
- Calculated separation distance
- Type of earth-termination system
- Details of the lightning equipotential bonding system to metal installations
- Details of electrical installations and existing equipotential bonding bars.

Fundamentals of inspection:

- Type
- Drawings of the LPS
- Standards and regulations at the time of installation
- Area classification drawings.

Inspection results and findings:

- Modifications to the structure or to the LPS
- Deviations from the applicable standards, regulations, requirements, and application guidelines applicable at the time of installation
- Earth resistance or loop resistance at the individual test joints, with information on the measuring method and the type of measuring device
- Total earth resistance (measurement with or without protective conductor and metal building installation).

9.10   CLOSING REMARKS

Lightning protection is crucial, especially in industries with hazardous areas. Careful attention and an inspection programme including arresters, down conductors, earthing systems, and SPD, combined with adequate training, are essential measures to protect hazardous areas against the dangers of lightning.

Lightning-induced fires and explosions in industries that process flammable materials result in significant loss of life, property damage, and substantial financial impact. The current state of the LPS must be determined in a site survey, taking into account the structural conditions, the existing documents, and possible requirements of property insurers.

A risk analysis needs to be performed in cooperation with the operator to define the protection measures required to prevent the destructive effects of lightning strikes and surges. With this information, designers use approved regulations and the recommendations from past accidents to design a complete protection concept, ensuring the safety of industrial operations and also the protection of surrounding communities.

BIBLIOGRAPHY

Adekitan, A. and Rock, M.—The undesirable interaction of lightning strike and floating roof tanks. In: *XVI International Symposium on Lightning Protection (SIPDA)*, Colombo, Sri Lanka, 2021, pp. 1–8.

API RP 2003—*Protection against ignitions arising out of static, lightning, and stray currents.* American Petroleum Institute, 2015.

API STD 650—*Welded tanks for oil storage.* American Petroleum Institute, 2021.

Directive 2014/30/EU—Directive of the European Parliament and of the Council of 26 February 2014, on the harmonisation of the laws of the Member States relating to electromagnetic compatibility (recast).

Durham, R. A., Durham, M. O. and Gillaspie, T. W.—Petrochemical facility lightning protection primer: Part 1: History and basic science. *IEEE Industry Applications Magazine*, vol. 27, no. 2, Mar./Apr. 2021, pp. 47–56.

Guthrie, M., Zipse, D. W. and Sanders, M. K.—NFPA 780 standard for the installation of lightning protection systems 2007 edition. In: *IEEE/IAS industrial & commercial power systems technical conference*, Edmonton, 2007.

Hossam-Eldin, A. A. and Houssin, M. I.—Design and assessment of lightning protection systems in petroleum structures. In: *11th International Middle East Power Systems Conference*, El-Minia, Egypt, 2006, pp. 468–475.

IEC 61643-11—*Low-voltage surge protective devices—Part 11: Surge protective devices connected to low-voltage power systems—requirements and test methods.* International Electrotechnical Commission, 2011.

IEC 61643-12—*Low-voltage surge protective devices—Part 12: Surge protective devices connected to low-voltage power systems—selection and application principles.* International Electrotechnical Commission, 2020.

IEC 62305-1—*Protection against lightning, Part 1: General principles.* International Electrotechnical Commission, 2024.

IEC 62305-2—*Protection against lightning—Part 2: Risk management.* International Electrotechnical Commission, 2024.

IEC 62305-3—*Protection against lightning—Part 3: Physical damage to structure and life hazard.* International Electrotechnical Commission, 2024.

IEC 62305-4—*Protection against lightning—Part 4: Electrical and electronic systems within structures*. International Electrotechnical Commission, 2024.

Lightning Protection Guide—3rd edition, Dehn International, 2014. Available at: www.dehn-international.com/sites/default/files/media/files/lpg-2015-e-complete.pdf

NFPA 780—*Standard for the installation of lightning protection systems*. National Fire Protection Association, 2023.

Parise, G., Allegri, M. and Parise, L.—A simplified method for the risk analysis of protection against lightning: Theory and a case study. *IEEE Industry Applications Magazine*, vol. 29, no. 5, Sept./Oct. 2023, pp. 36–43.

Rando, Ricardo—*Aterramento em atmosferas explosivas: Práticas recomendadas*. Blucher, 2021.

Rangel Jr., Estellito.—Garantindo a segurança nas instalações em áreas classificadas. In: *I Seminário Regional sobre Atmosferas Explosivas, Descargas Atmosféricas e Proteção contra Incêndio*, S. José dos Campos, SP, 12 Nov. 2004.

Vallejo, J. M., Batista, J. A. H. and Shipp, D. D.—Lightning and grounding issues impacting safety and performance of liquified natural gas-supplied power generation plants: Lessons learned to correct commonplace misunderstandings of requirements. *IEEE Industry Applications Magazine*, vol. 29, no. 5, Sept./Oct. 2023, pp. 65–75.

Zipse, D.—Advancement of lightning protection and prevention in the 20th century. *IEEE Industry Applications Magazine*, vol. 14, no. 3, May/Jun. 2008, pp. 12–15.

10 Electrical System Attention Points

10.1 INTRODUCTION

Protective devices are required by IEC 60079 series of standards for safe operation of electrical equipment to be installed in hazardous locations (also known as "Ex equipment"), and they can be located inside or outside the hazardous locations. In the EU, protective devices for Ex equipment fall within the scope of the ATEX Directive 2014/34/EU.

As a first rule, electrical protection should be set with its operating level as close to the normal operating levels as possible, without producing a situation in which nuisance trips become a problem, and should cover overload, short circuit of the supply, and the earth faults. The electrical protection should have the same properties, whatever the Zone of risk, but extreme care must be taken in respect of any circuits which enter Zones 0 or 20 due to the fact that the explosive atmosphere is assumed always to be present and that significant duration of overloads, faults, short-circuit faults or earth faults is not acceptable in these circumstances. In addition to the aforementioned, electrical protection/isolation facilities which isolate all phase conductors and the neutral conductors (but not the potential equalization or protective conductors) should be present in non-hazardous areas to permit the isolation of circuits entering the hazardous areas.

The neutral isolation by the switch isolating the phase conductors is recommended but not essential in these circumstances. The minimum requirement is for a separate link in the neutral conductor, which may provide manual isolation.

Where, however, additional isolation facilities in the hazardous area are required, the isolator must isolate the neutral at the same time as the phase conductors. This is because the neutral may be carrying fault current from other sources, even when the phase conductors are isolated, and in these circumstances, isolating the neutral with a manual link could produce an ignition-capable spark.

High-voltage circuits, such as those used for the supply of large rotating machines, should be treated in the same way as other supply circuits, but isolation is particularly critical here and must be as rapid as possible. Isolation of earth faults is particularly important as it is this type of faults which is most likely to cause unprotected sparking, and this isolation must be as near instantaneous as possible.

The foregoing requirements apply to all circuits entering the hazardous area except intrinsically safe circuits and non-incendive circuits, where voltage and current in the circuit are so limited as to prevent ignition-capable sparking, but in these cases, the electrical protection requirements generally apply to the apparatus in the non-hazardous area generating those circuits.

DOI: 10.1201/9781003500001-10

10.2 FIRES AND EXPLOSIONS DUE TO ELECTRICAL MALFUNCTIONS

One of the primary culprits behind industrial fires is electrical hazards. They can be caused by overloaded circuits, faulty wiring, and the use of substandard electrical equipment. In industrial plants, where the demand for power is high, even a minor electrical issue can lead to a significant event like business interruption or property damage.

From 2017 to 2021, the United States fire departments responded to an estimated annual average of 36,784 fires at industrial or manufacturing properties (including utility, defence, agriculture, and mining properties). The associated annual losses from these fires included 22 civilian deaths, 211 civilian injuries, and US$1.5 billion in direct property damage.

This annual average of 36,784 fires can be broken down into the following categories:

- 25,021 (68%) outside or unclassified fires
- 8,077 (22%) structure fires
- 3,687 (10%) vehicle fires.

Structure fires more commonly occurred in manufacturing or processing properties (63%) than in industrial properties, including utility, industrial, defence, agriculture, and mining properties (37%), and they accounted for the largest shares of civilian injuries (73%) and direct property damage (66%).

Equipment, or heat source, failure was a leading cause of structure fires in industrial and manufacturing properties. Electrical distribution, lighting, and power transfer equipment was identified as the leading equipment involved in ignition in industrial properties.

Electricity can generate fires in the following situations:

Arc flash:

An arc flash is a type of electrical phenomenon that occurs when a powerful electric current arcs through the air. It can be triggered by equipment failure, improper work procedures, or a lack of proper maintenance. Arc flashes release intense energy capable of causing fires, explosions, and severe burns, making them one of the most dangerous risks in industrial environments.

Flammable liquids and gases:

These substances can generate explosive atmospheres with high flammability potential. An electrical spark, or defective electrical equipment, can act as an ignition source, triggering an explosion. Oil and gas, chemical, pharmaceutical, paint, and other industries have such risks.

Combustible dusts:

Dust may seem harmless, but it can form clouds in the air, which can be ignited when their concentration is above their Lower Explosive Limit (LEL), and an electrical spark with enough energy comes as ignition source. Industries that produce metal, wood, coal, or grain dust are particularly at risk.

Faulty electrical equipment:

Industrial equipment can become a fire hazard due to overheating or because of sparks generated from defective insulation.

Static electricity:

This is a less commonly recognized but equally a dangerous source of ignition. Static electricity can cause sparks, igniting flammable atmospheres.

10.3   RESPONSIBILITIES

There are limits of responsibilities between the manufacturer and the end user. The manufacturer is responsible for product safety and for delivering installation and maintenance instructions. In the EU, the Directive 2014/35/EU guides the manufacturer to certify the product and its production line.

The end user is responsible for ensuring that the product is installed, maintained, and operated in a way that does not pose any risk of fire or explosion. In the EU, the Directive 89/391/EEC guides end users to use certified products and to elaborate risk analysis, operation instructions, training programmes, and safety procedures for operation and maintenance.

10.4   REQUIREMENTS FOR ELECTRICAL SYSTEMS

The electrical system of installations in hazardous locations must follow the requirements defined in standards as IEC 60079–14 and national codes, as National Electrical Code (NEC) in the United States, as described later in this chapter.

10.4.1   Wiring

The wiring systems include cables, cabling accessories, junction boxes, and raceways or conduits, which should be installed in positions that prevent exposure to mechanical damage, chemical effects, and heat. Cables are able to withstand quite high conductor temperatures, provided these are of very short duration and only occur very infrequently. The short-circuit conductor temperatures given in manufacturers' data can be used to determine the combinations of current and time which a conductor can carry. Actually, a quantity I^2t is used which is proportional to the energy liberated during the fault and, hence, to the conductor temperature rise. A similar relationship can be determined for the capability of protective conductors to carry earth-fault currents, based on the maximum temperatures that cable insulation or covering materials can tolerate, as informed by the manufacturer.

There are two main references on wiring systems for hazardous locations: The IEC 60079–14 standard and the NEC (US).

10.4.1.1   IEC Requirements

Where aluminium is used as the conductor material, it shall be used only with suitable connections and, with the exceptions of intrinsically safe and energy-limited installations, shall have a cross-sectional area of at least 16 mm^2.

Under IEC directions, where there is a risk of damage and exposure occurring, some form of mechanical protection is necessary, which can be provided by different ways, as examples, using cables with Steel-Wire Armouring (SWA) which have higher costs or metallic conduit. Suitable cable glands must prevent flame propagation or vapour entrainment through conduit systems. The metal conduit becomes part of the equipotential bonding system, and, therefore, measures must be taken to ensure good electrical continuity through it.

In IEC 60079–14, the general requirements for cables for fixed installations are the following:

a) Sheathed with thermoplastic, thermosetting, or elastomeric material (rubberlike). They shall be circular and compact. Any bedding or sheath shall be extruded. Fillers, if any, shall be non-hygroscopic.
b) Mineral-insulated, metal-sheathed (mineral-insulated cables shall be sealed)
c) Special, e.g. flat cables with appropriate cable glands. They shall be compact and any bedding or sheath shall be extruded. Fillers, if any, shall be non-hygroscopic.

And IEC 60079–14 requirements for flexible cables in fixed installations in hazardous areas (excluding intrinsic safety (IS) circuits) are:

• Ordinary tough rubber sheathed
• Ordinary polychloroprene sheathed
• Heavy tough rubber sheathed
• Heavy polychloroprene sheathed
• Flexible cables, plastic insulated, equally robust construction, and coated with resistant rubber.

All cables for fixed installations shall have:

• Flame propagation characteristics which enable them to withstand the tests according to IEC 60332–1 - 2 or IEC 60332–3-22 as appropriate
• Other protections against flame propagation (e.g. laid in sand-filled trenches)
• Cables entering hazardous areas shall be installed with a barrier to prevent flame propagation from a non-hazardous area into a hazardous area.

10.4.1.1.1   Cables for Dusty Environments

For hazardous areas, due to combustible dusts, all common types of cable can be used if they are drawn into screwed, solid-drawn, or seam-welded conduits. It is also possible to use cables that are inherently protected against mechanical damage and are impervious to dust, e.g.:

- Thermoplastic or elastomer-insulated, screened, or armoured cable, with a polyvinyl chloride (PVC), polychloroprene (PCP), or similar sheath overall
- Cables enclosed in a seamless aluminium sheath with or without armour
- Mineral-insulated cables with metal sheath
- Cables that are externally protected or not at risk of mechanical damage, and that are thermoplastic or elastomer-insulated with a PVC, PCP, or similar overall sheath, are permitted.

Note: Cables may need to be de-rated to limit surface temperature.

In dusty environments, the following requirements apply:

- Cable runs shall be arranged so that they are not exposed to the friction effects and build-up of electrostatic charge due to the passage of dust.
- Cable runs shall be arranged as far as possible so that they collect the minimum amount of dust and are accessible for cleaning. Wherever possible, cables that are not associated with the hazardous areas shall not pass through.
- Where layers of dust are liable to form on cables and impair the free circulation of air, consideration shall be given to reduce the current-carrying capacity of the cables, especially if low ignition temperature dust is present.
- When cables pass through a floor, partition, or a ceiling that forms a dust barrier, the hole that is provided shall be made good to prevent the passage or collection of combustible dust.
- When a metal conduit is used, care should be taken to ensure that no damage might occur to the connecting points, that they are dustproof, that the dust proofing of connected equipment is maintained, and that they are included in the potential equalization.

10.4.1.1.2   Earthing of Conductor Screens

Where a screen is required, it shall be electrically connected to earth at one point only, normally at the non-hazardous area end of the circuit loop. This requirement is to avoid the possibility of the screen carrying a possibly incentive level of circulating current in the event that there are local differences in earth potential between one end of the circuit and the other.

Special cases:

- If there are special reasons for the screen to have multiple electrical connections throughout its length, the arrangement of Figure 10.1 may be used, provided that:

- The insulated earth conductor is of robust construction (normally at least of area 4 mm², but 16 mm² may be more appropriate for clamp type connections).
- The arrangement of the insulated earth conductor plus the screen is insulated to withstand a 500 Vac rms or 700 Vdc as an applicable insulation test from all other conductors in the cable and any cable armour.
- The insulated earth conductor and the screen are only connected to earth at one point, which shall be the same point for both the insulated earth conductor and the screen and would normally be at the non-hazardous end of the cable.
- The insulated earth conductor is installed in a position that will prevent it being exposed to mechanical damage, corrosion, chemical influences, effects of heat, and the effects of UV radiation.
- The inductance/resistance ratio (L/R) of the cable installed together with the insulated earth conductor shall be established and shown that it is not adversely impacted by external electric or magnetic fields such as from nearby power lines or heavy current-carrying single-core cables.
- If the installation is impacted and maintained in such a manner that there is a high level of assurance that potential equalization exists between each end of the circuit (i.e. between the hazardous area and the non-hazardous area), then, if desired, cable screens may be connected to earth at both ends of the cable and, if required, at any interposing points, as shown in Figure 10.1.

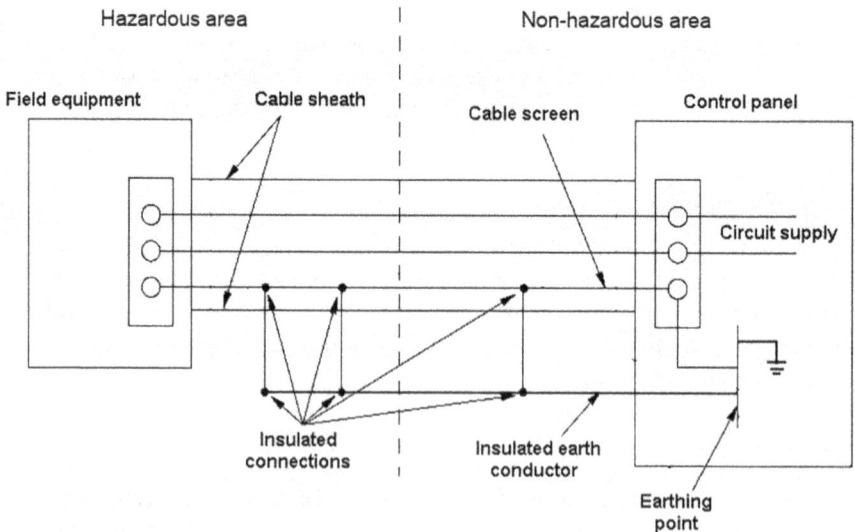

FIGURE 10.1 Earthing at interposing points only is accepted only if potential equalization is assured.

Note: Multiple earthing through small capacitors (e.g. 1 nF, 1,500 V ceramic) is acceptable, provided that the total capacitance does not exceed 10 nF.

10.4.1.1.3   Cable Armour Bonding
Armour should normally be bonded to the equipotential bonding system via the cable entry devices or equivalent, at each end of the cable run. Where there are interposing junction boxes or other apparatus, the armour will normally be similarly bonded to the equipotential bonding system at these points. In the event that armour is required not to be bonded to the equipotential bonding system at any interposing point, care should be taken to ensure that the electrical continuity of the armour from end to end of the complete cable run is maintained.

Where bonding of the armour at a cable entry point is not practical (typically when using plastic enclosures) or where design requirements make this not permissible, care should be taken to avoid any potential difference which may arise between the armour and the equipotential bonding system, giving rise to an incendive spark. In any event, there shall be at least one electrical bonding connection of the armour to the equipotential bonding system. The cable entry device for isolating the armour from earth shall be installed in the non-hazardous area or in Zone 2/22.

10.4.1.2   NEC Requirements
In the United States, the NEC defines the types of cables that can be used in hazardous areas, as shown in Table 10.1. The NEC requirements differ from IEC ones when defining particular types of cables, instead of just listing summarized characteristics, as IEC does.

The NEC covers in different articles, the approved wiring methods for hazardous areas: 501 for Class I locations, where the presence of flammable gases or vapours is foreseen; 502 for Class II locations, where the presence of combustible dust is

---

TABLE 10.1
NEC-Approved Wiring for Hazardous Locations

Class (Art. NEC)	Division	NEC Requirements
I (art. 501)	1	• Any suitable type of wire or cable if installed in threaded metallic conduit • Type MI (mineral-insulated) cable • Any suitable type of wire, or cable, if installed in non-metallic conduit encased in concrete and buried under at least 2 feet of earth • In certain industrial establishments, Type MC-HL (Metal Clad—Hazardous Locations) or Type ITC-HL (Instrumentation Tray Cable—Hazardous Locations) continuously corrugated welded armour (CCW) sheathed cable. Note: All of above must be terminated with approved fittings (end seals). • Optical fibre cable types (OFN, OFC, etc.) installed in raceways in accordance with NEC 501.10(A) and 501.15

TABLE 10.1 (*Continued*)
NEC-Approved Wiring for Hazardous Locations

Class (Art. NEC)	Division	NEC Requirements
	2	• All wiring methods permitted for Class I Div. 1 • Type PLTC or PLTC-ER cables installed in accordance with Article 725 including cable trays • Type ITC or ITC-ER installed as permitted in Article 727.4 • Type MC, MV, TC, or TC-ER cables installed in cable trays Note: All of above must be terminated with approved fittings (end seals). • Optical fibre cable types (OFN, OFC, etc.) installed in raceways in accordance with NEC 501.10(A) and 501.15.
II (art. 502)	1	• Any suitable type of wire or cable if installed in threaded metallic conduit • Type MI (Mineral-Insulated) cable • In certain industrial establishments, type MC-HL continuously corrugated aluminium cable, having an overall jacket and listed Note: All of above must be terminated with approved fittings (end seals). • Optical fibre cable types (OFN, OFC, etc.) installed in raceways in accordance with NEC 501.10(A) and 501.15
	2	• All wiring methods permitted for Class II Div. 1 • Any suitable type of wire or cable if installed in metallic conduit • Type MC or MI cables • Type PLTC or PLTC-ER cables installed in accordance with Article 725 including cable trays • Type ITC or ITC-ER installed as permitted in Article 727.4 • MC, MI, TC, or TC-ER cables installed in cable trays Note: All of above must be terminated with approved fittings (end seals). • Optical fibre cable types (OFN, OFC, etc.) installed in raceways in accordance with NEC 501.10(A) and 501.15
III (art. 503)	1 and 2	• Any suitable type of wire or cable if installed in metallic or PVC conduit • Type PLTC or PLTC-ER cables installed in accordance with Article 725 including cable trays • Type ITC or ITC-ER installed as permitted in Article 727.4 • Type MC, MI, TC, or TC-ER cables installed in cable trays Note: All of them must be terminated with approved fittings (end seals).

foreseen; and 503 for Class III locations where the presence of combustible fibres is foreseen.

It is important to note that some of NEC articles have precautions on situations where wiring and/or equipment outside the classified area could provide ignition of a flammable atmosphere by, as an example, releasing ignition-capable particles that could communicate to the adjacent classified areas. Although the details of these

FIGURE 10.2   An MC-HL (Teck90) cable.

FIGURE 10.3   An ITC-HL cable.

possible hazards vary greatly from process to process, and from electrical system to electrical system, it is important to know the requirements established in NEC articles 511 through 516. Additional examples can be also found in documents such as NFPA 55, which, as example, does not allow the installation of hydrogen systems beneath electric power lines.

The types MC-HL, in Figure 10.2, and ITC-HL, in Figure 10.3, are permitted to be used in Class I Division 1 areas, provided they are used only in industrial establishments with restricted public access, where the conditions of maintenance and supervision ensure that only qualified persons service the installation. They must be installed as per the requirements of NEC and must be terminated with fittings listed for the application.

10.4.1.2.1   Characteristics of Some NEC Cable Types

a) Cable MC-HL

A metal clad cable is defined as "a cable in which the conductors are enclosed in a corrugated metal sheath or interlocking metal tape, insulated, and the entire cable is put together at the factory". MC-HL cables must have a separate equipment of a full-size earthing conductor. It features one of several types of armour sheathing:

Aluminium Interlocked Armour (AIA)

This spirally wound cable is used in industrial applications, including pulp and paper, chemicals, mining operations, and power. AIA, when jacketed with PVC, is flame-retardant, impervious to water, and sunlight resistant and is best suited for utility and industrial areas where there is a potential risk of high-heat or fire and in tight spaces.

Continuously Corrugated Copper Weld (CLX)

This medium-voltage cabling is used for controls, instrumentation, and other low-voltage applications. The corrugated metal sheath can be made from aluminium, copper, steel, or bronze and has a wavy look to it. CLX cable is impervious to water and has uses in petro-chemical, transportation, pulp and paper, and high-tech applications.

Steel Wire Armouring (SWA)

Also known as Served Wire Armor, this cable armour uses strands of served galvanized steel wire. It is typically used for underground applications and in power and auxiliary control cable installations.

Galvanized Armor

This metal clad cable is protected by galvanized steel and a PVC jacket, making it well-suited for wet and oily applications. It is direct burial rated and can be used in mining, pulp and paper, and petroleum.

Teck90

Teck90 cable has smooth aluminium armour and can be used in industrial applications, including petro-chemical. It is impervious to water, sunlight resistant, and direct burial rated and can be installed in hazardous environments. These cables must have a gas/vapour tight continuous corrugated metallic sheath with an overall jacket of suitable polymeric material.

b) Cable ITC-HL

Type ITC (Instrumentation Tray Cable) is a factory assembly of two or more 300 V insulated copper conductors, no. 22 through 12 AWG, with or without earthing conductor(s), and enclosed in a non-metallic sheath, usually without armour, as defined by NEC Article 727. This cable is suitable for use in Class I Zones 1 and 2, and installation details are given in NEC Article 505.

The ITC-HL-ER (Instrument Tray Cable—Hazardous Locations—Exposed Run) allows installation out of conduit, without protection, as shown in Figure 10.4. It is oil resistant, sun resistant, and is suitable for direct burial for underground applications.

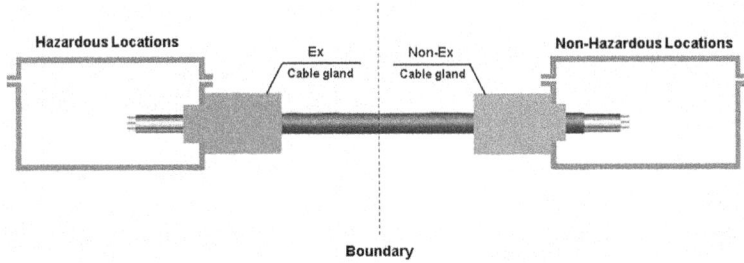

FIGURE 10.4 Installation of ITC-HL cable.

c) **Cable MI**

Mineral-Insulated cable consists of copper conductors inside a continuous copper sheath, which provides corrosion protection, and it also serves as an equipment-earthing conductor. The copper conductors inside the copper sheath are separated and insulated from one another by a dry magnesium oxide powder that is introduced during the manufacturing process. Type MI cable can be used in many applications in conditions where it is embedded in plaster, concrete, fill, or other masonry, whether above or below grade; in hazardous (classified) locations where specifically permitted by NEC; and where it is exposed to oil and gasoline.

Type MI cable shall not be used under the following conditions:

- In underground, unless protected from physical damage, where necessary
- Where exposed to conditions that are destructive and corrosive to the metallic sheath, unless additional protection is provided.

10.4.1.3 Intrinsically Safe Cabling

Only insulated cables whose conductor-earth, conductor-screen, and screen-earth test voltages are at least 500 Vac shall be used in IS circuits. If multi-stranded conductors are used in the hazardous area, the ends of the conductor shall be protected against separation of individual strands, for example by means of core-end sleeves (ferrules). The diameter of individual conductors within the hazardous area shall not be less than 0.1 mm. This applies also to the individual wires of a finely stranded conductor.

Installations with intrinsically safe circuits shall be erected in such a way that their IS is not adversely impacted by external electric or magnetic fields such as from nearby overhead power lines or heavy current-carrying single-core cables. This can be achieved, for example by the use of screens and/or twisted cores or by maintaining an adequate distance from the source of the electric or magnetic field.

In addition to the cable requirements in IEC 60079–14 (where cables with sheaths of a tensile strength lower than 2.5 N/mm² (in case of PVC) or 15.0 N/mm² (in case of polyethylene, polychloroprene, chlorosulphonated, and similar polymers) are not allowed), the following requirements apply:

(a) Intrinsically safe circuit cables are separated from all non-intrinsically safe circuit cables.
(b) Intrinsically safe circuit cables are so placed as to protect against the risk of mechanical damage.
(c) Intrinsically safe or non-intrinsically safe circuit cables are armoured, metal-sheathed, or screened.

Conductors of intrinsically safe circuits and non-intrinsically safe circuits shall not be carried in the same cable, and it is necessary to maintain the separation of IS cables from non-IS circuits. No segregation is required if metal sheaths, or screens, are used for the intrinsically safe or non-intrinsically safe circuits, as shown in Figure 10.5. If conductors of intrinsically safe circuits and non-intrinsically safe circuits run in the same duct, they shall be separated by an intermediate layer of insulated material, or by an earthed metal partition, as shown in Figure 10.6.

The electrical parameters C_C and L_C or C_C and L_C/R_C should be known, or the worst case values specified by the manufacturer should be assumed for all cables used.

10.4.1.3.1 SELV and PELV Systems

The level of voltage used in electrical systems associated with IS protection is categorized in IEC standards and specifies precautions to be taken depending on the circumstances.

- Extra-low voltage: Normally not exceeding 50 Vac or 120 Vdc ripple-free, whether between two conductors or to earth.

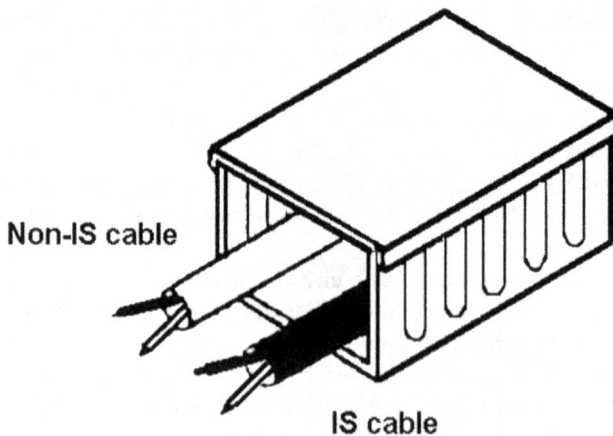

FIGURE 10.5 When an IS cable has a metal sheath, it can run in the same duct of non-IS cables.

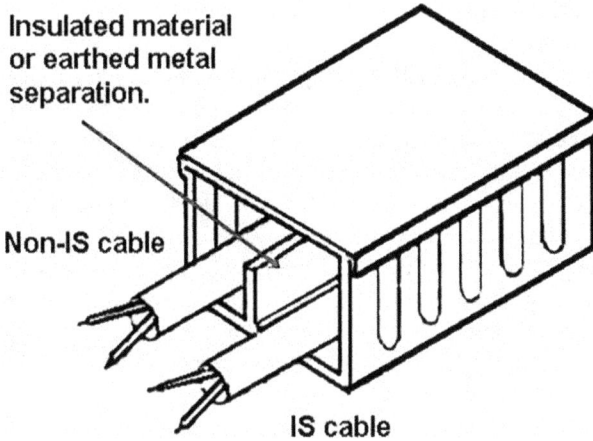

FIGURE 10.6 When using IS cables without sheaths with non-IS cables in the same duct, a separation is required.

- Safety extra low voltage (SELV): It is electrically separated from earth and from other systems in such a way that a single earth fault cannot give rise to the risk of electric shock.
- Protective extra low voltage (PELV): It is not electrically separated from earth but which satisfies all the requirements for SELV.
- Low voltage: Exceeding extra-low voltage but not exceeding 1,000 V_{AC} or 1,500 V_{DC} between conductors or 600 V_{AC} or 900 V_{DC} between conductors and earth.

The IEC 60079–14 standard recognizes the use of extra-low voltage systems as applicable to instrumentation. Circuits may or may not be floating with respect to earth, provided that they cannot give rise to the risk of shock. It therefore accepts the use of galvanic-isolating interfaces in IS circuits. Preferences are expressed towards a floating circuit being earthed at one point, or being monitored with respect to earth, using Earth Leakage Detection (ELD) systems.

10.4.1.3.2 Earthing of IS Circuits

Intrinsically safe circuits may be either isolated from earth (when galvanic isolators are used) or connected at one point to the equipotential bonding system (when safety barriers are used), if this exists over the whole area in which the intrinsically safe circuits are installed. The installation method shall be chosen with regard to the functional requirements of the circuits and in accordance with the manufacturer's instructions.

More than one earth connection is permitted on a circuit, provided that the circuit is galvanically separated into sub-circuits each of which has only one earth point. In intrinsically safe circuits that are isolated from earth, attention shall be paid to the danger of electrostatic charging. A connection to earth across a resistance of between

0.2 MΩ and 1 MΩ, for example for the dissipation of electrostatic charges, is not deemed to be earthing.

Intrinsically safe circuits shall be earthed if this is necessary for safety reasons, for example in installations with safety barriers without galvanic isolation. They may be earthed if necessary for functional reasons, for example with welded thermocouples. If intrinsically safe apparatus does not withstand the electrical strength test with at least 500 V to earth according to IEC 60079–11, a connection to earth at the apparatus is to be assumed.

Where the equipment is earthed and a bonding conductor is used between the equipment and the point of earth connection of the associated apparatus, such situations should receive careful consideration by a competent person, so as to avoid danger from circulating fault currents. Particular care should be taken where the requirements of EPL Ga apparatus have to be met.

If bonding conductors are employed, they should be adequate for the situation, have a copper cross-sectional area of no less than 4 mm^2, be permanently installed without the use of plugs and sockets, be adequately mechanically protected, and have terminals which conform to the requirements of type of protection "e" with the exception of the IP rating.

In intrinsically safe circuits, the earthing terminals of safety barriers without galvanic isolation (e.g. Zener barriers) shall be:

- Connected to the equipotential bonding system by the shortest practicable route
- For TN-S systems only, connected to a high integrity earth point in such a way as to ensure that the impedance from the point of connection to the main power system earth point is less than 1Ω. This may be achieved by a connection to a switch room earth bar or by the use of separate earth rods. The conductor used shall be insulated to prevent invasion of the earth by fault currents which might flow in metallic parts with which the conductor could come into contact (e.g. control panel frames). It shall also be given mechanical protection in places where the risk of damage is high.

The cross-section of the earth connection shall be either:

- There must be at least two separate conductors, each of which is rated to carry the maximum possible continuous current and has a minimum cross-sectional area of 1.5 mm^2 of copper, or
- At least one copper conductor with a minimum area of 4 mm^2.

Note: The provision of two earthing conductors should be considered to facilitate testing. If the prospective short-circuit current of the supply system connected to the barrier input terminals is such that the earth connection is not capable of carrying such current, then the cross-sectional area shall be increased accordingly or additional conductors used.

10.4.1.3.3   Verification of IS Circuits

When installing intrinsically safe circuits, including cables, the maximum permissible inductance, capacitance or L/R ratio, and surface temperature, shall not be exceeded. The permissible values shall be taken from the associated apparatus documentation or the marking plate.

The temperature classification of the equipment mounted in the hazardous area shall be determined from the label or documentation of that apparatus. The apparatus may have different classifications for different conditions of use (usually dependent on ambient temperature or input parameters U_i, I_i, and P_i).

10.4.1.3.4   Installations to Meet EPL Ga or Da

Intrinsically safe circuits shall be installed in a way that the power limitation at the safe level of the circuit is not exceeded, even when short circuits or earth faults occur. In installations with intrinsically safe circuits for Zone 0, the intrinsically safe equipment and the associated apparatus shall comply with IEC 60079–11, Ex ia. The circuit (including all simple apparatus, intrinsically safe apparatus, associated apparatus, and the maximum allowable electrical parameters of inter-connecting cables) shall be of level of protection "ia". In installations to meet the requirements of EPL Da, the intrinsically safe apparatus and the intrinsically safe parts of associated apparatus shall comply with IEC 60079–11, for Group III and at least to the level of protection "ia".

Associated apparatus with galvanic isolation between the intrinsically safe and non-intrinsically safe circuits is preferred. As only one fault in the equipotential bonding system, in some cases, could cause an ignition hazard, associated apparatus without galvanic isolation may be used only if the earthing arrangements ensure that the conductor used is insulated (to prevent invasion of the earth by fault currents, which might flow in metallic parts with which the conductor could come into contact), and any mains-powered apparatus connected to the safe area terminals is isolated from the mains by a double-wound transformer, the primary winding of which is protected by an appropriately rated fuse of adequate breaking capacity.

The circuit (including all simple components, simple electrical apparatus, intrinsically safe apparatus, associated apparatus, and the maximum allowable electrical parameters of inter-connecting cables) shall be of category "ia". If the intrinsically safe circuit is divided into sub-circuits, the sub circuit(s) in locations requiring EPL Ga, including the galvanically isolating elements shall have the level of protection "ia", but sub-circuits not in locations requiring EPL Ga, need only have the level of protection "ib" or "ic".

Note: Galvanic isolation can be achieved via the associated apparatus or via galvanically isolating apparatus within an intrinsically safe circuit in EPL Gb, Db, Gc, Dc, or non-hazardous locations.

If earthing of the circuit is required for functional reasons, the earth connection shall be made outside the locations requiring EPL Ga or Da but as close as is reasonably practicable to the EPL Ga or Da equipment. If earthing of the circuit is inherent in the circuit operation, as for example with an earthed tip thermocouple or a conductivity probe, this should be the only connection to earth, unless it can be demonstrated that no fault condition can arise as a result of the presence of more than one earth connection.

If part of an intrinsically safe circuit is installed in locations requiring EPL Ga or Da such that the apparatus and the associated apparatus are at risk of developing hazardous potential differences within the locations requiring EPL Ga or Da, e.g. through the presence of atmospheric electricity, a Surge Protection Device (SPD) shall be installed between each non-earth bonded core of the cable and the local structure as near as is reasonably practicable, preferably within 1 m, to the entrance to the locations requiring EPL Ga or Da.

Examples of such locations are flammable liquid storage tanks, effluent treatment plants, and distillation columns in petrochemical works. A high risk of potential difference is generally associated with a distributed plant and/or exposed equipment location, and the risk is not alleviated simply by using underground cables or tank installation.

The SPD shall be capable of diverting a minimum peak discharge current of 10 kA (8/20 μs impulse according to IEC 60060–1, ten operations). The connection between the protection device and the local structure shall have a minimum cross-sectional area equivalent to 4 mm^2 of copper. The sparkover voltage of the SPD shall be determined by the user and an expert for the specific installation.

The use of one or more low-voltage SPDs in an intrinsically safe circuit modifies the way in which that circuit is considered to be earthed. This shall be taken into account in the design of the intrinsically safe system.

The cable between the intrinsically safe apparatus in the locations requiring EPL Ga or EPL Da and the SPD shall be installed such that it is protected from lightning.

10.4.2 INSTALLATION OF CABLES

Cables must be circular and compact. Circular, because the cable entries have circular rubber seals, which assure the necessary sealing of the cable; and they must be compact, because a less-compact stranded line will not withstand the contact pressure of the cable gland's rubber ring.

In this case, the static friction between the rubber ring and cable will not be sufficient to ensure the necessary compressive strength. Especially when introducing the cable directly into the flameproof enclosure (Ex d), the cable's circular shape and compactness are major technical properties ensuring the cable safety.

As the IEC 60079–14 put the focus on the protection against damage of the cables, there are few main requirements, as described next.

- Cables with low tensile strength sheaths (commonly known as "easy tear" cables) shall not be used in hazardous areas, unless installed in conduit.
- Single insulated wires, excluding intrinsically safe circuits, shall not be used for live conductors, unless they are installed inside switchboards, enclosures, or conduit systems.

- Cable systems and accessories should be installed, so far as is practicable, in positions that will prevent them being exposed to mechanical damage, to corrosion, or chemical influences (e.g. solvents), to the effects of heat, and to the effects of UV radiation. For cables in hazardous areas exposed to ultraviolet (UV) rays, the UV resistance is a very important characteristic.
- Where an exposure of this nature is unavoidable, protective measures, such as installation in protecting conduit, shall be taken or appropriate cables selected (e.g. to minimize the risk of mechanical damage, armoured, screened, seamless aluminium sheathed, mineral-insulated metal-sheathed, or semi-rigid sheathed cables could be used).
- Where cables are subject to other conditions, e.g. vibration or continuous flexing, they shall be designed to withstand that condition without damage.
- Precautions should be taken to prevent damage to the sheathing or insulating materials of cables when they are to be installed at temperatures below $-5°C$.
- Where cables are secured to equipment, or cable trays, the bend radius on the cable should be in compliance with the cable manufacturer's data, or be at least eight times the cable diameter, to prevent damage to the cable;
- The bend radius of the cable should start at least 25 mm from the end of the cable gland.

10.4.2.1 Marking of Cables

Cables containing intrinsically safe circuits shall be marked, except if all intrinsically safe circuit cables, or all cables of circuits which are not intrinsically safe, are armoured, metal-sheathed, or screened. If sheaths or coverings are marked by a colour, the colour used shall be light blue, and such colour shall not be used for other purposes.

Alternative marking measures shall be taken inside measuring and control cabinets, switchgear, distribution equipment, etc., where there is a risk of confusion between cables of intrinsically safe and non-intrinsically safe circuits in the presence of a blue neutral conductor. Such measures include:

- Combining the IS wires in a common light blue harness
- Labelling
- Clear arrangement and spatial separation.

10.4.2.2 Potential Equalization

Potential equalization is required for installations in hazardous areas. For TN, TT, and IT systems, all exposed and extraneous conductive parts shall be connected to the equipotential bonding system. The bonding system may include protective conductors, metal conduits, metal cable sheaths, steel-wire armouring, and metallic parts of structures but shall not include neutral conductors.

Connections shall be secure against self-loosening and shall minimize the risk of corrosion which may reduce the effectiveness of connection.

An internal earth continuity plate may be fitted, for example, to allow for the use of metallic cable glands without the use of separate individual earthing tags. The material and dimensions of the earth continuity plate should be appropriate for the anticipated fault current.

If the armour or screens of cables are only earthed outside the hazardous area (e.g. in the control room), then this point of earthing shall be included in the potential equalization system of the hazardous area. If the armour or screen is earthed only outside of the hazardous area in TN system, there is a possibility that dangerous sparks may be created at the ending of the armour or screen in the hazardous area. Therefore, this armour or screen should be treated like unused cores.

Exposed conductive parts need not be separately connected to the equipotential bonding system if they are firmly secured to and are in conductive contact with structural parts or piping which are connected to the equipotential bonding system. Extraneous conductive parts which are not part of the structure or of the electrical installation, for example frames of doors or windows, need not be connected to the equipotential bonding system, if there is no danger of voltage displacement.

The minimum size for bonding conductors for the main connection to a protective rail shall be 6 mm², and that for supplementary connections shall be a minimum of 4 mm². Consideration should also be given to using larger conductors for mechanical strength.

Metallic enclosures of intrinsically safe or energy-limited apparatus need not be connected to the equipotential bonding system, unless required by the equipment documentation or to prevent the accumulation of static charge.

10.4.2.3 Cable Entry Devices

Neither the manufacturers nor the testing laboratory are obliged to determine in detail, what kind of cable gland has to be used. Moreover, for the certification of flameproof devices (Ex d), or increased safety (Ex e) devices, there are no defined requirements on cables and cable glands which have to be used, and neither the cable itself nor the cable glands are determined, much less tested.

Cables and lines are not included in the scope of the ATEX Directive 2014/34/EU and, therefore, cannot be certified in accordance with it. If an improper cable, or cable gland, is selected, the entire protection system can become unsafe.

This can cause problems as the following: A flameproof device (Ex d) generates 10 bar of explosion pressure; however, the selected combination of cable and cable gland can withstand only 6 bar. As a result, such a device is not safe!

The selection of suitable cable glands under IEC is described in IEC 60079–14, and the following rules are applied for direct Ex d entries:

- Use certified cable glands, considering the equipment explosion protection parameters and the ambient characteristics.
- If all the parameters are fulfilled and the cable is at least 3 m long, cable glands with rubber ring seals can be used.
- If the cable is shorter than 3 m, use barrier glands.

The requirements for the entries in Zones 20 and 21 (ATEX category 1D and 2D, dust explosion protection equipment) are basically the same as for increased safety. The only difference is the IP rating: IP6X for Zone 20 and Zone 21.

10.4.2.4   Fibre-Optic Cables

In hazardous areas, fibre-optic cables, especially those directly inserted into flame-proof boxes, are considered potentially more critical than copper wires. In this case, it is not relevant how much energy is transported but rather what longitudinal tightness can be achieved by the cable.

In practice, neither classic fibre-optic cables, with or without cable dividers, nor "breakout cables" are known which comply with the criteria for longitudinal tightness in accordance with Annex E of IEC 60079–14. Classic fibre-optic cables with cable dividers, or splice fields, require a lot of space. For such an installation, it is advisable to use an Ex e terminal enclosure.

This reliably prevents flame transmission through the cable; however, it does not exclude the risk of Zone entrainment. Breakout cables can be quite easily inserted into Ex d enclosures. They are already pre-fabricated for plug-in assembly.

For fibre-optic cables, the following rules stood the test of the practice:

- For cable glands into flameproof boxes, verify the requirements to use barrier glands.
- Ensure mechanical robustness and circularity of the cables.
- If in doubt, ask the manufacturer of the cable glands whether they are compatible with the fibre-optic cable selected.

10.4.2.5   Flexible Cables

In hazardous areas, generally also, flexible cables can be used for stationary devices as many times; drag-chain cables are required for applications. Nevertheless, it is sensible to examine closely the application. The critical questions at this moment are:

- How many cycles has the cable to withstand?
- And at what bending radius and ambient temperature?

The recommendation is to discuss the application requirements with the cable manufacturer. The IEC 60079–14 is vague, as it can be used from a robust plastic insulated cable, up to a light rubber hose cable, as long as it is ensured that the cable cannot be damaged. When planning cabling for portable and mobile devices, it will be necessary to check the applicable standard regarding the maximum voltages, currents, cross-sections, as well as the earthing requirements.

10.4.3   SPECIAL REQUIREMENTS

10.4.3.1   Emergency Switch-Off

IEC 60079–14 defined that for emergency purposes, at a suitable point outside the hazardous area, there shall be a single (or multiple) means of switching off electrical supplies to the hazardous area, except for those electrical equipment which must continue to operate in order to prevent additional danger. These shall be on a separate circuit. And, to allow work to be carried out safely, suitable means of isolation (adequately labelled) shall be provided for each circuit, or group of circuits, to include all circuit conductors (live and neutral).

10.4.3.2   Electrical Isolation

A means of isolation shall be provided to isolate all live conductors, including the neutral, to allow electrical work to be carried out safely. Where all conductors are not isolated by the same device, the means of isolation of other conductors shall be clearly identified.

The preferred means of isolation is by a device that operates in all relevant conductors at the same time. The means of isolation may include fuses and neutral links where relevant. Labelling shall be provided immediately adjacent to each means of isolation to permit rapid identification of the circuit or group of circuits thereby controlled.

There shall be effective measures or procedures to prevent the restoration of supply to the equipment, while the risk of exposing unprotected live conductors to an explosive atmosphere continues.

10.4.3.3   Circuits Traversing Hazardous Areas

Uninsulated conductors, including partially insulated crane conductor rail systems and low and extra-low voltage track systems, should not be installed above hazardous areas. Where circuits traverse a hazardous area in passing from one non-hazardous area to another, the wiring system in the hazardous area shall be appropriate to the EPL requirements for the route.

Where overhead wiring with uninsulated conductors provides power, or communication services, to a hazardous area, it shall be terminated in a non-hazardous area and the service continued into the hazardous area with cable or conduit. Openings in walls for cables and conduits between different hazardous areas, and between hazardous and non-hazardous areas, shall be adequately sealed.

There shall be effective measures, or procedures, to prevent the restoration of supply to the equipment, while the risk of exposing unprotected live conductors to an explosive atmosphere continues.

Cable routing should be arranged in such a way that the cables accumulate the minimum amount of dust layers while remaining accessible for cleaning. Where trunking, ducts or pipes, or trenches are used to accommodate cables, precautions should be taken to prevent the passage or collection of dusts in such places. Where layers of dust are liable to form on cables and impair the free circulation of air, consideration shall be given to derating the current-carrying capacity of the cables, especially if low Minimum Ignition Temperature (MIT) dusts are present.

10.4.3.4   Protection of Electrical Circuits

Electrical circuits and equipment shall be protected against:

- Short circuits
- Overloads
- Earth faults
- Loss of phase (of polyphase electrical equipment).

It is important to note that intrinsically safe and energy-limited circuits have specific requirements for electrical protection. Lower values of disconnection time than those stated in IEC 60364–4 - 41 may be required for installations in areas requiring EPL Ga, Gb, Da, and Db. In cases where automatic disconnection of the electrical equipment may introduce a safety risk, a warning alarm may be used as an alternative to automatic disconnection, provided that operation of the warning alarm prompts remedial action to be taken.

10.5   CONDUITS

Where there is a risk of damage the cable, some form of mechanical protection is necessary. The form of that protection varies from one country to another. As an example the German standard VDE 0100 allows the use of the very practical toughened outer sheath, while other European and non-European countries demand "steel-wire armouring" (SWA) cables. In the United States, metal conduit has been used, almost exclusively, to protect standard cables and also the data communications types.

Usually, the metal conduit becomes a part of the equipotential bonding system, and, therefore, measures must be taken to ensure good electrical continuity through it. Where the conduit system is used as the protective earthing conductor, the threaded junction shall be suitable to carry the fault current which would flow when the circuit is appropriately protected by fuses or circuit breakers.

Cable conduits are widely used in the United States and Canada. The NEC in the United States is based on the requirement to protect cables using all methods of protection and does not currently single out IS cables for special treatment.

Conduit systems do have both advantages and disadvantages, which must be considered before their use on an installation. They are costly to install, difficult to modify, or expand once in place, but they are highly reliable and durable. Combinations of metals that can lead to galvanic corrosion shall be avoided. The installation of metal conduits requires the installation of conduit seals (also known as sealing units) to prevent flame propagation through conduit systems, as shown in Figure 10.7.

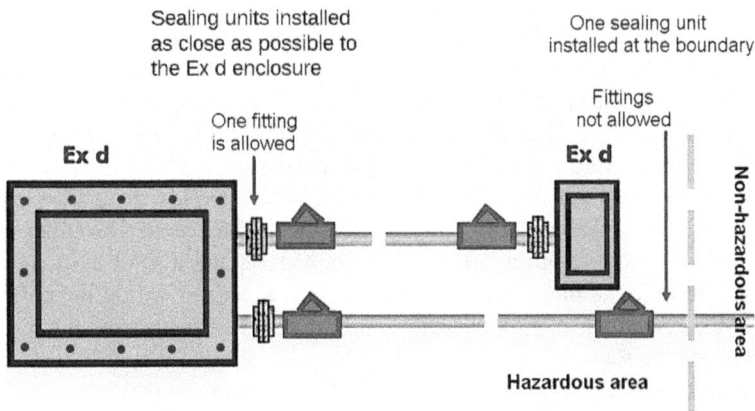

FIGURE 10.7   Sealing units in metallic conduits.

It is necessary to know the instructions given by the compound manufacturers, because temperature limits are usually specified, as examples: The minimum temperature at which the compound is allowed to be mixed, the minimum temperature at which it can be poured into fittings, and the time the seals need to cure when exposed to the minimum temperature.

10.5.1   NEC Requirements for Conduit Seals

Each conduit entry into an explosion-proof enclosure shall have a conduit seal where either of the following conditions apply:

1) The enclosure contains apparatus, such as switches, circuit breakers, fuses, relays, or resistors that may produce arcs, sparks, or temperatures that exceed 80% of the auto-ignition temperature, in degrees Celsius, of the gas or vapour involved in normal operation.

    Exception: Seals shall not be required for conduit entering an enclosure under any one of the following conditions:
    • The switch, circuit breaker, fuse, relay, or resistor is enclosed within a chamber hermetically sealed against the entrance of gases or vapours.
    • The switch, circuit breaker, fuse, relay, or resistor is immersed in oil.
    • The switch, circuit breaker, fuse, relay, or resistor is enclosed within an enclosure; identified for the location; and marked "Factory sealed", "Seal not required", or equivalent.
    • The switch, circuit breaker, fuse, relay, or resistor is part of a non-incendive circuit.
2) The entry is metric designator 53 (trade size 2) or larger, and the enclosure contains terminals, splices, or taps.

Attention points:

• An enclosure, identified for the location and marked "Factory sealed", or "Seal not required" or equivalent, shall not be considered to serve as a seal for another adjacent enclosure that is required to have a conduit seal.
• Conduit seals shall be installed within 450 mm from the enclosure or as required by the enclosure marking. Only threaded couplings or explosion-proof fittings such as unions, reducers, elbows, and capped elbows that are not larger than the trade size of the conduit shall be permitted between the sealing fitting and the explosion-proof enclosure.
• A conduit seal shall be required in each conduit run, leaving a Division 1 location. The sealing fitting shall be permitted to be installed on either side of the boundary within 3.05 m of the boundary, and it shall be designed and installed to minimize the amount of gas or vapour within the portion of the conduit installed in the Division 1 location that can be communicated beyond the seal.
• The conduit run between the conduit seal and the point at which the conduit leaves the Division 1 location shall contain no union, coupling, box, or other fitting except for a listed explosion-proof reducer installed at the conduit seal.

- Where the seal is located on the Division 2 side of the boundary, the Division 1 wiring method shall extend into the Division 2 area to the seal.
- As an exception, for underground conduit installed where the boundary is below grade, the sealing fitting shall be permitted to be installed after the conduit emerges from below grade, but there shall be no union, coupling, box, or fitting, other than listed explosion-proof reducers at the sealing fitting, in the conduit between the sealing fitting and the point at which the conduit emerges from below grade.
- Non-sheathed insulated single or multi-core cables may be used in the conduits. However, when the conduit contains three or more cables, the total cross-sectional area of the cables, including insulation, shall be not more than 40% of the cross-sectional area of the conduit.
- Long runs of conduits shall be provided with suitable draining devices to ensure satisfactory draining of condensate. In addition, cable insulation shall have suitable water resistance.

10.5.2 COMPARISON OF CABLE TRAYS X CONDUITS

In Brasil, the first experience with hazardous areas installations began with the oil refineries in the 1940s, using metallic conduits and Ex d components as per NEC requirements.

Following the introduction of jacketed offshore oil platforms in the early 1980s, work began on designing electrical installations using armoured cables, cable trays, and cable glands. This approach was based on the philosophy used for similar oil platforms in the North Sea and complied with IEC requirements. Those offshore installations also used fluorescent Ex d luminaires, with their bodies made with cast aluminium, as shown in Figure 10.8.

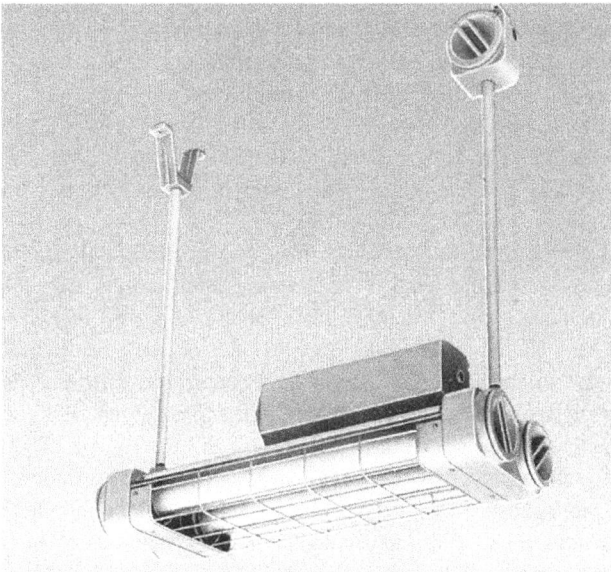

FIGURE 10.8   A typical Ex d fluorescent luminaire.

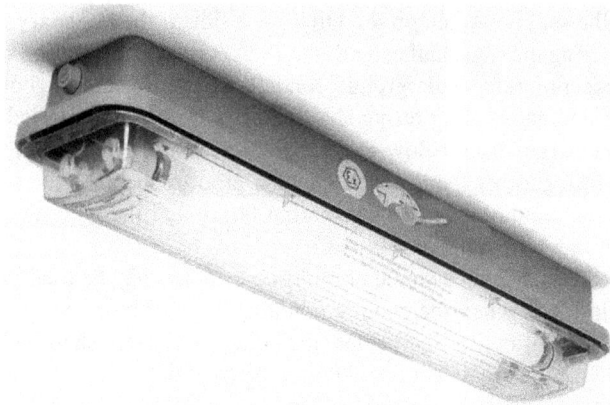

FIGURE 10.9   A typical Ex ed fluorescent luminaire.

TABLE 10.2
Comparison between Ex Luminaires in the 1990s

2 × 40 W	Ex d	Ex ed
Weight [kg]	26	12
Lighting level [lux]	340	530
Price [US$]	432.00	629.00

Corrosion due to the marine environment caused high costs of maintenance for those luminaires. In the early 2000s, the maintenance people started to use fluorescent Ex ed luminaires, made of fibre glass reinforced polyester, with polycarbonate lens, as shown in Figure 10.9.

Ex ed luminaires were lighter, with greater luminous efficiency, easier maintenance, and corrosion-free characteristics. A comparison of these two types is shown in Table 10.2.

Table 10.3 shows the Brasilian costs for two installation alternatives of an oil pumping station located in a Zone 1 area, with four oil pumps, a distribution panel, and four push button control stations, whose sketch is shown in Figure 10.10. The cables' costs are included.

It is necessary to consider that Ex ed components were more expensive, because they were imported and high taxes were applied, while the metallic accessories and Ex d components were produced in Brasil.

Using armoured cables, it cost 270% more than using non-armoured ones. However, the civil works costs, including the concrete encasement of conduits, were 153% higher than using the cable tray alternative. The installation using metallic conduit

TABLE 10.3

Cost Comparison in U.S. Dollars

Description	Conduits	Cable Trays
Panel, control stations, luminaires	9,665.91	16,752.82
Raceway, accessories	1,679.86	1,652.71
Civil works	8,451.96	5,580.78
Cables	281.83	760.55
Total cost (US$)	20,079.56	24,746.86

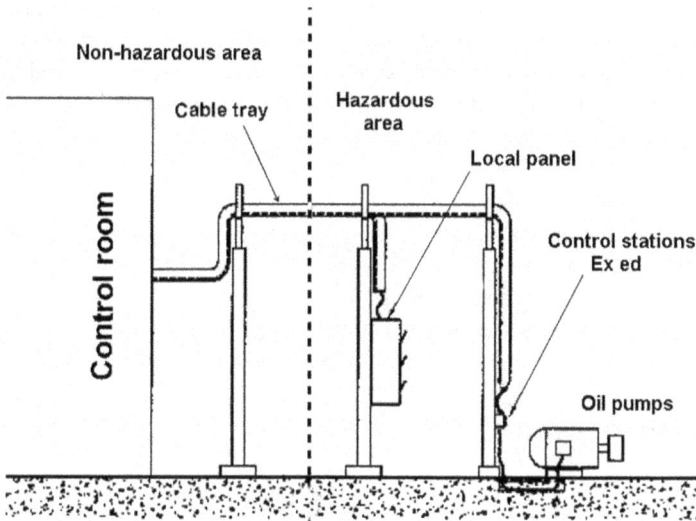

FIGURE 10.10   Installation of 4 × 20 hp oil pumps, 480 V, 3Ø, using a 6 m high cable tray, instead of buried metallic conduits.

was cheaper, but to know the whole scenario, it would be necessary to include the maintenance costs, and the better alternative depends on the particular characteristics of each facility's human and financial resources.

10.6   EX MOTORS' PROTECTION

In selecting a motor with protective devices for explosive atmospheres, the motor manufacturer's instructions and recommendations must be followed. As only the motor can be installed in a potentially explosive atmosphere, with the protective devices always kept in a safe area, the instructions are intended to prevent the motor from overheating or creating any sparks. To ensure safe operation, certain details need to be considered when selecting a motor together with its protective devices.

10.6.1 OVERLOAD

For Ex electric motors, especially the increased safety (Ex e), the overload protection device plays a very important role. The IEC 60079–14 stipulates that the thermal overload protection of rotating electrical machines (e.g. the bimetallic relay) must not be set above the motor's rated current. At 1.2 times the rated current, the device must operate within 2 hours. In Ex e motors, the protection must cover, in addition to continuous operation, the failure—considered foreseeable—of a locked rotor.

During the tests carried out by the certification body, the protection time is determined so that the temperatures in the stator windings and in the housing remain below the ignition temperature of the explosive atmosphere with a safety margin. The time t_E determined in this way is categorized according to the Temperature Classes.

The designer must then select a protective device that meets these shutdown conditions. In general, this requirement is met via combinations of I_a/I_n and t_E that are made available by the thermal relay manufacturers, as shown in Figure 10.11 for a insulation class F, T3 (Temperature Class) Ex motor, where:

A—maximum ambient temperature;
B—temperature in normal service;
C—Temperature Class limit;
1—temperature elevation in service;
2—temperature elevation with locked rotor; and
t_E—5 seconds, when the stator temperature reaches the T3 limit.

However, compatibility with the motor data must be checked on a case-by-case basis.

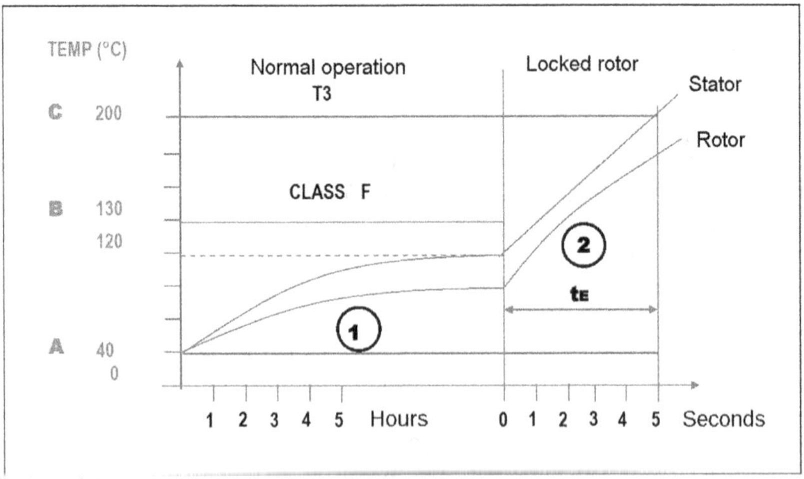

FIGURE 10.11 Time t_E of Ex e motors.

For motors with type of protection "e" terminal boxes, when using converters with high frequency pulses in the output, care should be taken to ensure that any overvoltage spikes and higher temperatures which may be produced in the terminal box are taken into consideration.

Additionally, for machines with type of protection "e", the following should be provided:

- Monitoring the current in each phase
- Close overload protection to the fully loaded condition of the motor.

Inverse-time overload protection relays may be acceptable for motors of duty type S1, which have easy and infrequent starts. Where the starting duty is arduous, or starting is required frequently, the protection device should be selected so that it ensures limiting temperatures are not exceeded under the declared operational parameters of the motor. Where the starting time exceeds $1.7t_E$, an inverse-time relay would be expected to trip the motor during start-up.

Thermal overload relays are electromechanical protection devices for the main circuit. Electronic overload relays offer reliable protection for motors. Both can make up a compact starting, together with contactors.

10.6.2   Phase Loss

The danger of three-phase squirrel-cage motors in the event of a phase failure is highlighted in IEC 60079–14, which states that "Precautions must be taken to prevent the operation of three-phase motors, if there is a loss of one or more phases, from causing overheating".

In the case of Ex e motors, there are additional recommendations:

The characteristics of machines with closed delta windings, in the event of a phase loss, must be considered. Unlike star-connected machines, the loss of a phase may not be detected during operation. The effect will be an imbalance in the currents in the machine supply and an increase in the motor heating. A delta-connected motor with a low torque load at start-up may also be capable of starting in this fault condition, which may remain undetected for a long period.

Therefore, for Ex machines with delta windings, protection against phase imbalance must be provided to prevent excessive heating in the event of this fault. The properties of delta wound motors in the case of the loss of one phase should be specifically addressed. Unlike star wound motors, the loss of one phase may not be detected, particularly if it occurs during operation.

The effect will be current imbalance in the lines feeding the motor and increased heating of the motor. A delta wound motor with a low torque load during start-up might also be able to start under this winding failure condition, and, therefore, the fault may exist undetected for long periods. Therefore, for delta wound motor, phase imbalance protection shall be provided, which will detect motor imbalances before they can give rise to excessive heating effects.

10.6.3 FREQUENCY CONVERTERS AND SOFT STARTING

Motors supplied by frequency and voltage converters require that either:

- The motor has been type-tested for this duty as a unit, in association with the converter, or soft start, specified in the descriptive documents according to IEC 60079–0 and with the protective device provided; or
- if the motor has not been type-tested for this duty as a unit, in association with the converter, or soft start, then means (or equipment) for direct temperature control by embedded temperature sensors specified in the motor documentation, or other effective measures for limiting the surface temperature of the motor housing, shall be provided.

The effectiveness of the temperature control shall take into account power, speed range, torque, and frequency for the duty required and shall be verified and documented. The action of the protective device shall be to cause the motor to be electrically disconnected.

The selection and installation of motors fed by frequency converters must take into account items that can reduce the voltage at the motor terminals. Other hazards must also be taken into account.

Note 1: A filter at the converter output can cause a voltage drop at the machine terminals. The reduced voltage increases the motor current and slip and thus increases the temperature of the motor stator and rotor. This temperature increase may be more noticeable under constant rated load conditions.

Note 2: Additional information on the application of motors fed by a converter can be found in IEC/TS IEC TS 60034–25. The main concerns include the frequency spectra of the voltage and current, as well as their additional losses, the effects of overvoltages, bearing currents, and high-frequency grounding.

10.6.4 OVERTEMPERATURE

The use of embedded temperature sensors to control the limiting temperature of the motor is only permitted if such use is specified in the motor documentation. The time t_A specifies the response time of the temperature sensors and has to be verified.

Thermistor motor protection relays monitor the winding temperature of motors which have PTC temperature sensors installed. This direct temperature measurement enables the thermistor motor protection relays to evaluate various motor conditions such as overheating, overload, and insufficient cooling. Check if it has a suitable approval for the use in hazardous areas.

10.6.5 SHORT CIRCUIT

Manual motor starters, also known as Motor Protection Circuit Breakers (MPCB) or Manual Motor Protectors (MMP), are electromechanical protection devices for the main circuit. They are mainly used to switch motors ON/OFF manually and to provide fuseless protection against short-circuit, overload, and phase failures. Fuse-less protection saves costs and space and ensures a quick reaction under short-circuit condition by switching the motor off within milliseconds. Starter combinations are set up together with contactors.

10.6.6 SURGES

IEC 60079–14 ed. 5 alerts on its clause 5.11.5.1 that switching overvoltages can occur if vacuum circuit breakers, or vacuum contactors, are used when a high-voltage motor is switched off. These transients depend on various installations system and design factors such as arc-extinguishing principle of the contactor or switch, size of the motor, length of the power supply cable, system capacitance, among others.

From IEC 60079–14:

> In some cases, multiple restrikes can result in switching overvoltages which are too high for the insulation of the motor stator winding leading to insulation deterioration and incendive sparks. In practice, this generally occurs when high-voltage motors with starting currents $I_A > 600$ A are disconnected during start-up or during a stalled or overload condition.
>
> Vacuum circuit breakers or vacuum contactors are commonly associated with high voltage transients. Surge suppressors should be installed in the switchgear, between the circuit breaker and the motor cable termination, for each of the three conductors to earth.

Although the IEC 60079–14 text seems to be only a suggestion for installation of surge suppressors, this should be a requirement, as even multiple restrikes are not a pre-condition for the occurrence of overvoltages on insulation. Older the insulation, higher will be the possibility of a spark.

IEC 60079–14 also says that:

> The peak voltages which arise as a result can damage the winding insulation, which can lead to insulation deterioration and incendive sparks. If vacuum circuit breakers or vacuum contactors are used for motor switching, the motor installation design should consider using an appropriate surge suppressor, such as a zinc oxide varistor with spark gap.

In order to verify the suitability to the clause 5.11.5.1 of IEC 60079–14 ed. 5, an Ex e induction motor, four poles, three-phase, 60 Hz, insulation class F, $\Delta T = 80$ K, frame 355, 400 kW, 4160 V, certified in accordance with IEC 60079–7, was subjected to a locked rotor test in an explosive mixture atmosphere composed by air and $(21 \pm 5)\%$ of hydrogen.

Note: With a maximum coolant temperature of 40°C, the maximum permitted over-temperature in the thermal class 130 (B) is $\Delta T = 80$ K.

On Tables 10.4 and 10.5, the test results are shown, where flame propagation hap-pened, confirming that a surge suppressor protection is recommended.

TABLE 10.4

Locked Rotor Test of Ex e Motor in an Explosive Mixture of Hydrogen and Air

Test n°	O₂ %	H₂ %	T_amb °C	rh %	Voltage (Φ-N) (kV)			Current (A)			Locked Rotor Time (ms)	Flame Propagation	
					V_U	V_V	V_W	I_U	I_V	I_W		Yes	No
1	16.70	20.50	26.3	84.9	2.41	2.43	2.43	542	611	562	1,177		X
2	16.70	20.50	26.3	84.9	2.33	2.38	2.43	568	636	578	1,177		X
3	16.70	20.50	23.8	95.6	2.33	2.38	2.38	568	628	578	1,177		X
4	16.70	20.50	23.8	95.6	2.33	2.38	2.43	568	619	565	1,177	X	

Note: Minimum voltage test shall be 90% of the nominal value.

In this case, the system was calculated for V = 2.162 kV

(V_{nom} = 4.160 / √3 = 2,402 V × 0.9 = 2.1612 kV)

TABLE 10.5
Currents and Voltages

Voltage		RMS	Peak	Current		RMS	Peak
Phase-neutral	V_U	2.33	–	Current (A)	I_U	568	–
voltage (kV)	V_V	2.38	–		I_V	619	–
	V_W	2.43	–		I_W	565	–
Average phase-		2.43		Average current (A)		584	
neutral voltage				Duration time (ms):			
(kV)				1,177			

Note: Date—Time: 2008/11/29–15:11:08

10.6.7 VERIFICATION

Checking the tripping of the overload protection, and also the tripping of the over-current protection with locked rotor, must be included in the periodic inspections. Measuring the tripping values is recommended, although IEC 60079–17 leaves this point to the discretion of the maintenance personnel, both in the initial and periodic inspections. The actual tripping time should be within a tolerance, depending on the device specification.

Medium and large motors can also become stuck if they start with the phase sequence reversed and may suffer damage.

For safety reasons, the test must be performed phase-by-phase, and relays in this condition are expected to operate in a shorter time than when operating with three-phase power. If the tripping times are longer than specified, there is a suspicion that the characteristic curve is seriously altered.

10.7 ELECTRIC HEATERS

Heaters, unless installed as part of another certified assembly (e.g. electric motor anti-condensation heater), shall have the following protection, in addition to overcurrent:

- In a TT or TN type system, a residual current device (RCD) with a rated residual operating current not exceeding 100 mA shall be used in order to limit the heating effect due to earth-fault and earth-leakage currents. Preference should be given to RCDs with a rated residual operating current of 30 mA.
- In an IT system, an insulation-monitoring device shall be used to disconnect the supply, whenever the insulation resistance is not greater than 50 ohms per volt of rated voltage.

For short-circuit calculations, the load current of the complete trace heating circuit should be taken into consideration.

The resistance heating device shall be prevented from exceeding the limiting temperature when energized by one of the following means:

- A stabilized design using the temperature self-limiting characteristic of the resistance heating device
- A stabilized design of a heating system (under specified conditions of use).
- A safety device by sensing based on the temperature of the resistance heating device, or the surrounding temperature, with additional parameters, like:
  - in the case of liquids, the heating device can be covered by at least 50 mm of liquid by means of a level monitor (dry run protection).
  - in the case of flowing media (such as gas and air), the minimum throughput can be ensured by means of a flow monitor; or
  - for the heating of work pieces, the heat transfer can be ensured by the fixing of the heating device or with auxiliary agents (heat-conducting cement).

For locations where EPL Gb or Db equipment are required, the safety device shall de-energize the resistance heating device, or unit, either directly or indirectly. For locations where EPL Gc or Dc equipment are required, the safety device shall de-energize the resistance heating device, or unit, either directly or indirectly, or provide an output for an alarm intended to be located in a constantly attended location.

Reset shall only be manual with the aid of a tool, and only after the previously defined process conditions have returned, except when the information from the safety device is continuously monitored. In the event of failure of the sensor, the heating device shall be de-energized before the limiting temperature is reached. The adjustment of the safety devices shall be locked and sealed and shall not be capable of being subsequently altered when in service. Thermal fuses should be replaced only by parts specified by the manufacturer.

10.7.1   ELECTRICAL TRACE HEATING SYSTEMS

In electrical trace heating systems, the outer metallic covering, metallic braid, or other equivalent electrically conductive material of the trace heater shall be bonded to the earthing system to provide an effective earthing path. In applications where the primary earthing path is dependent on the metallic covering, metallic braid, or other equivalent electrically conductive material, the chemical resistance of the material shall be taken into account, if the exposure to corrosive vapours or liquids might occur.

Stainless steel type braids and sheaths typically have high resistance and may not provide effective earthing paths. Consideration should be given to alternative earthing means or supplemental earthing protection.

There are two designs for electrical heating systems:

Stabilized design:

The electrical trace heating system is designed in such a way that, under foreseeable conditions, the surface temperature of the electrical resistance trace heater does not exceed the limits of the Temperature Class, or the maximum surface temperature, minus 5 K for temperatures lower than or equal to 200°C or minus 10 K for temperatures higher than 200°C. A stabilized design for EPL Gb or Gc does not normally need additional protection against excessive temperatures.

Controlled design:

Controlled design applications require the use of a temperature control device to limit the maximum surface temperature. The temperature limiting device operates independently from the temperature controller. A protective device, such as a temperature limiter, de-energizes the system and prevents the temperature from exceeding the maximum permissible surface temperature. In the event of a fault or damage to a sensor, the heating system is de-energized in order to replace the defective equipment.

10.8 CLOSING REMARKS

Industrial fire and explosion incidents due to electrical system malfunctions highlight the critical importance of rigorous maintenance, inspection, and upgrade practices for electrical systems. Each of these incidents result in significant loss of life, property damage, and substantial financial impacts. Subsequent investigations provided valuable recommendations for preventing future accidents. These recommendations emphasized the need for fire detection systems' fault detection sensors, surge protection, regular safety audits, and comprehensive training for workers. Implementing these recommendations is essential to ensure the safety of industrial operations and the protection of the surrounding communities.

BIBLIOGRAPHY

Anantasate, S., Chokpanyasuwan, C., Pattaraprakorn, W. and Bhasaputra, P.—Application of hazard & operability study to safety evaluation of electrical design for a major power system upgrade of oil & gas production plant. In: *6th International Conference on Electrical Engineering/electronics, Computer, Telecommunications and Information Technology*, Chonburi, Thailand, 2009.

Crow, Daryld Ray and Mohla, Daleep—Auditing your electrical safety management program: A method to identify key gaps in electrical safety at your site. In: *2023 IEEE IAS Petroleum and Chemical Industry Conference (PCIC)*, New Orleans, LA, USA, 2023.

Directive 89/391/EEC—Council Directive 89/391/EEC of 12 June 1989 on the introduction of measures to encourage improvements in the safety and health of workers at work, 1989.

Directive 2014/34/EU—Council of the European Parliament of 26 February 2014 on the harmonisation of the laws of the Member States relating to equipment and protective systems intended for use in potentially explosive atmospheres, 2014.

Directive 2014/35/EU—Council of the European Parliament of 26 February 2014 on the harmonisation of the laws of the Member States relating to the making available on the market of electrical equipment designed for use within certain voltage limits, 2014.

Floyd, H. Landis—A practical guide for applying the hierarchy of controls to electrical hazards. In: *2015 IEEE IAS Electrical Safety Workshop*, Louisville, KY, USA, 2015.

IEC TS 60034-25—*Rotating electrical machines—Part 25: AC electrical machines used in power drive systems—application guide*. International Electrotechnical Commission, 2022.

IEC 60060-1—*High-voltage test techniques—Part 1: General definitions and test requirements*. Ed. 3.0. International Electrotechnical Commission, 2010.

IEC 60079-0—*Explosive atmospheres—Part 0: Equipment—General requirements*. International Electrotechnical Commission, 2017.

IEC 60079-11—*Explosive atmospheres—Part 11: Equipment protection by intrinsic safety "I"*. Ed. 7.0. International Electrotechnical Commission, 2023

IEC 60079-14—*Explosive atmospheres—Part 14: Electrical installation design, selection and installation of equipment, including initial inspection*. Ed. 6.0. International Electrotechnical Commission, 2024.

IEC 60079-17—*Explosive atmospheres—Part 17: Electrical installations inspection and maintenance*. International Electrotechnical Commission, 2023.

IEC 60332-1-2—*Tests on electric and optical fibre cables under fire conditions—Part 1–2: Test for vertical flame propagation for a single insulated wire or cable—procedure for 1 kW pre-mixed flame*. International Electrotechnical Commission, 2004.

IEC 60332-3-22—*Tests on electric and optical fibre cables under fire conditions—Part 3–22: Test for vertical flame spread of vertically-mounted bunched wires or cables—category A*. International Electrotechnical Commission, 2018.

IEC 60364-4-41—*Low voltage electrical installations—Part 4–41: Protection for safety—protection against electric shock*. International Electrotechnical Commission, 2005.

McClung, L. Bruce and Hill, Dennis J.—Electrical system design techniques to improve electrical safety. In: *2010 IEEE IAS Electrical Safety Workshop*, Memphis, TN, USA, 2010.

McGree, Tucker—*Fires in industrial and manufacturing properties*. NFPA Research, 2023.

NFPA 55—*Compressed gases and cryogenic fluids code*. National Fire Protection Association, 2023.

NFPA 70—*National Electrical Code (NEC)*. National Fire Protection Association, 2023.

Rangel Jr., Estellito—Brasil moves from divisions to zones. In: *49th Petroleum and Chemical Industry Conference*, New Orleans, Conference Record, IEEE, 2002, pp. 23–29. Available at: http://ieeexplore.ieee.org/document/1044981/

Rangel Jr., Estellito—Eletricidade estática. *EMEx Section, Eletricidade Moderna Magazine*, São Paulo, ed. 374, May 2005, p. 178.

Rangel Jr., Estellito—Eletricidade estática. *EMEx Section, Eletricidade Moderna Magazine*, São Paulo, ed. 526, Jan. 2018, p. 58.

Rangel Jr., Estellito—Proteção de motores Ex. *EMEx Section, Eletricidade Moderna Magazine*, São Paulo, ed. 551, Feb. 2020, p. 62.

Rangel Jr., Estellito—Riscos da eletricidade estática. *EMEx Section, Eletricidade Moderna Magazine*, São Paulo, ed. 443, Feb. 2011, p. 160.

Rangel Jr., Estellito—Riscos da eletricidade estática II. *EMEx Section, Eletricidade Moderna Magazine*, São Paulo, ed. 561, Sep./Oct. 2021, p. 66.

VDE 0100-100—*Low-voltage electrical installations. Part 1: Fundamental principles, assessment of general characteristics, definitions.* Verband der Elektrotechnik, 2009.

11 Inspection and Maintenance

11.1 INSPECTION

11.1.1 THE IMPORTANCE OF INSPECTIONS

In the UK, an Ex private Inspection Company issued their results of four years of inspections in the "close" grade, which involved more than 71,000 items in several companies in the segments of gas production, oil refining, and storage and distribution of chemical products, which revealed a very worrying picture:

- Only 35% of the Ex equipment could be considered safe for use, without any non-conformities
- Defects capable of causing ignition were detected in 14% of the items
- Other defects that compromised the safety of the unit were presented in 27% of the items
- Minor defects, which did not compromise safety, were identified in 24% of the items.

The most common defects found were:

- Equipment unsuitable for the classification of the area
- Cable entry devices with corrosion and inadequate installation.

It is worth noting that these results were got in the UK, where there is a specific professional certification system (voluntary) for workers who perform Ex installations, called CompEx. In countries without such resources, probably the same Ex inspections could present even worse results, mainly due to the lack of specific technical training for installers and maintainers.

11.1.2 CONCEPTS OF AN INSPECTION PLAN

When we talk about "inspection", the first image that many facilities' managers associate with it is that it should be something merely bureaucratic, carried out after a few years of operation of the plant. They may even consider it as something unnecessary, as by their point of view, "everything is working perfectly", and when a problem occurred, the maintenance team has always been acting quickly.

This image, which is very common, derives from the focus on operational continuity, but, in the context of explosion safety, it is necessary to go further, because it

 DOI: 10.1201/9781003500001-11

is not enough for the equipment to be just working; it has to be working safely. And safety will only be guaranteed if the equipment maintains its original characteristics, or, if something needed to be modified, from that moment on, it offers a level of safety equal to the original.

In fact, the first inspection of the electrical installation in a hazardous location must be done before it comes into operation, and within a comprehensive scope. The first step on planning the inspection plan is to define their grades and types of inspection.

11.1.3 GRADES OF INSPECTION

The grade of inspection may be visual, close, or detailed.

Visual

This grade of inspection is very limited, as it does not need the inspector standing beside the equipment and does not involve the opening of the enclosure or the use of any tools. It can be done with the equipment energized, and the focus is on obvious external problems, such as corrosion or damage, and which will not normally require isolation of the installation, unless any potential problems are identified.

The recommended minimum requirement of a visual inspection of all equipment is carried out at least every two years, but, depending on the rate of equipment deterioration (based on the results of the visual inspection), the interval between inspection periods may need to be reduced.

Close

This grade of inspection can reveal more details than can be observed in a visual inspection, as it requires the inspector to be next to the equipment and allows the use of simple tools such as a ruler and a magnifying glass. Close inspections can be also performed with the equipment energized.

Detailed

This grade of inspection will require the equipment to be isolated, because it will include a complete verification within the equipment.

11.1.4 TYPES OF INSPECTION

There are four types of inspection described in IEC 60079–17, as follows in the next subsections.

Initial

It is a fully detailed inspection, to be done before the facility starts it operation, or when some Ex equipment has just been replaced. The initial inspection is done under the detailed grade.

Periodic

It may be visual or close grade. A visual or close periodic inspection may lead to the need for a further detailed inspection. The intervals between periodic inspections must not exceed three years.

Sample

Some companies perform sampling inspections, but this alternative should be understood as having a limited spectrum, as it will not reveal important faults, such as loose connections, in the uninspected components. They may be visual, close, or detailed.

The size and composition of all samples are mainly determined by environmental factors. As it only covers part of the installed equipment, if maintenance is required for the sample, all Ex equipment of the same type or function will need to be assessed to ensure the same corrective measures are applied.

Continuous Supervision

This is a type of inspection provided for in IEC 60079–17, which is based on the idea that the plant operator himself (and not a member of the maintenance team) performs an inspection (visual or close, depending on the ease of access to the equipment), while carrying out his work routine. In practice, this scheme only works to identify the most glaring flaws visually, not only because the training and experience of the operators are focused on the plant's process parameters, but also because there is no way to establish defined controls of periodicity and criticality, as the idea would be to carry out such inspections in the "operators' free time".

11.1.5   ESTABLISHING AN EX INSPECTION PLAN

The risks of explosions in industries that process flammable materials, due to electrical installations, are highly controlled with the investment in maintenance plans and in the training of professionals.

As provided for in the IEC 60079–17, all electrical and electronic equipment intended to operate in potentially explosive atmospheres must be included in an Inspection Plan that compasses:

- A detailed initial inspection before the plant goes into operation, when it must be verified that the equipment is intact, correctly installed, and suitable for the characteristics of the location
- Periodic inspections that issue records and trigger corrective maintenance when non-conformities are found.

Each maintenance—preventive or corrective—performed on Ex equipment must be recorded in their respective individual files containing their data (model, manufacturer, serial number, date of purchase, date of installation, etc.). This control, in addition to enabling the evaluation of the performance of the Ex equipment, will also allow for a rapid identification in the event of a recall by the manufacturer.

The following steps are to be followed in the inspection plan:

1) Get the area classification documents.
2) Select the inspection team.
3) Obtain data on all installed Ex equipment.
4) List all equipment in inadequate conditions.
5) Eliminate non-conformities immediately.
6) Verify the necessity to review the frequency of inspections.

Let us explain each item in detail:

1) **Get the area classification documents**

It is a prerequisite for the inspection process to have the updated documents of the area classification study on hands, because the first step to be taken is to verify whether the equipment meets the requirements for the classified area in which it is installed.

2) **Select the inspection team**

Skilled personnel shall be provided with sufficient training to enable familiarity with the installation which they attend. This training shall include any plant, equipment, and operational or environmental conditions which can relate to the needs of the explosion protection.

In addition, refresher training must be provided regularly. Therefore, a worker who only has experience in maintaining industrial equipment in non-classified areas cannot be assigned to this activity.

The skilled personnel will need to be capable of identifying any alterations on installation, based on the updated documents and equipment characteristics. If there are no properly trained personnel at the facility, it will be necessary to hire a specialized company to carry out the inspection.

3) **Obtain data on all installed Ex equipment**

It is necessary to know the inventory of installed Ex electrical equipment, especially on their types, Equipment Groups, Temperature Classes, and categories. The certificates of conformity for each Ex electrical and electronic equipment must be kept available for the inspection personnel. Caution is required if a letter suffix ("X" or "U") appears after the certificate number, as this means that special conditions of safe use apply.

Each preventive or corrective maintenance performed on Ex equipment must be recorded in its respective individual file, which contains information such as model, manufacturer, serial number, date of purchase, and date of installation. This control, in addition to enabling the evaluation of the performance of the Ex equipment, will also allow for rapid identification of it in the event of a recall issued by the manufacturer.

4) **List all equipment in inadequate conditions**

The inspection sheets available in IEC 60079–17 can be used to identify the non-compliant equipment. It is important to note that if the integrity of the Ex equipment is compromised, the safety of the plant is at risk. Repair work on Ex equipment must be only carried out by competent workshops and personnel.

5) **Eliminate non-conformities immediately**

All equipment detected with their Ex integrity impacted must be identified, and the necessary repairs must be carried out without delay. Repair work on Ex equipment must only be carried out by trained personnel, in a equipped workshop, and using the manufacturer's documentation and the conformity certificates as basis.

6) **Adjust the frequency of periodic inspections**

The frequency of inspections should be determined on the basis of the scenario encountered during the inspection of the equipment. A finding that the integrity of the Ex equipment has been compromised, for example due to corrosion, should lead to a more frequent inspection so that protective products can be applied. On the other hand, if the equipment maintains its integrity unchanged between inspections (e.g. electrical contacts with imperceptible changes in cleanliness), the interval between periodic inspections may be extended.

11.1.6 WHAT TO CHECK IN AN EX INSPECTION?

The IEC 60079–17 provides a checklist for various types of Ex protection in its annexes. Table 11.1 shows the model suggested by the standard for the Ex i and Ex iD types of protection.

The grades of inspection are designed as: D—detailed, C—close, and V—visual.

TABLE 11.1
Checklist for Ex i and Ex iD Types of Protection

Check that		Grade of Inspection		
		Detailed	Close	Visual
A	EQUIPMENT			
1	Circuit and/or equipment documentation is appropriate to the EPL/Zone requirements of the location	X	X	X
2	Equipment installed is that specified in the documentation—fixed equipment only	X	X	
3	Circuit and/or equipment category and group are correct	X	X	
4	Equipment Temperature Class is correct	X	X	
5	Installation is clearly labelled	X	X	
6	Enclosure, glass parts, and glass-to-metal sealing gaskets and/or compounds are satisfactory	X		

TABLE 11.1 *(Continued)*
Checklist for Ex i and Ex iD Types of Protection

Check that	Grade of Inspection		
	Detailed	Close	Visual
7 There are no unauthorized modifications	X		
8 There are no visible unauthorized modifications		X	X
9 Safety barrier units, relays, and other energy limiting devices are of the approved type, installed in accordance with the certification requirements and securely earthed where required	X	X	X
10 Electrical connections are tight	X		
11 Printed circuit boards are clean and undamaged	X		
B INSTALLATION			
1 Cables are installed in accordance with the documentation	X		
2 Cable screens are earthed in accordance with the documentation	X		
3 There is no obvious damage to cables	X	X	X
4 Sealing of trunking, ducts, pipes, and/or conduits is satisfactory	X	X	X
5 Point-to-point connections are all correct	X		
6 Earth continuity is satisfactory (e.g. connections are tight, conductors are of sufficient cross-section) for non-galvanically isolated circuits.	X		
7 Earth connections maintain the integrity of the type of protection	X	X	X
8 Intrinsically safe circuit earthing and insulation resistance are satisfactory	X		
9 Separation is maintained between intrinsically safe and non-intrinsically safe circuits in common distribution boxes or relay cubicles	X		
10 As applicable, short-circuit protection of the power supply is in accordance with the documentation	X		
11 Specific conditions of use (if applicable) are complied with	X		
12 Cables not in use are correctly terminated	X		
C ENVIRONMENT			
1 Equipment is adequately protected against corrosion, weather, vibration, and other adverse factors	X	X	X
2 No undue external accumulation of dust and dirt	X	X	X

Additionally, important items for safety of the installation need to be included in a dedicated inspection scheme. As an example, ventilation systems intended to prevent the formation of explosive atmospheres and the associated monitoring systems must be checked by a competent person against the intended duty before they are first brought into service and should be examined at regular intervals.

Ventilation systems with adjustable items (e.g. regulators, baffles, variable-speed fans) should be examined every time the settings are changed. It is desirable for such items to be locked against interference. Where ventilation systems are adjusted automatically, the examination should cover the entire range of settings.

11.1.6.1   Fundamental Items

The main inspection items applied to equipment of all types of protection are highlighted in the following questions:

- Is the equipment suitable for the classified area?
  - Verify: EPL, Equipment Group, and Temperature Class
- Is any component damaged?
  - Verify if:
    - A deterioration of enclosures, fixings, and cable entries due to corrosion, happened
    - Physical damage to enclosures and/or cables occurred.
- Is the equipment subjected to an environmental attack?
  - Verify if:
    - Chemicals, vapours, and/or accumulation of condensates have caused damages in the form of change in colour or surface roughness alteration.
    - Undue accumulations of dust and dirt, particularly on the outside of the enclosure and on cable trays, could harbour corrosive liquids and solvents.
    - There is an excessive vibration at the point of installation, which could cause deterioration of connections.

For fixed and floating offshore petroleum facilities, API RP 14F has recommendations for the routine and detailed inspections on electrical equipment. It is highlighted that the investment in refresher training on hazardous locations' installation requirements needs to be considered also for operational teams, in order to prepare them to quickly identify non-conformities at site and to ask the maintenance team for a prompt repairing action.

11.1.6.2   Verification Items for Particular Types

Some types of protection have particular verification items, as follows.

11.1.6.2.1   Type Ex d
Attention points for this type of protection:

a) **Unauthorized modifications**

The assembly of electrical panels using Ex d enclosures requires, in addition to knowledge of the technical requirements regarding the functionality of the circuits, that a careful analysis be carried out by an accredited laboratory, aiming at its certification for the whole assembly. The IEC 60079–17 states that inspection should look for signs of unauthorized modification, but, in practice, this is difficult, as it is necessary to check the original design of the equipment.

It is not unusual to find a "vacant space" inside the certified Ex d panel, and, after a while, there is a need to expand the circuit to add a new component, as example, an auxiliary transformer. At this point, workers without the required knowledge can think that it seems very convenient to add the new component exactly in that "vacant space inside".

But, adding new components inside an Ex d panel can put the installation at risk. Let us know the reasons why. First, because the certified panel had its surface temperature assessed on the basis of the thermal load defined by the original components. In other words, the installation of a new internal component may cause the panel to change its Temperature Class and become incompatible for safe use with the gas processed in the plant, thus compromising its safety. Second, the Ex d panel is tested to withstand any internal explosion, and the arrangement of the internal components has a significant influence on this.

A change in the internal geometry may allow the occurrence of the "pressure piling" phenomenon, as a result of sub-division of the interior of a flameproof enclosure. An explosion at one side of an obstacle pre-compresses the flammable mixture in the direction of the opposite side, resulting in a secondary explosion that can reach a pressure value around three times that of the initial explosion. See Figures 11.1 and 11.2.

The pressure piling can also be propagated through the conduit to another Ex d enclosure connected to it. This results in a new explosion inside the second Ex d

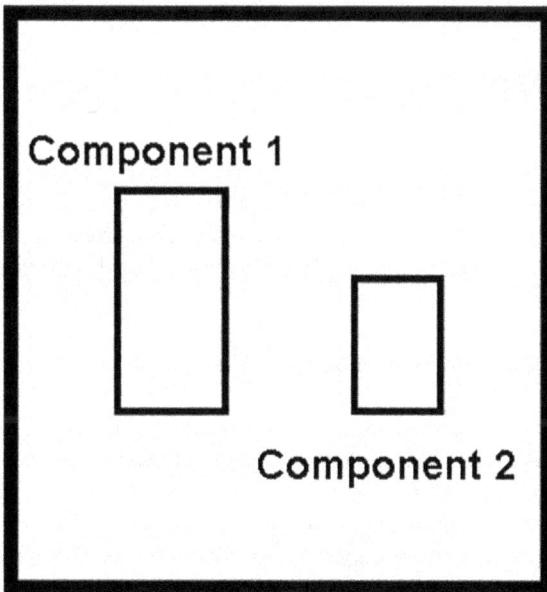

FIGURE 11.1 Original interior of a certified Ex d panel, with two components.

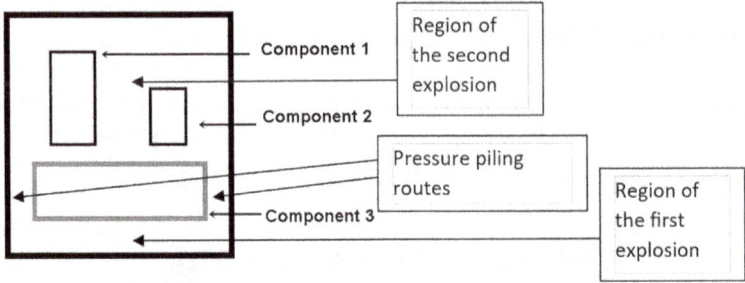

FIGURE 11.2 Adding a new, third component can allow the "pressure piling" phenomenon.

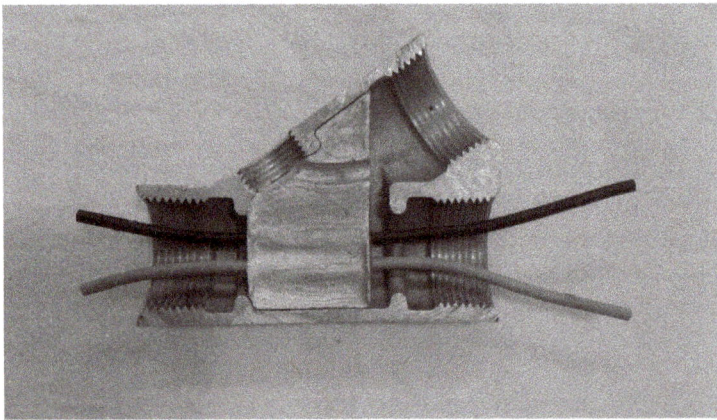

FIGURE 11.3 Sectional view of an Ex d conduit seal unit, showing the draining channel.

enclosure, many times higher than the first one, which can collapse the latter. This so-called "pressure piling" can cause the equipment to explosively rupture. So, the main function of a conduit sealing fitting is to prevent the propagation of an explosion from the inside to the outside of the "Ex d" flameproof enclosure through the metallic conduits of the electrical system.

b) **Absence of the sealing compound**

The Ex d installations with metallic conduits need conduit sealing fittings, in order to prevent the passage of gases, vapours, or flames from one portion of a conduit system to another. They also restrict large amounts of ignitable gases or vapours from accumulating to confine explosive pressure. As shown in Figure 11.3, conduit sealing fittings for Ex d installations are explosion-proof, dust-ignition proof, and raintight also.

The standards are specific about sealing conduits as directed by manufacturer's instructions using approved damming fibre and sealing compound. Inferior sealing methods, such as filling a conduit seal with silicone or electrical sealing putty, are unacceptable.

These remedial methods fall short of standard compliance and achieving functional objectives. The seal fitting installation is not complete if the seal is not dammed and poured. So, if the sealing fitting chamber is not filled with an approved sealing compound, the explosion propagation between adjacent Ex enclosures would not be stopped, and this is considered a dangerous non-conformity.

c) **Integrity of the flamepaths**

Flamepaths and internal parts of glasses and glass/metal seals are not rare to be found damaged. The flanges and internal parts of glass mountings and glass/metal seals should be examined for damage or corrosion effects. This is an internal inspection, requiring the panel is isolated before the removal of covers.

It is important to ensure that flameproof joints shall not be painted. Flameproof enclosures may be used in dusty environments and in combined dust and gas/vapour locations if additional precautions are adopted. An enclosure that is specifically designed as flameproof/weatherproof shall be used where a flameproof enclosure is exposed to the weather or installed in wet conditions as available.

Weatherproofing is usually achieved in this type of enclosure by gasketed joints, which are additional to and separate from the flamepaths. The weatherproofing of other flameproof enclosures may be achieved by the use of a suitable grease, or flexible non-setting compound in the flame path, provided that chemicals with which they may come into contact do not adversely impacted these.

Gaskets shall not be inserted in flamepaths. IEC 60079–14 on clause 10.3 allows enclosure painting by the user (a maintenance task) after the complete assembly of the Ex d enclosure. But, on the same paragraph, it warns that in this case, the ink penetration into the gap is possible to occur, which can compromise the hot gases' cooling when they pass through the flamepaths.

The NFPA 70B also stresses that grease, paint, and dirt shall be cleaned from machined joints.

As safety cannot be compromised, the inspector needs to ensure that flameproof joints are kept clean.

d) **Artificial IP improvement**

When reassembling flameproof enclosures, all joints shall be thoroughly cleaned and may be lightly smeared with a suitable grease to prevent corrosion and to assist weatherproofing. Until the early 2000s, Ex d method of protection had no defined requirements for a high IP rating. It was customary to use silicone grease on the flanges of the Ex d enclosures to "increase" the IP rating, which was supported by IEC 60079–14.

The use of silicone grease in the assembly activity can prevent seizure in the cylindrical joints and in the threaded joints, facilitating the coupling, considering the tight tolerances involved. A grease must be used that does not "age" and that does not contain solvents that can evaporate and corrode.

However, the idea of improving the IP rating by applying a grease-bearing textile tape is controversial (see Figure 11.4) because in the event of an internal explosion, the tape acts as a solid obstacle, unlike silicone grease which is expelled by the force of escaping gases. IEC 60079–14 specifies the minimum distance that solid obstacles must be located from flameproof flange joints, depending on the Gas Subdivision of the hazardous area (see Table 11.2).

Although the use of grease-bearing textile tape, employed outside of a straight flanged joint, under the conditions established in IEC 60079–14 (with gases allocated to Group IIA, that the tape should be restricted to one layer surrounding all parts of the flange joint, as recommended by an old ERA report), is allowed, a test carried out at CEPEL (Brasilian Ex accredited lab) with a tape from an English manufacturer, wrapping an Ex d box certified for Group II B, revealed two problems when the propagation test was done:

- Occurrence of a propagation of the explosion to the external environment once, as the hot gases could ignite the flammable grease (flash point = 290°C) embedded in the tape
- Overpressure of 16% inside the enclosure, which was not recorded in the certification documents.

FIGURE 11.4   Application of greased tapes over the Ex d flange joint.

TABLE 11.2
Minimum Distance from the Flameproof Flange
Joints to Solid Obstacles

Equipment Group	Minimum Distance [mm]
IIA	10
IIB	20
IIC	30

This greased-textile tape is used as a corrosion protection for buried metal pipe-lines when they emerge from the soil, and has been only recommended for this application by the British Gas Company. As it is not approved for use on Ex d enclo-sures, aggravated by the fact that the grease is flammable, such use is considered a non-conformity. The recommendation is to verify if the flange joint is in a perfect condition, where the use of a thin layer of grease would be used just for corrosion protection; if not, proceed with a flange repair, or replace the Ex d enclosure.

e) Cable entries

As Ex d enclosures need to contain a possible internal explosion, it is essential that the cable entry devices are certified and correctly installed. In Figure 11.5, from the left to the right, Ex d entry devices for use with unarmoured cables, armoured cables, conduits containing various cables, and conduits containing a single multi-cable are shown.

f) Corrosion

Where a flameproof enclosure is exposed to corrosive conditions, its safety features may be impaired by corrosion of the enclosure, cement, or other sealing materials. It shall therefore be suitably protected by, for example, painting external surfaces (including the cement) and the greasing of flanges.

Consideration shall also be given to increasing the frequency of maintenance. A recommendation is that the interval between inspection periods shall not exceed three years, without seeking expert advice.

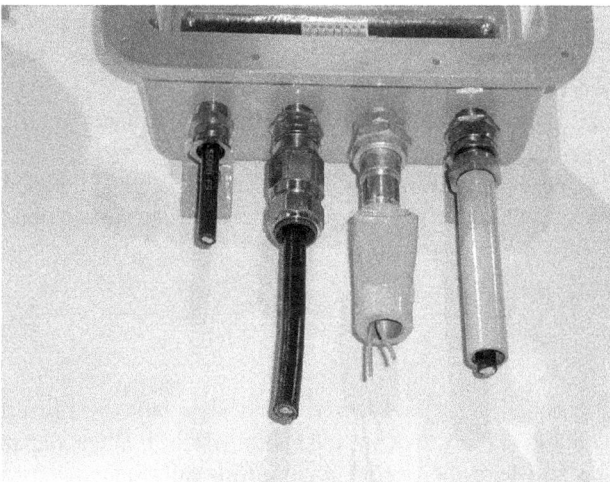

FIGURE 11.5   Different Ex d cable entry devices.

11.1.6.2.2   Type Ex p

In this type of protection, the inspection sheet should include the following verifications:

a) **Integrity of ducts**

Verify if inlet and exit ducting are undamaged. This is intended to ensure that there is no damage to the ducting which could cause gas leakage into or out of the ducting as this could contaminate the enclosure or exhaust area dependent upon the operating system.

b) **Design pressures**

Verify if the pressure and flow of pressurizing gas are adequate. The internal pressure needs to be kept above the minimum and below the maximum specified pressure for the enclosure. The pre-energization purge period should also be checked.

11.1.6.2.3   Type Ex o

This type of protection relies on the immersion of conductive parts in an isolated liquid, so it is necessary to verify if there is no leakage of oil by observation of the sight glass. If electrical conductors are exposed, above the oil level, then, equipment isolation and prompt repair are necessary.

11.1.6.2.4   Types Ex i and Ex iD

These types of protection have the following specific points to be inspected:

a) **Documentation**

Check if the documentation informs the following:

- Manufacturer, equipment type and certificate numbers, level of protection, and, in case of gases, Equipment Group, Temperature Class, and, in case of dust, the maximum surface temperature
- Electrical parameters such as capacitance and inductance, length, type, and route of cables
- Special requirements of the equipment certificate(s) and detailed methods by which such requirements are met in the particular installation.

b) **Cables**

Verify if:

- The cables comply with the documentation, especially regarding multi-core cables containing more than one intrinsically safe circuit, and to the protection afforded where cables containing intrinsically safe systems and other cables, run in the same conduit, duct, or cable tray.

- Cable screens are earthed in accordance with the appropriate documentation.
- The integrity of the connections between intrinsically safe circuits and the earth point is kept.

c) **Barriers**

Verify if:

- The correct barriers are fitted.
- The connection of the potential equalization (bonding/earth) bar and its isolation from the enclosure are correctly made.

d) **Loop impedances**

There is no requirement to measure the earth loop impedance of mains-powered equipment associated with intrinsically safe circuits, other than that required for normal control room instrumentation to protect against electric shock.

As, in some equipment, the intrinsic safety earthing is internally connected to the equipment frame, any impedance measurements (such as between the earth pin of the plug and the equipment frame, or the equipment frame and the control panel) shall be made using a tester specifically designed for use on intrinsically safe circuits.

e) **Circuits' segregation**

Verify if segregation is maintained in multi-circuit junction boxes. This is an internal inspection, but isolation is usually not necessary, unless the box contains non-intrinsically safe circuits at potentials which constitute an electric shock risk. Segregation between intrinsically safe and non-intrinsically safe circuits should be maintained as defined in IEC 60079–11 and shown in Figures 11.6 and 11.7.

In brief, when separation is accomplished by distance, then, the clearance between bare conducting parts of terminals shall be at least 50 mm. When separation is

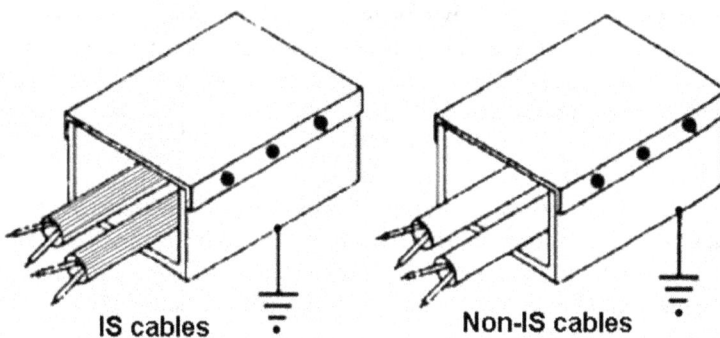

FIGURE 11.6 IS and non-IS cables can run in different earthed metallic cable trays.

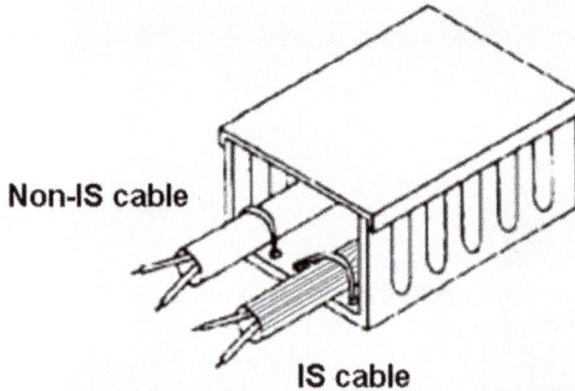

Non-IS cable

IS cable

FIGURE 11.7 IS and non-IS cables can be run in the same duct, provided they are separated by at least 50 mm and adequately tied.

accomplished by locating terminals for intrinsically safe and non-intrinsically safe circuits in separate enclosures or by the use of either an insulating partition or by an earthed metal partition between terminals with a common cover, IEC 60079–11 establishes particular requirements.

f) **Multi-core cables**

Multi-core cables with several IS circuits should not be used in Zone 0 without first studying the possible fault combinations. Only multi-core cables with additional external protection against mechanical damage, exclusively with IS circuits (< 60 Vp) running in adjacent cores, can be considered as not subject to faults.

g) **Frequency**

The recommendation is that in extremely adverse conditions, the interval between periodic inspections may be as low as three months but should not normally exceed two years. However, in extremely good, stable environmental conditions, the interval between inspections may be extended.

11.1.6.2.5 Ex Motors

Visual inspections are necessary to identify possible physical damage, the need for cleaning, signs of corrosion due to air condensation, or other aggressive agents. If the motor is not running for a long time, it is important to monitor parameters such as ambient temperature and relative humidity, especially in humid climates, to check the need to use internal heaters, in order to prevent low insulation in the windings from compromising immediate start-up when necessary.

The recommendation is that the interval between inspections in motors shall not exceed one year. Also verify that the protective devices for the current ratio I_A/I_N do not exceed the time tE stated on the motor's marking plate. Tripping times shall also be measured by current injection at initial inspection.

11.1.7 INSPECTION SOFTWARE PROGRAMMES

As the number of Ex equipment in the plant increases, the volume of information and the number of service orders require increasingly greater resources from the inspection team. In these cases, it may be interesting to adopt dedicated software for the inspection of Ex equipment, which automatically manages inspection periods, records pending issues, and indicates necessary corrective maintenance.

Corrective maintenance can be planned using a repair priority classification, based on the Zone, the risk of ignition, and the type of defect. There are some software programmes specialized in Ex inspections, with features such as:

- Flexibility to define the risk level of each piece of equipment (assigning a score that allows the creation of risk group categories and scheduling activities related to the equipment in each category)
- Capability of assessing legal compliance and asset integrity
- Fully customizable capability, allowing each client's inspection strategy to be introduced in order to automatically generate the relevant service orders, adapted to the repair priorities defined by the client
- Capability of extracting various types of reports that can be consulted at any time via internet.

In order to enjoy all the benefits of the software, it is not enough to simply purchase it and install it on the company's internal network. A planned implementation, with the appropriate assistance from a specialized consultancy, will take care of verifying the Ex experience of the inspection team, in order to locate the right persons on the right tasks.

11.2 MAINTENANCE

Regular servicing is required to maintain the safety of electrical systems in hazardous areas. This requires human and material resources, as well as adherence to safety procedures specifically tailored to site conditions.

11.2.1 THE IMPORTANCE OF EX MAINTENANCE

Hazardous areas are found in small and large facilities such as chemical processing plants, oil production platforms, flammable products' storage tank farms, oil refineries, grain silos, and many others, which require careful management of the explosion and fire risks.

Unfortunately, many companies that have classified areas do not take proper care of their electrical installations, and as a result, they can experience occurrences of small fires and explosions. Usually, employees in these companies do not receive the necessary training, and these small fire events are treated as "isolated cases", where their causes are never clarified, and preventive measures are rarely implemented.

The UK's Health and Safety Executive—through the Control of Major Accident Hazards Programme (COMAH)—carried out an inspection of the electrical installations of 12 chemical and petrochemical industries in the UK in 2002. The idea was to obtain a more realistic assessment of the extent to which the COMAH guidelines were being respected.

The adopted strategy was to first interview the plant management to find out what measures the company was adopting to control risks and, then, carry out an on-site inspection to confirm whether these measures were effectively implemented. The summary of the report revealed the following:

- Of the 12 sites inspected, only 1 was considered to have adequate electrical installations. This meant that 92% of the installations were non-compliant! "Adequate" meant the proper equipment specification and maintenance.
- Six plants had such poor electrical installations that the government agency issued Improvement Notices, which are legal notices to require the operator to bring the equipment up to standard or face prosecution.
- Of the five remaining non-compliant plants, three were in a position to receive a legal summon to make urgent improvements, but considering the specific conditions of the plants at that time, such as renovation work already being contracted, the regulatory body signed a Conduct Adjustment Agreement, a less severe penalty.
- Most of the 11 non-compliant companies acknowledged that their plants had hazardous areas but did not have any documentation regarding this.
- Most of the 11 non-compliant companies did not have professionals, either by their own or hired, with the necessary engineering training to coordinate a risk management policy in this area.
- The training given to workers was considered poor at five of the sites visited. Although the companies claimed to employ trained workers, only one of them had gone through a refresher programme for training in explosive atmospheres.
- The maintenance history of Ex equipment was non-existent in the smaller companies. Therefore, it was not possible to demonstrate that the equipment had maintained its integrity since its installation.

With this example, it is clear that it is necessary for business owners to understand that the risk involved is: Explosion! It can take lives, destroy properties, shut down the company's activities, and even, ruin the economy of a city. Companies that process flammable substances should consider training, risk management plans, and audits carried out by competent professionals as part of their daily routine. Beware of those who offer "low" prices but do not even know the applicable standards. This is vital for the business to flourish.

Considering that if poorly maintained, electrical installations can act as a source of ignition in hazardous atmospheres, a maintenance plan is highly recommended as a tool to monitor the integrity of the installation, guaranteeing the safety of goods and people. It is important to emphasize that only after the required repairs have been completed, will the safety of the installation be restored.

11.2.2 REFERENCE STANDARDS

The publication IEC 60079–19, on Ex equipment maintenance and repairs, has caused concern for some users, rather than providing peace of mind. Why? Because in addition to establishing requirements for the repairer, in order to ensure that the repaired equipment is returned to the customer with the same original safety characteristics, it also establishes responsibilities and requirements for the user!

Analysing the scope of the standard, it can be seen that its main focus is on repairs to Ex motors, which historically have been repaired more frequently than other devices for explosive atmospheres. With this standard, users have become aware that it is not enough, for example, that an Ex motor returns from repair "operational". To ensure the safety of the installation, the characteristics of the motor in relation to the Ex requirements must be respected and verified.

And, to guarantee such characteristics, the standard established requirements on the technical training of repairers also. This explains the concern of users with many questions that have arisen since then, such as:

- "Did that small workshop, which always offered the lowest price to repair our Ex motors, know and respect the requirements applicable to each motor type of protection?"
- "What procedure did they adopt to repair that imported motor that is in the most dangerous area of the facility?"
- "Did they know the manufacturer's drawing tolerances?"
- "Do their employees know the applicable standards?"

In fact, repair services for Ex equipment have several special requirements. Particularly in the case of motors, it is worth noting that, even if the equipment is "operational" when it returns from a repair, it may no longer meet its original Ex characteristics, which compromises the safety of the installation, as the risk we are dealing with is an explosion. Among the requirements of the IEC 60079–19, the following points stand out for the repairer:

- Repairers must have a Quality Management System that meets the requirements of the ISO 9001.
- The repairer must obtain all information from the manufacturer and the user for equipment repair services. This may include information regarding previous repairs, revisions, or modifications. Among the data that must be available for the repair are:
  - Technical specifications
  - Drawings

- Disassembly and assembly instructions
- Certificate limitations (special conditions for safe use), when specified
- The list of spare parts.
- The repairer must have access to and be familiar with the standards that relate to the types of protection that are required for explosive atmospheres.
- Upon the completion of the work, the service report must be submitted to the user, containing, among other information, descriptions of the faults detected, complete details of the repairs and revisions, and a list of the parts replaced or recovered.

For the end users, the following items stand out:

- It is recommended that the equipment design and other referenced documents be obtained as part of the original purchase contract.
- Records of any previous repairs, overhauls, or modifications should be kept by the user and made available to the repairer.
- The user should, whenever possible, notify the repairer of the failure and/or nature of the work to be carried out.
- The user should ensure that the repairer can demonstrate compliance with the relevant requirements of IEC 60079–19.

As it can be seen, the IEC 60079–19 also pointed out responsibilities to the user. This is a completely new fact, which requires users to integrate and structure their purchasing and maintenance departments, so that they have important information at hand for future repairs of Ex equipment.

It is also worth highlighting the identification of repaired equipment, with the placement of a symbol that reflects the condition of the equipment after the repair has been carried out, as follows:

a) *Equipment repaired in accordance with the manufacturer's specification and conformity certification*: The symbol to be placed on the equipment is the letter R inscribed in a square, as shown in Figure 11.8.

This marking should only be used when the repair is in accordance with the standard, and the repairer has sufficient evidence of full compliance with the manufacturer's certification documents and/or specifications.

b) *Equipment repaired in accordance with the requirements of the standards but not with the certification documents*: The symbol to be placed on the equipment is the letter R inscribed in an upside-down triangle, as shown in Figure 11.9.

c) *Other situations: If, after repair, the equipment is not in accordance with item a) or b), the original* certification markings must be removed to provide a clear indication that the equipment can no longer be associated with that considered certified.

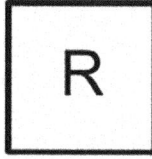

FIGURE 11.8   Sign of a repaired equipment that met the requirements of the standard and the conformity certificate.

FIGURE 11.9   Sign of a repaired equipment that met only the standard's requirements.

IEC 60079–19 also established requirements of technical knowledge that the repairer must have, emphasizing, therefore, that the maintenance of Ex equipment cannot be done by just anyone. To ensure users have real peace of mind, a recommendation is to conduct a technical audit, carried out by a professional with proven competence, to identify the repaired equipment in relation to the situations provided for in the standard. In this way, the actual safety situation of the plant could be mapped and assessed, and criteria for a repairer selection policy could be developed and implemented.

In the United States, NFPA 70B requires the implementation of an electrical maintenance program, which includes training, accountability, and documentation of maintenance activities. In essence, while NFPA 70B does not specifically target hazardous equipment in isolation, it provides the framework for maintaining all types of electrical equipment, including those with inherent hazards, to minimize risks and ensure safety.

11.2.3   Recommendations

When performing Ex maintenance services, it is important to note the following:

- Remedial action is to be carried out where necessary. Caution is expressed in that safety must be preserved prior to and during remedial work.
- Flexible cables are particularly prone to damage and may require special consideration in that more frequent inspection may be necessary.
- If any equipment is taken out of service, the disconnected cables must be dealt with by correct termination in an appropriate enclosure, being correctly isolated and insulated or being correctly isolated and earthed.
- Where special bolts and other fastening, or special tools, are required, these items shall be available and shall be used.
- Environmental conditions including ambient temperature and chemical attack effects are required to be examined and verified for suitability.

- All parts of installations shall be kept clean, particularly where accumulations of substances can cause heat build-up.
- Equipment should be undamaged. Weatherproofing applied to equipment should be maintained effectively.
- Care shall be taken in respect of anti-condensation devices operating, vibration not loosening fixings, and the generation of static electricity.

11.2.4   MAINTENANCE PLAN

To ensure the integrity of Ex equipment, its maintenance plan cannot be limited to a generic checklist. Each installation has its own characteristics, and the responsible professional needs to elaborate the Ex maintenance plan considering them, because the industrial processes and their environmental conditions can impact the integrity of the Ex installation. The maintenance plan of Ex equipment has some distinctive topics that require special attention, as explained in the next subsections.

11.2.4.1   Attention Items

a) Environmental conditions

Electrical equipment in classified areas, from all types of protection, can be adversely impacted by environmental conditions such as corrosion, ambient temperature, exposure to ultraviolet radiation, water ingress, and dust accumulation. In the case of equipment installed in coastal regions, known to be impacted by the characteristic salinity of sea water and atmosphere, whether cleaning with high-pressure water jets, additional precautions must be taken to mitigate possible impacts on corrosion.

Corrosion of metal parts can impact the integrity of the equipment's protection type, and if the casing or a component is severely corroded, its replacement should be considered. Therefore, it is recommended that metal enclosures in corrosive environments be protected with an appropriate coating.

Plastic enclosures also require care, as they may present surface cracks due to UV rays, capable of impacting the integrity of the enclosure.

b) Documentation

For maintenance to be performed correctly, at least the following must be available:

a) Area classification drawings, indicating the extents of the Zones.
b) Equipment characteristics such as Temperature Class, type of protection, and degree of protection (IP).
c) Technical documents with specific information for maintenance such as list of spare parts, certificates of conformity, and the previous inspection reports.

c) Qualified personnel

IEC 60079-17 establishes that inspection and maintenance of Ex installations must be performed only by qualified personnel. A qualified person is one who has received training and has demonstrated skills and knowledge in the construction and operation of electric equipment and installations and the hazards involved.

Whether an employee is considered to be a "qualified person" will depend upon various circumstances in the workplace. For example, it is possible and, in fact, likely for an individual to be considered "qualified" with regard to certain equipment in the workplace but "unqualified" as to other equipment type.

In the United States, OSHA 1910.332(b)(3) established training requirements that specifically apply to qualified persons. Workers must be formally authorized by the responsible professionals, who will use their training records on specific technical knowledge as a basis for individual authorization to perform the tasks.

The legislation of some countries considers that an employee who is undergoing on-the-job training and who has demonstrated the ability to perform duties safely at his or her activity level will be able to perform the tasks as long as he or she is supervised by the team coordinator. Training must include instruction on the basic concepts of area classification, and:

- Types of protection and their properties
- Functional characteristics of the equipment
- Interpretation of the information given by the Ex certificate of conformity
- Safety procedures for performing services in classified areas
- Knowledge of the operation and maintenance manuals' orientations.

In addition, refresher training must be provided regularly. Therefore, a worker who only has experience in maintaining industrial equipment in non-hazardous areas cannot be assigned to repair Ex equipment.

d) Modifications

The standard IEC 60079–14, which defines the requirements for electrical installations in explosive atmospheres, addresses the issue of internal changes to Ex equipment:

> Alteration of the internal components of the equipment is not permitted without re-evaluation of the equipment, because conditions may inadvertently be created which lead to pressurepiling, change in temperature class or other issues that may invalidate the certificate.

This is an interesting point, as apparently "harmless" modifications may compromise the characteristics of the equipment in the event of an explosion. In turn, the standard IEC 60079–19, which provides the requirements for maintenance and repairs, addresses the topic also:

> No modification shall be made to certified equipment unless such modification is permitted in the certificate or approved in writing by the manufacturer.

So, in Ex installations, there must be a clear understanding and commitment from the user to the original design. This is justified by the fact that the test approval was obtained on those characteristics. In "common" electrical equipment, there is no great concern regarding modifications, because the focus is limited to functionality;

but, in Ex equipment, the focus goes beyond simple functionality, as safe operation is required.

d) Services in Zone 2

Although many professionals give little importance to Zone 2, and in some cases even consider it as a "non-classified" region, the IEC 60079–17 defines specific safety precautions when performing maintenance services in these locations. According to the IEC 60079–17, essential work involving exposure to live parts may be carried out under a safe work procedure that demonstrates the proposed work on energized circuits won't cause ignition, provided the necessary precautions are applied.

This is extremely difficult to assess, considering the usual operating conditions of a chemical industrial unit, for example.

e) Traceability

Ex equipment are usually subjected to legal requirements, as an example, to have a certificate of conformity. Therefore, when installed, data such as model, brand, purchase date, and serial numbers must be recorded, so that, in the event of manufacture defects, if a recall is issued, the identification of impacted units will be easier.

In this case, the user only will be able to identify which components are included in the recall, if he has the maintenance equipment file complete and updated. Some recalls are well known on Ex lighting fixtures and cable glands. By replacing these devices, consumers were able to maintain the safety of their installations in classified areas.

f) Repairs

In the maintenance of Ex equipment, there are cases in which the manufacturer no longer exists, or the edition of the standard used at the time of manufacture of the equipment is not available. In these situations, IEC 60079–17 recommends that an expert evaluates whether the equipment and points that do not comply with the current provisions can be accepted, without compromising the unit's safety.

11.2.5 Special Requirements

11.2.5.1 Ex i Systems

Ex i systems offer many advantages but require careful maintenance by properly trained personnel.

The design of intrinsically safe systems aims to ensure that any short circuits, and even open circuits in the field circuit, cannot ignite a gas atmosphere.

Maintenance takes advantage of this resource to perform services with the circuits energized. However, major repairs cannot be performed (e.g. repairing printed circuit boards in classified areas).

In practice, the possible actions are limited to the available tools, and, therefore, deciding what can be done would not be difficult. IEC 60079–17 restricts the actions that can be performed with the circuits energized, as described next:

- Disconnection, removal, or replacement of devices and cables
- Adjustment of controls necessary to perform calibration
- Removal and replacement of plug-in components

- Use of test instruments specified in the documentation. If these are not specified, only those that cannot impact the intrinsic safety of the circuit may be used.
- Other specifically documented and authorized maintenance activity.

These requirements are in line with standard maintenance practice for field-mounted equipment and, therefore, nothing new. Work on associated devices installed in non-classified areas must only be carried out if the electrical equipment, or the relevant part of the circuit, is disconnected from the one located in the hazardous area. It is similarly restricted, but there is greater freedom to operate on terminals belonging to non-classified areas.

The most recent interfaces operate at 24 V, with no risk of electrocution. However, it is not uncommon to find interfaces with relay outputs that switch higher voltages — and therefore are with a risk of shock. Where this risk exists, warning labels must be placed and procedures for maintenance activities drawn up.

When special installation and/or maintenance precautions are required, they usually are specified in the certificates of conformity and in the equipment manuals. This information must be made available to technicians if they do not have easy access to the certificate and/or instructions. Although Ex equipment requiring special precautions has an "X" after the certificate number, this often goes unnoticed by maintenance personnel.

11.2.5.2   Ex i Maintenance Workshops

Repair of intrinsically safe equipment and associated devices may only be carried out under favourable conditions by trained technicians using suitable tools. IEC 60079–19 provides guidance on such repairs. There are practical limitations that must be assessed. For example, intrinsic safety barriers are invariably encapsulated and non-repairable.

Interfaces are usually in boxes that are difficult to open and coated with resin. If the technician does not know the circuit parameters and does not have special test equipment, nothing can be done.

In general, replacing a component with an identical one is preferable for economic and safety reasons. Some repairs can be carried out without impacting the safety of the equipment, when observation of the damage allows and quickly determines the extent of the repair. For example, damage to the boxes generally does not directly impact the intrinsic safety of the devices, and, consequently, it is possible to restore the enclosure to its original IP rating.

Repairing printed circuit boards is often considered impractical. Removing components without damaging the board is difficult, restoring the protective varnish during reassembly requires care, and maintaining the original creepage and insulation distances may not be possible.

And there are additional complicating factors: For example, if the manufacturer used lead-free soldering, a repair with ordinary lead soldering will often result in faulty connections.

A record of all repairs should be kept in the equipment folder. The use of "before" and "after" photographs of repairs, included in the maintenance records, is highly recommended.

11.2.5.3 Ex d

The MESG—Maximum Experimental Safe Gap—is one of the parameters used to assess the integrity of an Ex d enclosure, and its maximum value needs to be verified when a maintenance service is done. Its basic principle establishes that, during an internal explosion, the pressure developed must be contained safely and the propagation of the same explosion to the external explosive atmosphere must be prevented by means of pressure relief provided by specially designed openings. The definition of the MESG is "maximum gap of a 25 mm long joint that prevents any transmission of an explosion during ten tests carried out under conditions specified in IEC 60079–1".

The explosion-proof joint gap is defined as the distance between the corresponding surfaces of an explosion-proof joint when the electrical equipment enclosure is assembled. Regarding the Ex d enclosures, they usually are made of aluminium, cast iron, and stainless steel.

Comparing with the cast iron, aluminium has the advantage of being much lighter and, thus, facilitates both the assembly and the maintenance of the system; in respect to the stainless steel, it has lower cost. Furthermore, aluminium has a better corrosion resistance than cast iron, which must be galvanized and varnished protected.

In the past, aluminium enclosures needed special coatings to resist highly corrosive atmospheres such as offshore plants or chemical plants with presence of strong acids. Recently, it was found out that such phenomenon was caused by the incorrect use of aluminium-copper alloys, certainly suitable for their mechanical characteristics and for the easy machining, but, they had the disadvantage of not being completely resistant to corrosion, because of the presence of an electrolyte. Nowadays, the improved corrosion resistance is achieved by the pure aluminium.

Silicon alloys are also highly resistant to corrosion in marine environments, and also in the presence of strongly acidic vapours, which normally characterize hazardous areas.

The aluminium-magnesium alloys also have high resistance to corrosion, but, these alloys, however, cannot be used for the construction of flameproof enclosures, or of any component that is used in areas with potentially explosive atmospheres, because they may cause sparks when rubbed with metal tools. IEC 60079–0 allows for aluminium alloys with a magnesium content of up to 6%. Most manufacturers currently use aluminium-silicon alloys, which contain between 5% and 13% silicon.

Copper is present only as an impurity, and the primary alloys used today contain copper to a maximum of 0.05% in ingods and about 0.1% in the casting. Such alloys ensure perfect protection against corrosion in any environment.

In the past, aluminium alloys with a copper content of at least 0.3% were used. In the best conditions, the quantity of copper was six times higher than it is today. This required the use of resins as corrosion-protective coatings, usually epoxy resin, inside and outside the Ex d enclosure. Figure 11.10 shows this. The inner part is a rigid resin compound and the outer part is flexible.

FIGURE 11.10 Flanged joints of two Ex d enclosures with corrosion-resistant coatings.

Given that each flammable gas or vapour behaves differently during an explosion, even when subjected to the same conditions, the different flammable gases and vapours are classified in relation to the MESG as follows:

* Group I: Mines (involving methane and coal dust).
* Group II: Relating to surface installations, with the following sub-groups:
  - IIC, for gases whose MESG is less than 0.5 mm
  - IIB, for gases whose MESG is between 0.5 and 0.9 mm.
  - IIA, for gases whose MESG is equal to 0.9 mm.

A recurring question when presenting the MESG is: "What should be the IP protection rating for an Ex d enclosure?"

IEC 60079–1 does not establish a single IP rating for Ex d enclosures. In fact, the design of this equipment aims to prevent the external explosive atmosphere from igniting if an internal explosion occurs. Manufacturers of Ex d enclosures that have invested in the technological updating of their production processes are nowadays offering to the market certified Ex-d enclosures with high protection levels such as IP6X.

11.2.5.4 Ex p

The pressurized system requires a careful maintenance, not only because of the possibility of the entry of dirty and/or humid air, but also because the functionality of the pressurization controller must be periodically tested and documented.

When accessing the internal parts of the enclosure, it is necessary to wait 10 to 20 minutes before opening the covers to allow the hot internal parts to cool down.

Factors such as the installation cost, annual air consumption costs, and, sometimes, the unpleasant noises of the pressurization controllers tend to limit the use of this type of protection on big motors and turbomachinery enclosures, usually.

11.2.5.5 Electric Motors

Ex electric motors are widely used in industry, but they require special care during the installation, commissioning, operation, and maintenance activities. In order to ensure the safe operation of these motors in hazardous areas, periodic inspections must be carried out by duly trained professionals, as well as correct maintenance and repair activities, when necessary.

11.2.5.5.1 Parameters

Periodic maintenance on motors must include the monitoring of electrical characteristics such as insulation resistance and polarization index, among others.

11.2.5.5.2 Protection Devices

Thermal protection can provide useful information for both preventive actions and maintenance procedures. The alarm and shutdown temperature settings must be in accordance with the manufacturer's recommendations.

Other protection devices such as surge protectors and vibration sensors must follow the manufacturer's recommendations. Associated Ex i equipment must comply with the parameters defined in the certificate of conformity.

11.2.5.5.3 Repair Workshops

In the event of repairs being necessary, there are actions that must be provided by both the repair company and the owner of the motor, as recommended in IEC 60079–19, and, among them, are:

- The repairer must have a quality management system (ISO 9001).
- The repairer must have a technical manager, duly registered in the local professional engineering council.
- The owner of the motor must provide its certificate of conformity and inform the operational conditions of use.
- The repairer must obtain, either from the user or from the manufacturer, the necessary documentation, including drawings and instructions for disassembly.
- The repairer must obtain the spare parts necessary for the service, preferably from the manufacturer.

11.2.5.5.4 Avoiding Bad Experiences

Cases of motors that were sent to repair workshops without adequate resources in terms of materials and labour, and that failed soon after returning to operation, are not uncommon.

Among the most common causes for this occurrence are:

- Replacing components with "similar" ones that do not have Ex conformity certificate.
- Assembling bearings without respecting the maximum tolerance provided in the documentation.

It is recommended to verify which conditions apply in your country (established by legislation or insurance companies), regarding actions taken by plant personnel (with motor experience but no Ex personal certification), on tasks such as disassembling the motor to replace bearings. In the United States, if a motor manufactured for use in hazardous location has any sign of being dismantled, (match-marks, prick punches, scratched paint, etc.), and it does not have the certifier's stainless steel re-built tag riveted to it, denoting the repair shop's active file number and other info related to its Class/Groups, and so on, the motor is deemed no longer suitable for use in hazardous locations, and this can impact the insurance company requirements.

The Occupational Safety and Health Administration (OSHA) defines what is an "acceptable" installation:

An installation or equipment is acceptable to the Assistant Secretary of Labour, and approved within the meaning of this subpart S:

- If it is accepted, or certified, or listed, or labelled, or otherwise determined to be safe by a nationally recognized testing laboratory recognized pursuant to § 1910.7; or
- With respect to an installation, or equipment of a kind that no nationally recognized testing laboratory accepts, certifies, lists, labels, or determines to be safe, if it is inspected or tested by another Federal agency, or by a State, municipal, or other local authority responsible for enforcing occupational safety provisions of the NEC, and found in compliance with the provisions of the NEC as applied in this subpart; or
- With respect to custom-made equipment, or related installations that are designed, fabricated for, and intended for use by a particular customer, if it is determined to be safe for its intended use by its manufacturer on the basis of test data which the employer keeps and makes available for inspection to the Assistant Secretary and his authorized representatives.

In a brief, an installation is "accepted" by OSHA if it has been inspected and found by a nationally recognized testing laboratory to conform to specified plans or to procedures of applicable codes.

If the motor is not correctly repaired, it may suffer damage when it returns to operation or later soon, which will inevitably lead to more costs, including loss of production for the industry, which will be much more expensive than having the service performed by a repairer that is properly equipped and has personnel trained to correctly perform repairs on Ex motors.

Remember that if there is an explosion, or other incidents involving Ex equipment, there will undoubtedly be an investigation. If the investigation reveals that there was compromising work, compensation will be paid for the damages caused. If the work performed was competent and followed legal standards and requirements, there should be no evidence of intent.

11.2.5.6 Luminaires

Luminaires' enclosures are more readily opened than is the case for other equipment, because of the needs of relamping, etc. and should also carry a label warning against opening while energized, unless they are interlocked. This is necessary as it is normal in standard industrial circumstances to carry out such practices as relamping live, and this must not be the case in hazardous areas.

In addition, it is necessary to check the time required for accumulated energy to dissipate, or for the internal temperature to drop, before opening the covers. This time should be indicated on the luminaire label.

11.2.5.7 Gas Alarm Systems

The performance of gas alarm systems must be verified by a competent person after installation and at suitable intervals, taking care of any pertinent national regulations and manufacturer's instructions.

Where hybrid mixtures could arise, the detectors must be suitable for both phases and be calibrated against the possible mixtures.

11.2.5.8 Non-Electrical Equipment

In order to prevent explosions, the Maintenance and Inspection (MI) programme shall include non-electrical equipment, also. It is important that all inspections and testing be tailored to the specific equipment and environment of the process equipment.

General industry standard procedures will not be applicable in all situations and may be lacking necessary hazard information. As such, employers must tailor site-specific MI procedures that address their unique Process Safety Management (PSM) covered process.

Employers must train each employee involved in maintaining the ongoing integrity of process equipment in an overview of that process and its hazards and in the procedures applicable to the employee's job tasks, to ensure that the employee can safely perform the job task.

As examples of failures on inadequate site-specific inspection in petroleum refineries, the following can be pointed:

- Failure of inspection and test H_2S monitors on a regularly scheduled basis and failure on inspecting and calibrating flow alarms, temperature alarms, flame detectors, auxiliary burners, level alarms, shutdown alarms, and control valve alarms. The worst practice is to verify monitoring devices and alarms only when there was a malfunction.
- Failure to ensure that the pressure gauges of hazardous processes, as a butylene feed coalescer and recycle isobutane coalescer, were calibrated on a periodic basis.
- Failure to adequately inspect and test process piping such as a stream vapour, a stream draw, a diesel draw, a naphtha overhead line, a naphtha/crude exchanger inlet, and a stream reflux line.

- Failure to adequately inspect and test heat exchangers, resulting in nuts being loose, or missing, on anchor bolts.
- Failure to inspect earthing cables for deficiencies.

Employers must ensure that their inspection and testing procedures are specific to their covered processes; otherwise, equipment may be incidentally omitted from the MI programme. The first step of an effective mechanical integrity programme is to compile and categorize a list of process equipment and instrumentation for inclusion in the programme.

11.3   CLOSING REMARKS

Inspection and maintenance of installations in hazardous areas are essential and add value to safety. They must be carried out using appropriate schedules by trained personnel who must collect and analyse the relevant data.

The benefit obtained is especially reflected in the safety of the installation, provided that maintenance has the resources to carry out all necessary repairs in a short period of time. In many companies, workers who perform Ex equipment maintenance are not aware that a single non-conformity, whether in installation or maintenance, can put the safety of the entire industry at risk, as the possibility of explosions is at stake.

In the modern industrial context, with plants that process several flammable substances and have a high production capacity, it is necessary to consider that an explosion will have disastrous consequences not only for the unit but also for employees and the neighbouring community. It should be considered as a moral and legal commitment to keep them safe by implementing consistent Ex maintenance programmes. Therefore, the maintenance team must have the human and material resources to ensure that all necessary repairs are carried out immediately.

BIBLIOGRAPHY

API RP 14F—*Recommended practice on design, installation, and maintenance of electrical systems for fixed and floating offshore petroleum facilities for unclassified and class 1, division 1 and division 2 locations*. American Petroleum Institute, 2008.

Cottin, O. and Lefebvre, X.—Inspection, maintenance and repair of ATEX equipment. In: *PCIC Europe*, Paris, 2007.

Fotau, D., Colda, C., Buriam, S., Friedmann, M., Magyari, M. and Moldovan, L.—Aspects regarding the use and maintenance of electric equipment to be used in explosive atmospheres protected by flameproof enclosures. *Recent advances in Electrical Engineering*, 2013, pp. 113–116.

IEC 60079-1—*Explosive atmospheres—Part 1: Equipment protection by flameproof enclosures "d"*. Ed. 7.0. International Electrotechnical Commission, 2014.

IEC 60079-11—*Explosive atmospheres—Part 11: Equipment protection by intrinsic safety "i"*. International Electrotechnical Commission, 2023.

IEC 60079-14—*Explosive atmospheres—Part 14: Electrical installation design, selection and installation of equipment, including initial inspection*. Ed. 6.0. International Electrotechnical Commission, 2024.

IEC 60079-17—*Explosive atmospheres—Part 17: Electrical installations inspection and maintenance*. Ed. 6.0. International Electrotechnical Commission, 2023.

IEC 60079-19—*Explosive atmospheres—Part 19: Equipment repair, overhaul and reclamation*. Ed. 4.0. International Electrotechnical Commission, 2019.

ISO 9001—*Quality management systems—requirements*. International Organization for Standardization, 2015.

Mackay, Enrique Valera and Berna, Luis Andrades—Consideraciones de los ambientes offshore en el uso de cajas de distribución Ex e. In: *IV Petroleum and Chemical Industry Conference Brasil, presenter: Manuel Aros Gonzalez,*, Rio de Janeiro, 2012.

Maia, Pedro and Carvalho, Flaviano—Best practices in maintenance, repair and overhaul of flameproof motors. In: *PCIC Middle East*, 2022.

Murdoch, Peter—Inspections of installations in hazardous areas adds value to safety and maintenance strategies. In: *II HazardEx*, Coventry, UK, 2003.

NFPA 70B—*Standard for electrical equipment maintenance*. National Fire Protection Association, 2023.

OSHA 1910.332—*Electrical, training, occupational safety and health administration*. Occupational Safety and Health Standards, 1990.

Pearson, Jeff—Explosion protected equipment—UK law and good practice in maintenance. In: *II HazardEx*, Coventry, UK, 2003.

Rangel Jr., Estellito—Instalações elétricas em atmosferas explosivas: inspeção e manutenção são fundamentais para a segurança. In: *I IEEE ESW Brasil—International seminar of electrical safety in the workplace*, Guararema, Brasil, 2003.

Rangel Jr., Estellito, Queiroz, A. R. and Oliveira, M. F.—The importance of inspections on electrical installations in hazardous locations. *IEEE Transactions on Industry Applications*, vol. 52, no. 1, 2015, pp. 589–595. Available at: https://ieeexplore.ieee.org/document/7167696

Rangel Jr., Estellito—A manutenção Ex. *Section EMEx, Eletricidade Moderna Magazine*, São Paulo, ed. 371, Feb. 2005, p. 160.

Rangel Jr., Estellito—Alterações não-autorizadas. *Section EMEx, Eletricidade Moderna Magazine*, São Paulo, ed. 456, Mar. 2012, p. 148.

Rangel Jr., Estellito—Cuidados na manutenção Ex. *Section EMEx, Eletricidade Moderna Magazine*, São Paulo, ed. 518, May 2017, p. 56.

Rangel Jr., Estellito—Espaços disponíveis. *Section EMEx, Eletricidade Moderna Magazine*, São Paulo, ed. 422, May 2009, p. 300.

Rangel Jr., Estellito—Fitas engraxadas. *Section EMEx, Eletricidade Moderna Magazine*, São Paulo, ed. 461, Aug. 2012, pp. 168–169.

Rangel Jr., Estellito—Inspeção e manutenção Ex. *Section EMEx, Eletricidade Moderna Magazine*, São Paulo, ed. 482, May 2014, p. 124.

Rangel Jr., Estellito—The importance of inspections in classified areas. *Vector Magazine*, EE Publishers, South Africa, Oct. 2005, pp. 52–54.

Rangel Jr., Estellito—Manutenção Ex. *Section EMEx, Eletricidade Moderna Magazine*, São Paulo, ed. 431, Feb. 2010, p. 163.

Rangel Jr., Estellito—Manutenção de motores Ex. *Section EMEx, Eletricidade Moderna Magazine*, São Paulo, ed. 525, Dec. 2017, p. 66.

Rangel Jr., Estellito—Manutenção em sistemas Ex-i. *Section EMEx, Eletricidade Moderna Magazine*, São Paulo, ed. 470, May 2013, p. 138.

Rangel Jr., Estellito—MESG. *Section EMEx, Eletricidade Moderna Magazine*, São Paulo, ed. 511, Oct. 2016, p. 78.

Rangel Jr., Estellito—Reparos em equipamentos Ex. *Section EMEx, Eletricidade Moderna Magazine*, São Paulo, ed. 403, Oct. 2007, p. 218.

Rangel Jr., Estellito—Softwares para inspeção Ex. *Section EMEx, Eletricidade Moderna Magazine*, São Paulo, ed. 556, ano 49, Nov./Dec. 2020, p. 72.

12 Safe Working Practices

12.1 INTRODUCTION

In many countries, the sale of Ex equipment does not only depend on the conformity to technical standards, but also, it is subjected to legal regulations. The most common requirement is the presentation of the Ex certificate of conformity, which is the document that confirms the equipment has been assessed by an accredited body, it has passed all tests required for its type of protection, and therefore, it is considered safe to be installed in a hazardous area. So, purchasing Ex equipment needs additional information than purchasing electrical equipment for a non-hazardous area.

While purchasing equipment for use in a non-hazardous area can be done by choosing the functional characteristics described in the manufacturer's catalogue, with Ex equipment it is also essential to check whether the certificate of conformity confirms that the equipment is suitable for the hazardous area in question, and, also, whether there are any limitations on its use, which may be decisive in choosing the most appropriate option for the project.

It is important to note that there are differences between industrial equipment for non-hazardous and for hazardous locations. As an example two fluorescent lighting fixtures may have similar external appearance, but the one for hazardous locations has considerable differences in internal components (such as the ballast) and, also, cannot be allowed to install in locations with atmospheres of certain flammable gases.

In order to evaluate the commercial proposals, it is necessary to know the types of Ex protection and the conditions under which the Ex certificate of conformity was issued. And in the case of international projects, it is also necessary to check whether the Ex certificates of conformity issued in other countries can be accepted in the project, because legal requirements can be applied, as example:

> In European Union (EU): ATEX 2014/34/EU (ATEX 114), that established the requirements for Ex equipment to be sold in EU.
> In UK: the UK Conformity Assessed (UKCA) marking, also known as the UKEx approval, is the conformity marking for products in United Kingdom, that are intended to be used in potentially explosive atmospheres. The UKCA marking replaced the ATEX marking, and only UKCA-issued Ex certificates are accepted in the UK.

The description of the requirements established by some legislation regarding hazardous locations is given next.

12.2 ATEX DIRECTIVE

The acronym ATEX is derived from the French title of this European Directive: "Appareils destinés à être utilisés en ATmosphères EXplosibles". The origins of the ATEX

Directive, applicable on all countries of the European Union (EU), can be found in the Treaty of Rome, which stipulated free movement of goods and services in the EU.

The ATEX Directive consists of two parts: ATEX 114, which concentrates on the duties of the Ex manufacturer, and ATEX 153 (referred to the Article 153 of the Treaty on the Functioning of the European Union), which was previously known as ATEX 137 (referred to the ex-Article 137 of the EC Treaty), which focuses on the end user's obligations.

ATEX 114 does not apply to:

- Medical devices intended for use in the medical field
- Devices and protective systems where the risk of explosion is due solely to the presence of explosives or unstable chemicals
- Devices intended for use in a domestic, non-commercial atmosphere, where a potentially explosive environment rarely occurs, and only resulting from accidental gas leaks
- Personal protective equipment covered by European Council Directive 89/686/EEC
- Sea-going vessels and mobile offshore installations as well as the equipment on board these ships or installations
- Means of transport, i.e. vehicles and trailers thereof, intended solely for the carriage of persons in the air, by road or rail or on the water, and means of transport, insofar as they are designed for the carriage of goods in the air, by public road or rail network, or on the water. Non-excluded from the scope of this Directive are vehicles intended for use in places where there may be an explosion hazard.
- Devices falling within the scope of Article 346 (1) (b) of the TFEU.

ATEX 137 had established "The Protection of Worker at Risk from Potentially Explosive Atmospheres" (Directive 99/92/EC). ATEX 153 replaced it and it is known as the "Use" directive, requiring that employers draw up documented evidence of risk analysis for their site, and that area classification and site inspections must be carried out where potentially explosive atmospheres may develop.

ATEX 153 does not apply to:

- Areas used directly for medical treatment of patients
- Use of appliances burning gaseous fuels
- The manufacture, handling, use, storage, and transport of explosives or chemically unstable substances
- Mineral-extracting industries
- Transport by land, water, and air, except means of transport intended for use in a potentially explosive atmosphere.

Under ATEX, the parties responsible for preventing accidents due to explosive atmosphere are the manufacturer and the end user. When it comes to the safety of the plant, the end user alone is responsible.

In order to fulfil his legal responsibilities under ATEX 153, the safety manager must demonstrate proven competence and experience of the plant. In practice, this could result in imprisonment following an accident while further detailed investigations are carried out and faults are proven. If no safety manager has been formally nominated, the plant manager becomes personally responsible for explosion safety.

Even if the work is contracted out to a third party, the end user is responsible. ATEX 153 does allow outsourcing, but the competence within many organizations may need to be reinforced to check the quality of such work.

Repair shops, and other third parties, have to conform either to ATEX 114 (the manufacturer's directive) or ATEX 153 (the "use" directive, also called ATEX 153). In practice, it means the repair shop needs to align itself with either a manufacturer or the end user, who will take final responsibility to check the work and assess the repair shop.

ATEX 153 introduced a new inspection regime, which is also found in many countries outside the EU. In the past, equipment would have been inspected by engineers from a national authority on electrical items; ATEX 137 had shifted this responsibility to the Labour Office, or equivalent authority, and the inspections were carried out by personnel responsible for health and safety or fire safety. They are not, necessarily, engineers and probably cannot make a judgement on whether a particular piece of equipment is safe; they will verify if the relevant paperwork exists, and if it is updated.

12.2.1 RESPONSIBILITIES FOR EMPLOYERS

Essentially, the employer is required to take all reasonable measures to prevent the formation of an explosive atmosphere in the workplace. Where this is not possible, measures must be taken to avoid the ignition of any explosive atmosphere. In addition, the effects of any explosion must be minimized in such a way that workers are not put at risk.

The main obligations of employers are to:

- Classify the workplace into Zones
- Select Ex-certified equipment according to Zones
- Identify, using warning signs, locations where explosive atmospheres may occur
- Train the workers
- Identify possible ignition sources
- Establish a permit-to-work system
- Prepare the Explosion Protection Document (EPD).

Let us talk about these items in detail.

12.2.1.1 Classify the Workplace into Zones

The first non-conformity found in Ex installation audits is the lack of the area classification drawing, or when there is one, it presents details that are not related to the installation.

As seen in Chapter 3, area classification is an analysis process applied on a case-by-case basis. We can point out as non-conformity at this stage, the preparation of drawings without references to the industrial plant data.

Unfortunately, it is very common to find drawings that only reproduce "typical figures" found in some technical documents but completely dissociated from the installation. This is attributed to the technical inability of the professional in charge of the area classification study.

An electrical installation designer once asked if the API RP 505 had been correctly specified in the project contract, because that document "had no figures for natural gas installations". Then, it was clarified that the figures in that document were merely illustrative and that in several of them, there was a note warning that "the area classification analysis should be done with criteria, judgement since in many cases smaller or larger distances could be applied". So, the area classification for a natural gas installation should be modelled by the designer, as he is the only person responsible for the study, not the document.

Ex installations require to have up-to-date documentation, as it will be very important to guarantee the safety of workers, especially on maintenance interventions. Plant expansions, and the acquisition of new equipment, are activities that cause changes to the existing documentation, and, usually, they are not properly documented.

The IEC 60079–10–1 states: "Revisions must be carried out during the life of the plant". This recommendation is somewhat generic and does not provide any guidance to the user. An experienced professional can interpret the change in the operational process' conditions as a trigger for a review of the area classification drawings.

A preventive measure to avoid outdated drawings is to define that the area classification plan be submitted to a critical analysis every 3 years to verify the need for review, preferably by the same team that prepared the previous one. And, to preserve the register of the occurred changes, it is important to adopt the following orientation: "When updating area classification documents, the scope of the changes made must be included and described in detail. Generic descriptions such as: 'revised according to comments' or 'revised where indicated' are not permitted".

12.2.1.2 Select Ex Certified Equipment According to Zones

Many countries, including those in the European Union, have their own Ex certification system. As example, in Brasil, it was stipulated by INMETRO (Government Organism to Quality and Metrology issues) that imported Ex equipment must have a Brasilian certificate, issued by an authorized certification body, even if it has a valid foreign Ex certificate.

The certificate aims to give the user peace of mind that the equipment has passed the tests required by the standards and is, therefore, considered safe for use in the designated hazardous area. When purchasing Ex equipment, it is recommended to verify whether the products delivered by the supplier meet not only the technical specifications but also the local legal requirements.

It is not unusual that the purchaser of electrical equipment is unaware of the legal requirements for Ex equipment and acquires them in a way, similar to general-purpose electrical equipment by consulting manufacturers' catalogues. This practice of acquiring Ex equipment is considered a non-conformity, as there is the possibility that not only the catalogue is outdated but also that some equipment, although advertised as "flameproof", cannot be considered an Ex equipment if it does not present the respective Ex certificate of conformity.

12.2.1.3 Signage for Hazardous Areas

The signage must be in correspondence of the area classification drawings. Figure 12.1 shows a model of a sign for hazardous areas. It consists of a yellow triangle with black edges (as recommended by ISO standards), and the letters "Ex" inside, in black, on a red background (the colour recommended by ISO standards to indicate fire risks). This model has already been used by Brasilian companies.

It is important to note that we should not place text on warning signs (e.g. "danger", "do not smoke", etc.), because the brain reaction when we see a symbol is faster and more effective than reading texts, especially when they are written in a foreign language. That is why international traffic signs avoid texts.

The location of signs must correspond to the area classification drawings, usually placed in the entrances of the industrial plant's units, and/or close to the sources of release, adding effectively relevant information for the safety of workers: The Zone, Equipment Group, and Temperature Class permitted for fixed and portable equipment that may be used in the area.

This signage helps the inspection activity, allowing a quick check to be made whether the installed equipment is suitable in that hazardous area. It is also valuable because it allows checking whether the marking of the portable equipment brought by the maintenance service team is compatible with the recommendations mentioned in the permit to work and is complying with the requirements for the hazardous area. Also, including the area classification drawing number onto the sign, it allows quick reference to check the height and extent details of the hazardous area.

It is important to emphasize that it is not enough to simply install the signs; it is necessary to train employees and visitors about their meaning, and what to do when one is seen.

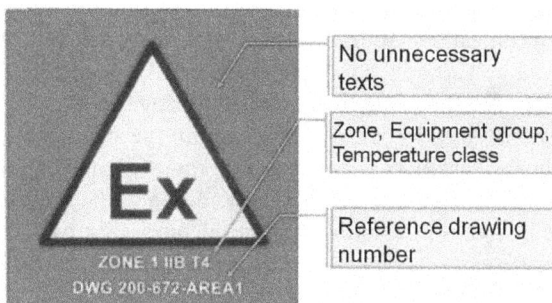

FIGURE 12.1 Warning sign model for hazardous areas, classified as Zone 1, based on the area classification plan and ISO recommendations.

12.2.1.4   Training of Workers

Employers must provide workers with training which informs them, in addition to the characteristics of the equipment and the facility's electrical system, the safety procedures that must be followed when carrying out services in hazardous areas. This training must explain how the potential explosive atmosphere arises, in what parts of the workplace it is likely to be present, and the safety procedures. The correct way of working with the equipment must be explained, and this also involves explaining the meaning of the warning signs in hazardous areas and knowing what mobile work equipment may be used there.

The duty to provide training also applies to the employees falling outside the contract.

Workers must receive safety training:

- On recruitment (before starting work)
- In the event of a workplace transference, or a job change.
- When a new equipment is installed and operated for the first time
- When new equipment or technology is introduced.

It is understood that the training must be carried out by a registered professional authorized by the company to do so. The documentation proving the qualification, and the authorization given to the worker for performing the tasks that were covered by the training courses, must be kept in the company's human resources department in the employee's record file.

This is aligned with IEC 60079–14, as the interventions on Ex equipment (installation, selection, and erection) are allowed to be carried out only by persons whose training has included instruction on the various types of protection and installation practices, relevant regulations, and on the interpretation of the area classification drawings. This is also a legal requirement in many countries, harmonizing that the competency of the worker shall be relevant to the type of task to be undertaken.

As IEC 60079–14 also highlights that "appropriate continuing education, or training, shall be undertaken by personnel on a regular basis"—a training plan that provides for updating with the newest editions of the standards for Ex installations, aims to eliminate time spent on rework, and effectively increases the safety level of the industrial plant. It is important to emphasize that electricians, holding a certificate of completion of a technical course in industrial equipment and systems for non-classified areas, cannot be authorized to work on equipment in classified areas without specific training on the equipment they will be working on.

As Ex installations need to be carried out within the standards' requirements, it is realized that hiring installers based only on previous experience is flawed. This is because new requirements emerge with each edition of the standard, and also the training received by the worker in his previous job, may not cover all risks, and even the equipment technology, encountered in the new worksite.

12.2.1.5   Identification of Possible Ignition Sources

The area classification documents show where the explosive atmospheres can occur, and with this information, the electrical and electronic equipment can be safely selected, depending on their types of protection, stopping them to act as sources of ignition.

However, electrical equipment is not the only source of ignition in these industries. For this reason, the ATEX 153 establishes that the EPD covers a risk assessment in order to identify possible sources of ignition, considering the processed products' ignition information taken from the Material Safety Data Sheet (MSDS) provided.

Among the possible sources of ignition, depending on the particular industrial process, the following can be present:

- Hot surfaces
- Flames and hot gases
- Mechanical sparks (especially in maintenance services)
- Static electricity
- Chemical reaction (thermal instability)
- Lightning
- Stray electric currents (from cathodic protection systems)
- Radiation (optical, ionizing)
- Ultrasonic
- Adiabatic compression and shock waves.

12.2.1.6 Permit to Work

The procedure was developed for the chemical industry, but the principle is equally applicable to the management of complex risks in other industries or situations. The permit-to-work procedure is used in potentially dangerous plant process, when employers are scheduled to work:

- In confined spaces
- In underground or in deep trenches
- Near live equipment or unguarded machinery
- In hazardous areas
- Near corrosive or toxic materials.

All the aforementioned situations are high-risk working situations that should be experienced only by the workers who have previously received the required technical training and are able to follow the permit-to-work conditions. Permits to work must adhere to the following principles:

- The Site Manager has overall responsibility for the permit-to-work, even though he may delegate its issuing.
- It must be recognized as the master instruction, which, until it is cancelled, overrides all other instructions.
- It applies to everyone on site, other trades, and sub-contractors.
- It must give detailed information on, for example, which piece of plant has been isolated, what work is to be carried out, and the time at which the permit comes into effect.
- It remains in force until the work is completed and is cancelled by the person who issued it.
- No other work is authorized on the same permit form. If the planned work must be changed, the existing permit must be cancelled and a new one issued.

The permit to work in a hazardous area is mandatory, and it is required to be documented and formalized. It is also permits the liberation to work in hazardous areas through the elimination of the flammable product that originated the classification of it.

Electrical installations need to be inspected, tested, maintained, and it is often necessary to search for defects in corrective maintenance activities, which imply the presence of energized circuits, machines, and tools that generate sparks. Under these conditions, if it is inadequate to apply de-energization, procedures must be adopted to suppress the risk agent, either by dilution of the flammable product or combustible dust, in the atmosphere, or by eliminating the presence of them, to ensure the safety of the operation and the worker.

Once the work has been finished, a check must be made to establish whether the plant is still safe or has been made safe again. All concerned must be informed when the work is finished.

The permit-to-work form should indicate at least, the following details:

- The location where the work is to be carried out
- Clear identification of the work to be undertaken
- Hazard identification
- Necessary precautions
- Personal Protective Equipment (PPE) needed
- When the work will begin and when it is expected to end
- Acceptance, confirming understanding.
- The marking of Ex equipment, and/or tools, suitable for the Zone
- On closing, report on any anomaly discovered during the work.

Where work may give rise to flying sparks (e.g. welding, flame cutting, grinding), suitable screening should be provided, and a fire sentry is recommended.

The people doing the work (to whom the permit is given) take on the responsibility of following and maintaining the safeguards set out in the permit, which will define what is to be done (no other work is permitted) and the time scale in which it is to be carried out.

The permit-to-work system must help the communication between everyone involved in the process or the type of work. Employers must train staff in the use of such permits, and, ideally, training should be designed by the company issuing the permit, so that sufficient emphasis can be given to particular hazards present and the precautions which will be required to be taken. Therefore, in accordance with IEC 60079–17, for safety in hazardous areas, it is necessary to eliminate the possibility of sparks during the execution of the services or guarantee the absence of an explosive atmosphere during the entire execution time of the task.

If there is a necessity to use electrical equipment to perform the job, the work permit must describe the requirements for them, in the forms of their EPL, Temperature Class, and Equipment Group, considering the area classification documents and the local signage.

12.2.1.7   The Explosion Protection Document

The Explosion Protection Document (EPD) required by ATEX 153 is intended to demonstrate that explosion risks have been determined and assessed, showing that adequate prevention and/or protection measures have been taken. The EPD is integrated with the safety and health policies of the company and needs to be reviewed every time a change is noted in procedures, equipment, or processes.

The EPD can be either a separate document or part of a combined safety report and, as a whole, must cover both technical and organizational aspects of explosion prevention and protection. Thus, for the majority of process plants, it is recommended that these two aspects be separated and dealt with in different documents.

It must be emphasized that the compliance with the minimum requirements set out in the Directive does not guarantee compliance with the appropriate national laws.

12.2.1.7.1   Technical Content

The technical content of the EPD includes:

- **The description of utilities and activities**

  This section should include a short description of the essential aspects of the process or plant, sufficient to identify the objectives, key operations, and major plant items.
- **Fire and explosion characteristics of materials**

  In order to carry out the assessment of a possible risk present in the process and, in particular, the likelihood of ignition, the required characteristics go beyond just whether a material is capable of forming an explosive atmosphere.
- **Hazardous area classification**

  ATEX 153 includes requirements for the classification of hazardous areas in terms of Zones.

  The concept of area classification has been used for many years for the selection of electrical equipment. However, the Directive's requirement for areas containing explosive dust atmospheres, together with the need to register the results of area classification in the EPD, will often lead to a reappraisal of the zone classification. This will be driven by the need specified in the Directive to only use equipment, both electrical and non-electrical, of a specific category in a particular Zone.
- **Identification of possible ignition sources**

  Ignition sources can arise not only from the equipment used but also from the mode or type of operation carried out. Thus, a careful assessment of the likelihood of them occurring during the process is needed. Static electricity, specifically mentioned in the Directive as a possible ignition source, needs also to be considered.

- **Selection of Ex equipment**

    Equipment and protective systems for all places in which explosive atmospheres may occur must be selected on the basis of the categories set out in Directive 94/9/EC. In particular, the following categories of equipment must be used in the Zones indicated, provided they are suitable for gases, vapours or mists, and/or dusts as appropriate:
    - In Zone 0 or Zone 20, category 1 equipment
    - In Zone 1 or Zone 21, category 1 or 2 equipment
    - In Zone 2 or Zone 22, category 1, 2, or 3 equipment.

    As currently with electrical equipment, the choice of suitable equipment also depends on ensuring that the maximum surface temperature remains below the ignition temperature of the probable explosive atmosphere.

    For gases and vapours, equipment is classified into classes T1 to T6 depending on the maximum surface temperature, and this can be compared to the autoignition temperature of the gas or vapour.

    For explosive dust atmospheres, the maximum allowable surface temperature is the lower of either two-thirds of the ignition temperature of the dust cloud or the Layer Ignition Temperature (LIT) minus 75 K.

- **Risk assessment of any work activities involving flammable substances**

    The Directive specifies that the EPD should include an assessment of the risks associated with the work being carried out. In the majority of cases, this needs to be a discussion on the likelihood of occurrence of an explosive atmosphere simultaneously with an effective ignition source and justification of the measures taken. These will need to take into consideration the ignition energy and ignition temperature required to ignite the particular explosive atmosphere.

- **Explosion protective measures in place**

    This section of the EPD should include a summary of the measures which have been applied to prevent the occurrence of an explosion or to protect against the effects of an explosion. Measures include avoiding ignition sources and preventing the formation of explosive atmospheres. One example of the latter is inert gas blanketing.

12.2.1.7.2   Organizational Content

The organizational content of the EPD is a part of the safety management system of the facility and describes the company's policies on:

- Safety, health and environmental policy/responsibilities
- Management of change (MOC)
- Permit to work
- Procedures for visitors
- Training.

The safety management aspects will not be covered in detail here, as it varies from company to company, but, some important points that should not be missed in order to guarantee the safety integrity of the plant are:

a) **Responsibilities:**

There may be circumstances in which it is necessary to ensure that safe environmental conditions are maintained at the time of carrying out the work, in which case it will be necessary to plan for supervision of these conditions before starting the work. The employer must establish an environmental supervision when necessary, as in the case of carrying out activities that, by their nature or by the equipment used, can generate or increase the risk of explosion.

Another case in which it would be necessary to carry out environmental supervision is when, in order to guarantee safety, the temperature or any other environmental parameter must be limited, and, therefore, it must be ensured that the limit is not reached. Environmental conditions must also be monitored whenever the conditions of areas susceptible to the presence of explosive atmospheres change.

b) **Management of Change**

Chemical processing enterprises should establish policies to manage deviations from normal operations. Systematic methods for managing change are sometimes applied to physical alterations such as those that occur when an interlock is bypassed, new equipment is added, or a replacement is "not in kind".

Management of change (MOC) policies should include abnormal situations, changes to procedures, and deviations from standard operating conditions. For an MOC system to function effectively, field personnel need to know how to recognize which deviations are significant enough to trigger further review. It is essential to prepare operating procedures with well-defined limits for process variables for all common tasks.

Once on-site personnel are trained on MOC policy and are knowledgeable about normal limits for process variables, they can make informed judgements regarding when to apply the MOC system. Once a deviation is identified that triggers the MOC system, it is the management's responsibility to gather the right people and resources to review the situation. The skills of a multidisciplinary team may be required to thoroughly identify potential hazards, develop protective measures, and propose a course of action.

c) **Permit to work**

One of the basic preventive measures of an organizational content, against the risk of explosion, should be the provision of specific written instructions to workers to inform about the precautions and work guidelines to follow in the activities to be carried out in hazardous areas, especially in those activities that may aggravate the risk of explosion. This was explained in Section 12.2.1.6.

d) **Procedures for visitors**

This policy should include special considerations for service providers hired to perform services in hazardous locations. Additionally to the normal visitor registration, where their names, contact information, purpose of visit, and other information are usually required, it is recommended to check if they have concluded training on safety procedures for services in hazardous locations. A special identification card can be provided to them, which will facilitate when they ask for working permits from the operational staff and also when accessing certain areas of the premises.

e) **Training**

The technical training on hazardous location topics, including concepts of area classification, types of protection, and safety procedures, is essential for professionals to correctly perform their tasks. It is necessary that the Operations Manager establishes a good relationship with Human Resources Manager, in order that when necessary to contract trainings, the HR does not treat this activity as a "product purchase".

As example, if a product, let us say a television set, of brand XYZ, model 123 is required, the purchaser can make a search in several retail stores and choose to buy where the lower price is, because he will receive the same product that was specified. However, when buying training, other parameters impact the price, such as the expertise of instructors and the quality of the handouts. More details were discussed in Section 12.2.1.4.

12.2.2   RESPONSIBILITIES FOR EQUIPMENT MANUFACTURERS

It is important to know the implications of the ATEX 114 on purchasing Ex equipment.

On April 20, 2016, the current version of ATEX, which governs the conditions for marketing protective equipment and systems (electrical or mechanical) intended for use in hazardous areas, came into force. The document covers both new products manufactured by a company established in the EU and new or second-hand products imported from countries outside the EU.

The following are some of the provisions that imposed additional responsibilities on manufacturers, distributors, and Ex product certification bodies:

- "Equipment" is any device capable of igniting an explosive atmosphere, whether under normal or faulty conditions.
- The EC declaration of conformity was replaced by the EU declaration of conformity, which in a single document brings together all the certificates of conformity required by the European directives applicable to the product.
- Concerns about products from non-member countries have led to additional requirements for importers, who must ensure that the products they place on the EU market comply with the requirements of the new directive and must also ensure that the marking and corresponding documentation of the equipment are available to the competent supervisory authorities in the member countries.
- For ten years from the date the product is placed on the market, importers must keep a copy of the EU declaration of conformity and corresponding technical documentation available to the supervisory authorities.
- Manufacturers must ensure that the product is accompanied by instructions and safety information in a language easily understood by end users and in accordance with the decision of the Member State. Such instructions and

safety information, as well as the labelling, must be clear, understandable, and intelligible.

* The impartiality of conformity assessment bodies must be guaranteed.
* Conformity assessment bodies must take out civil liability insurance, unless this liability is covered by the Member State.
* Member States must establish rules on the penalties applicable to infringements committed by the parties, as provided for in their national law, of the Directive and take the necessary measures to ensure their implementation. The penalties provided for must be effective, proportionate, and dissuasive.
* Certificates of conformity issued under the previous Directive, 94/9/EC, remain valid. If modifications are made to the product, a new certificate must be issued, meeting the requirements of the new Directive.

12.2.3 SPECIAL SITUATIONS

12.2.3.1 Used Ex Equipment

This is an interesting situation, as compulsory certification applies only to new Ex equipment. So, for temporary installations where, as example, gas compressors are rented, the principle adopted is that rented equipment must follow the ATEX 114 also, even if manufactured outside the EU, and intended for installation in the EU.

In case of reconditioned Ex equipment, which has been modified to:

1) Restore its external appearance
2) Restore its performance.

ATEX 114 would not apply if the modifications were not substantial. "Substantial modifications" were defined as those impacting one or more of the following requirements:

* Integrity of explosion protection
* Specification of materials
* Design or assembly
* Existence of ignition sources
* Safety devices
* Information given to the customer, i.e. the operating manual.

12.2.3.2 Plant Expansions

Parts of the new plant connected to the existing plant, for example by a conveyor belt—a typical case in storage expansions, when more silos are installed—would not be considered an expansion, but rather the addition of new equipment, and they would have to comply with ATEX 153.

An extension would be something directly added to the existing equipment, for example an increase in the size of a tank, which would change its behaviour in the event of an explosion. The volume added to the original vessel would not be relevant. The key point is whether the newly expanded plant would operate under the same parameters as the original part before the extension.

12.3 BEYOND SAFETY PROCEDURES

It is important to note that the safety procedures must be so worded that all workers can understand and apply them. If the establishment employs workers who do not have an adequate command of the language of the country, the operating instructions must be written in a language that they understand.

But, to have safety procedures is not enough to guarantee safety; it is necessary to check if they are been respected and if any obstacle to follow them exists. The best way to check this is programming periodic safety audits, verifying the compliance with technical and legal requirements.

12.3.1 THE IMPORTANCE OF EX AUDITS

As seen, explosion safety in hazardous areas depends on a great deal of care, from the design phase until the maintenance of the unit in operation. To verify the compliance with technical and legal requirements, it is recommended to do periodic audits, carried out by experienced personnel with proven experience.

Audits can cover, as example, the following topics:

a) **Purchase specifications**

The technical specification is a very important document, because in addition to describing the desired material, it will be used as a reference for the incoming inspection. Audits can check if technical and legal requirements are been included in the Ex equipment description in purchase orders.

b) **Documents update**

According to IEC 60079–10–1, area classification documents should be reviewed every three years. Audits can check whether this period has been adhered to.

c) **Maintenance procedures**

It is important to check if the safety measures are been followed, as an example:

- Carrying out work on live electrical systems and equipment is prohibited in hazardous areas. Work on intrinsically safe circuits is a permissible exception.
- In hazardous areas, earthing or short-circuiting is only permissible if there is no danger of explosion.
- In the case of all work carried out in hazardous areas, there must be no possibility of ignitable sparks or excessively hot surfaces occurring that may cause an explosion—a potentially explosive atmosphere.
- Replacing components in Ex equipment requires that the characteristics of the original components be respected. Audits can check if maintenance files are kept updated, with maintenance activities duly registered.

- During servicing, items of equipment or plant which could cause an explosion if inadvertently switched on during the work must, if possible, be mechanically and/or electrically isolated. For example, if open flame operations are carried out in a container, all pipes from which a potential explosive atmosphere may be emitted, or which are connected to other containers where such an atmosphere could be present, must be separated from the container and blinded off or closed by some comparable means.

d) **Training process**

The hiring of technical courses deserves special attention. Details such as instructors' resumes and assessment of the course plan (if it reflects the characteristics of the Ex equipment used in the facility) are more important than price because, a bad course can cause dissatisfied employees, services performed with rework, waste of material, and other issues. Audits can check whether the workers assigned to a service are qualified and, also, if services improved their results after training.

e) **Checklist**

Checklists are very important, allowing auditors to compose a dashboard where a plant's compliance index can be estimated, and corrective measures can be applied where necessary, aiming at explosion safety. An example of a checklist is shown in Table 12.1, where new items can be added, depending on the characteristics of the industrial facility.

12.3.2 Prepared for Recalls

Despite the mandatory certification requirements for equipment intended for use in hazardous areas, there have already been cases of "recalls", that is Ex equipment that were placed on the market with manufacturing non-conformities that would impact the safety of users.

The Field Operations Directorate 1–2017 bulletin from HSE (regulatory body in the UK), announced that a family of Hadar Ex luminaires (models HDL100, HDL106, HDL 206, and HDL109), despite already having Ex certificates of conformity, was considered unsafe for use in hazardous areas. The impact of this announcement on the market was significant, given the wide application of this luminaire family in the industry and the manufacturing period stipulated in the bulletin: From 2006 to 2016—ten years!

According to the bulletin, after tests were carried out on samples by a certifying body, these luminaires did not meet the requirements for impact tests, degree of protection, and safety items established in the ATEX 114 for the operating temperature range announced by the manufacturer. As highlighted in the bulletin, users must immediately inspect all their luminaires of those models, and those identified as inadequate must be replaced with others that meet the installation requirements. As the risk involved was explosion, it was recommended that until the replacement was completed (which could be done via relocation or new purchase orders), the luminaires considered inadequate in hazardous areas should be kept de-energized.

TABLE 12.1

A Suggestion for an Ex Audit Checklist

Item	Situation	Recommendations	Deadline
Are hazardous areas properly identified with warning signs?			
Has the area classification plan been periodically reviewed?			
Are the tools used in hazardous areas approved for such use?			
Are work permits issued after an on-site assessment?			
Is the documentation of hazardous areas accessible to all workers?			
Is the periodic inspection plan for hazardous areas being carried out?			
Are only qualified workers authorized to perform services in hazardous areas?			
Is the scope of authorization for workers trained in installations in hazardous areas, formally recorded?			
Is the refresher training plan for workers authorized to perform services in hazardous areas, being respected?			
Have the electrical equipment certified for use in hazardous areas, a record of their periodic inspections?			

So, companies with an Ex maintenance plan where each Ex equipment had its historic file including purchasing details, installing dates, and maintenance records could identify and isolate the involved items.

12.4 PRECAUTIONS ON MAINTENANCE SERVICES

Before any intervention, it is essential to carry out detailed planning and a comprehensive risk analysis, elaborating a safety work procedure for the task. Services in electrical installations must be planned and carried out in accordance with specific and standardized work procedures, with a detailed description of each task, signed by a legally competent person authorized by the employer.

Operating instructions for workplaces where there is a risk of an explosive atmosphere should cover whether special personal protective equipment must be worn, lockout-tagout procedures, evacuation procedures, incident communication procedures, and information on where explosive atmospheres are likely to occur, as shown on area classification drawings and local safety Ex signage.

The signs will alert workers on the type of Zone, the Equipment Group, and the Temperature Class, information required to allow the selection of electrical equipment and tools that can be used safely in the location, without acting as sources of ignition.

12.4.1 WATER JETS

The use of water jets can generate electrostatic charges that are potentially hazardous to personnel and act as ignition sources in hazardous locations. Several serious fires have been attributed to electrostatic discharge during water jet cleaning operations, and electrostatic arc discharge has been known to damage composite materials. The water jet industry's trend towards using higher velocity liquid jets and water of higher purity increases the potential of generating thousands of volts of electricity and has caused several fires from undesired ESD generation around flammable materials.

Usually, the liquid must be a poor conductor in order to build a substantial electrical charge. Plain water used for water jet cleaning is not normally considered a poor conductor of electricity, but, under certain conditions, it can be.

Some tests revealed the following:

- Clean seawater sprayed on oil contaminated walls produced a negative electrical charge
- Clean seawater sprayed on clean walls produced a positive electrical charge
- Water-containing detergents sprayed on either dirty or clean walls produced a negative electrical charge.

The conclusion was that the water spray created a charged cloud of water droplets, which was able to store sufficient energy to create an electrical discharge. The charged clouds were able to create between 2.2×10^4 and 1.0×10^6 V during the washing process. So, when liquids flow through any system, they rapidly build up an electrical charge. If the liquid is conductive, the liquid just as rapidly gives up the electrical charge. For non-conductive liquids, the time to dissipate the electrical charge may be substantial. For high-velocity water jets, the presence of a corona indicates that the potential may be as high as 30,000 V. Among the ways to control static electricity in water jet cleaning operations, are:

- Always bond all components together with an earthing cable and have a secure connection to earth
- Use conductive piping and hoses, verifying if the gaskets do not insulate the piping from the system
- Minimize fluid velocity by using larger diameter piping or hoses.

Any operations in hazardous areas, or inside tanks that contain flammable materials, should be evaluated by experienced engineers to make sure that such operations cannot generate sufficient energy to cause an ignition.

12.5 CONSIDERATIONS ABOUT PARTICULAR DEVICES

12.5.1 FACIAL MASKS

With the COVID-19 pandemic, we had to follow several procedures to hinder the spread of the coronavirus. In addition to frequent hand washing, the use of masks was essential.

But, in hazardous areas, not the mask filtration efficiency is taken in consideration alone. Some masks cannot be used in hazardous areas, because as the mask material is insulating, depending on the activities carried out, there may be a build-up of electrostatic charge, and so, as a precaution, their use is prohibited.

Considering that after some time, the masks become damp, permission for use them in Zones 1 and 2, in Groups IIA and IIB, can be granted, provided that:

• The activities do not generate a significant electrical charge.
• There are no metal components on the surface of the mask.
• The masks are put on and removed outside the hazardous area.

12.5.2 METALLIC TOOLS

In industrial environments, maintenance services often require the use of metallic tools, such as sledgehammers, which, when colliding with metallic parts of the installation, can produce sparks. One concern is whether the spark can ignite an explosive atmosphere, as this possibility must be eliminated in hazardous areas.

The risk of ignition of gases and vapours by impact involving light metals is well known. Several countries have imposed severe restrictions on the use of light metals in coal mines. Lloyds Register of Shipping has restricted the use of aluminium and aluminium paints in sea-going oil tankers. Various restrictions on the use of light metals are also common in many of the oil companies in the North Sea.

There are the "non-sparking tools" made by non-ferrous alloys, such as bronze and copper-beryl, for example, and as these tools are less resistant to impacts than those made of so-called "hard metals", such as steel, there are some records of accidents due to the tool breaking apart during use.

Some entities have published studies on the subject. The API—American Petroleum Institute—for example, published "Sparks from hand tools" in 1956, which stated that this type of tool should not be considered as explosion protected. The latest edition of that document, from 2004, titled "RP 2214—Sparks ignition properties of hand tools", kept practically the same text and conclusions. Another publication, the ISGOTT—International Safety Guide for Oil Tankers & Terminals establishes in item 4.5.2:

> The use of hand tools, such as picks and scrapers, for steel preparation and maintenance, may be authorized without a hot work permit. Their use must, however, be restricted to the deck area and facilities not connected to the cargo system. The work area must be degassed and free of combustible materials. The ship must not be in cargo, ballasting, bunkering, tank cleaning, purging or inerting operations. Tools made of non-ferrous metals are only less likely to create an igniting spark, and are not as efficient as those produced by equivalent tools made of ferrous material. Particles of sand, concrete and other substances can become embedded in the surface of the tool, and when struck against ferrous metals or other hard metals, they can produce sparks. The use of non-ferrous metal tools is therefore not recommended.

Today, entities such as FM Approvals (USA) and BAM—Federal Institute for Materials Research and Testing (Germany)—issue certificates for non-sparking

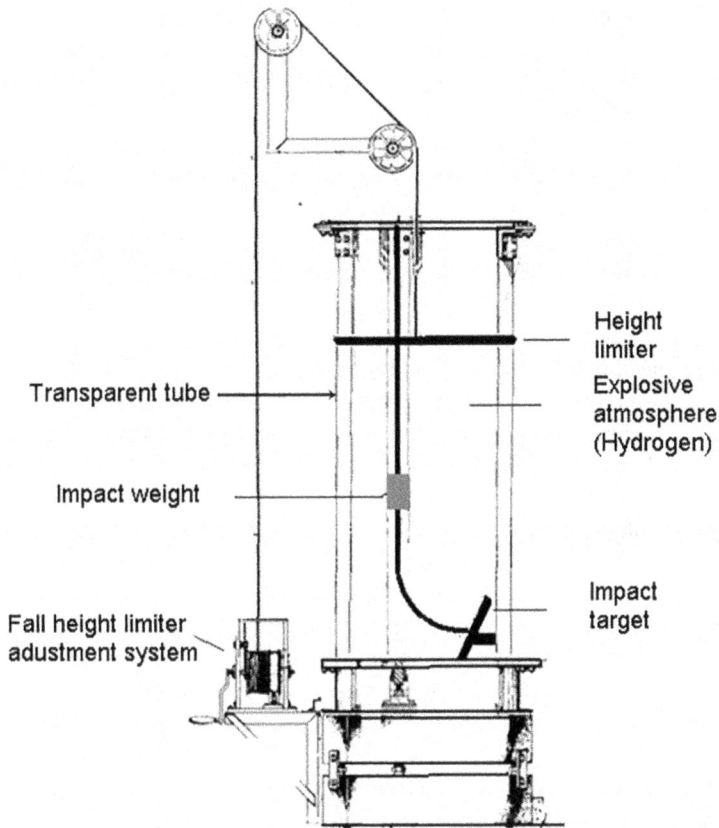

Height
limiter

Transparent tube

Explosive
atmosphere
(Hydrogen)

Impact weight

Impact
target

Fall height limiter
adustment system

FIGURE 12.2   Device used for the evaluation of a metallic tool ignition test.

tools, where their safety limits are explained in detail. Despite all literature about this topic, it was found that a Brasilian PhD thesis pointed, mistakenly, that "any type of metallic tools could be used in any explosive atmospheres without explosion risks". The thesis was based on experiments using the device shown in Figure 12.2.

The test was made using a device constituted of a plastic tube with a diameter of 300 mm and a length of 3,400 mm; a carbon steel "impact weight" of 3 kg that fell from a 3 m height sliding on a metal rail; and an explosive atmosphere of H_2. After 18 repetitions, as no ignition occurred, it was wrongly concluded that "no ignition risks" were there. But, a careful examination of the device construction reveals that the bend on the trail, the friction forces between the "impact weight" and the trail, and the inclination of the "impact target" were responsible for considerable energy losses, and thus the potential energy, which had been theoretically estimated by the ideal vertical free fall equation, was actually much smaller, which distorted the conclusion of the thesis.

This was confirmed by the observation in the Annex V of the thesis, that "during the assembly of the test device (in a room with clean atmosphere), a blow with a carbon steel hammer caused a visible spark", which demonstrated that the device did not reproduce the energy liberated in a real scenario, which should have disqualified the thesis. Promoting as safe the use of ordinary metallic tools in hazardous areas is considered "fake news".

The available literature studies show that the matter is not simple, and it is necessary to differentiate quite clearly between various types of impacts. For example, impacts involving light metals are not necessarily incendiary, as impacts between light metals and non-rusted materials do not produce visible sparks and usually are not able to ignite methane/air mixtures. However, impacts between light metals and rust can easily produce thermite mixtures and thermite reactions. If the light metals contain magnesium, more incendiary sparks are produced than by magnesium free alloys. Harder light metal alloys are generally more incendiary than softer alloys.

Sparks from iron and steel are not reported to ignite methane/air. However, steel sparks may ignite mixtures of hydrogen/air or acetylene/air. Impacts between iron or steel and various metals or minerals may ignite methane–air mixtures, but the ignition is then caused by the hot surfaces arising from the friction, and not by the metal sparks.

Impacts involving hard steels produce higher temperatures at the contact surfaces than softer steels. Because of the difference in heat conductivity, impacts of steel against minerals produce higher temperatures at the contact surfaces than impacts of steel against metals.

The probability of ignition of combustible gas in air mixtures by mechanical impacts is determined by the following parameters, characterizing the impact and the gas mixture:

- Chemical and physical properties of colliding bodies (hardness, grain structure, surface topography, etc.)
- Relative velocity of colliding bodies
- Contact force
- Contact area.

Experiments reported in the literature most often are not defined in the terms of these parameters but rather in terms of apparatus-dependent parameters.

It is difficult, and in most cases not possible, to use the existing knowledge to analyse the potential hazards in practical situations.

In conclusion, the use of metallic tools in hazardous areas, without any additional safety measure, is not recommended!

12.5.3   Mobile Phones

A video recorded by a security camera at a gas station in the Pinheiros district of São Paulo, Brasil, in November 2007, showed an explosion of a tanker truck while it was filling up at the gas station. According to the press, the source of ignition may have been the cell phone of the attendant, who was over the truck and died three days after the incident.

This fact reignited the debate about the risks of using mobile phones in hazardous areas. Many persons refuse to believe that such small phones could cause explosions.

What laypeople do not know is that the energy handled by a mobile phone depends on the energy of the installed battery. Therefore, a person cannot test a mobile phone of a certain brand and model and conclude that the same result would be obtained for others.

Additionally, given the explicit prohibition of mobile phone use in explosive atmospheres contained in users' manuals, the use of mobile phones in hazardous areas is permitted only for Ex-certified phones. If common mobile phones are to be used for short periods, this must be under the specific conditions described in safety procedures and work permits, and only when there is no explosive atmosphere present.

12.5.4 Computers, Tablets, and Cameras

Computers in hazardous areas follow the same requirements as electrical equipment: They must be certified according to a type of protection.

The first question to ask is whether there is a real need for the computer to be in the hazardous area. If the answer is "yes", there are some types of explosion protection available in the computer market, and some examples are given next:

Ex p:

The pressurized (Ex p) concept has as its principle to keep the flammable gas isolated from contact with any hot surface or sparking equipment. This is achieved by installing the computer in an enclosure with a minimum protection rating of IP-54 and pressurizing it with air or an inert gas.

The two methods of purge and pressurization utilized to place equipment in hazardous locations are "continuous flow" and "leakage compensation". In the continuous flow method, a steady, uninterrupted constant flow of air travels into the pressurized enclosure and is even maintained after the purge time has elapsed. During the purging phase, the flow of air serves to clear any potential gas from the enclosure. After purge phase, the flow of air maintains the enclosure at a pressure higher than the surrounding atmosphere.

Leakage compensation uses an initial high-flow rate purge to clear the enclosure of flammable gasses; it then maintains the minimum required pressure using a controller to regulate a smaller flow of air into the enclosure to compensate for any leaks in the enclosure system.

When the equipment is turned on, the controller ensures that a pre-determined number of air exchanges have occurred in the enclosure before powering up the computer. The internal pressure is constantly monitored to ensure that it remains higher than the external pressure, preventing the entry of any flammable gas. The controller compensates for any loss by admitting more air into the enclosure.

But does this type of protection have any drawbacks? The first and most common is the fact that many systems go out of operation due to leaks or even inadequate air supply. This compromises operational continuity, and, consequently, the maintenance team is often under pressure to bypass the safety interlocks required by standard.

As the enclosure is practically "sealed" to prevent excessive internal heat, the consumption of air or inert gas increases, which can become a problem, especially when using high-performance processors and large monitors with Cathode Ray Tubes (CRTs). The pressurized system requires careful maintenance, not only for leaks or the entry of dirty and/or humid air, but also because the functionality of the pressurization controller must be periodically tested and documented. When accessing the internal parts of the enclosure, it is necessary to wait 10 to 20 minutes before opening the covers to allow the hot internal parts to cool down.

Considering other factors such as the installation cost, annual air consumption costs, and the unpleasant noises of the pressurization controllers, maybe better options are available, using other types of Ex protection.

A factor that contributed greatly to the search for new solutions was the availability of larger Ex flat-screen monitors—18" and 24" LED Crystal Displays (LCDs) at affordable prices.

Ex d:

In this design, the monitor, keyboard, pointing device, and CPU are enclosed in one or more explosion-proof enclosures to allow them to be installed in the hazardous area. In most cases, as the only requirement of the production line personnel is that the operator is able to access the plant control system software, one option is to install a terminal in the area, and it can be connected to any standard computer located within about a 100 metres of the plant.

The only parts installed in the hazardous area in this case would be the monitor, keyboard, and pointing device. The connection of the video monitor, keyboard and pointing device outputs could be made via a transmission card using four-pair CAT5 or CAT6 cables.

The advantages of this option are easy network installation, simple maintenance, and relative protection from harsh plant conditions such as temperature, vibration, and humidity. The terminal connecting remotely to the computer allows the users to maintain their communication system, taking advantage of all the commercial and technical benefits of standardization.

Ex i:

A modern solution is the intrinsically safe computer terminal that has been adopted as the preferred choice by many chemical and pharmaceutical industries, as well as other industries such as paints, refineries, oil rigs, and distilleries. In the market, we already find intrinsically safe computer terminals certified, as an example ATEX 2 II G EEx ib IIC T4 (suitable for Zones 1 and 2), consisting of 15" or 18" flat-screen LCD monitors, installed on pedestals with rotation mobility up to 300°, and with trackball-type pointing devices. IP65 is optional.

In environments where there is also a need for easy-to-clean and corrosion-resistant enclosures, stainless steel plays a good role. The terminal can be installed up to 500 m away from the main computer, if necessary.

Ex n:

Another option is the Ex-type restricted breathing, Ex nR, which can be used in the manufacture of cabinets to accommodate monitors, CPUs, and keyboards.

An interesting feature of this option is that it allows the use of personal computers (except for the keyboard—which is of the membrane type—and the mouse, which is of the durapoint type, used in some notebooks), which can reduce the cost by up to 50%, compared to the Ex i and Ex p options.

There are some manufacturers offering this option, with ATEX certification (II 3G EEx nR II T6).

The cabinet is made of stainless steel.

12.5.4.1 Handheld Computers

Many industrial applications require data collection in the field, process control and even asset control, activities that are well performed by "handheld" units. Their use in hazardous areas has been growing, thanks to technological adaptations made to products originally launched for "clean industries", that is to say, those without hazardous areas—by companies specializing in original equipment manufacturers' (OEM) integration. Integration is the process of seamlessly incorporating third-party products, or components, into a manufacturer's own products, which resulted in customized solutions. As example, the circuits are generally re-engineered to result in a type of Ex i protection—intrinsic safety.

The market already offers handheld computers with ATEX certification (II 1 G EEx ia IIC T4), which are capable of running various applications in hazardous areas. The most common casing is made of reinforced plastic, resistant to drops of up to 2 m height onto concrete, and with IP-65, resistant to various solvents also, which makes them suitable even for offshore use.

12.5.5 DIESEL MOTORS

For use in hazardous areas, diesel motors have to follow specific requirements. If the equipment is used in an underground mining, a special design is required, regarding the risks of methane and the presence of coal dust.

It will be necessary to assess the points subjected to friction, high temperatures, and the possibility of sparks being emitted by the exhaust, in addition to the requirements for auxiliary electrical equipment of the diesel engine, such as alternators, starter motors, and electronic controls. For diesel engines to operate within hazardous areas, they require safety modifications that include water-cooled turbo and exhaust system, high-temperature sensors that link to engine shutdown systems, flame arrester, spark arrester, and air inlet shutdown valves.

These modifications are required to meet and comply with Class I Division 2 and Class I Zone II (NEC), ATEX Directive 94/9/EC Zone II hazardous locations with Category 3G (Zone 2), Equipment Group IIA, T3 (200°C) certification. In addition to these safety requirements, other considerations may be taken into account due to the possibility of explosive gases being admitted into the engine.

12.5.6 LITHIUM-ION BATTERIES

Li-ion batteries are widely used in various maintenance tools, such as drills, flashlights, and emergency luminaires. Among the reasons why lithium-ion batteries are gaining popularity and replacing the lead acid batteries in many applications (despite the fact that they are more expensive at initial purchasing) are:

- Lithium ion achieves an energy density of 125–600+ Wh/L versus 50–90 Wh/L for lead acid batteries. This means that for same energy consumption, a lithium-ion battery may last eight to ten times longer than a lead acid.
- The weight and size of lithium batteries are much lower in comparison to lead-acid batteries with the same capacity.
- Charging a lead-acid battery can take more than 10 hours, whereas lithium-ion batteries can take around 3 hours, depending on the size of the battery. So, lithium-ion batteries can charge quicker than lead acid ones.
- Depth of Discharge (DoD) indicates the maximum energy of a fully charged battery that can be used without recharging. Lithium batteries provide 85–100% of their stored capacity, regardless of the rate of discharge. Lead-acid batteries typically provide less usable energy with higher rates of discharge close to 50% only.
- The price of a lithium-ion battery is about two times higher than a lead-acid battery with the same capacity. However, lithium-ion battery lasts longer than a lead-acid battery.
- Usually, lithium-ion batteries possess a cycle life (number of charging and discharging cycles a battery can undergo without compromising its performance) of 5,000, whereas a lead-acid battery lasts for 300 to 500 cycles.

The only area where lead acid batteries perform better than lithium-ion batteries is in the locations of cold climate. Low temperature can cause significant reduction of stored capacity. The lithium-ion batteries are not sensitive to capacity reduction; however, they may face problem of recharging in sub-zero temperatures.

However, both types of batteries hide risks during usage. For example, in both lead-acid and lithium-ion batteries, overcharging may lead to an explosion. Furthermore, the sulphuric acid in the lead-acid battery is highly corrosive, and there is a chance of leakage. If overcharged, hydrogen and oxygen gases may evolve, leading to an explosion.

Lithium-ion and lead acid batteries use the same technology to store and provide energy. The primary difference lies in the material used as cathode, anode, and electrolyte. In a lead-acid battery, lead is used as the anode, and lead oxide is used as a cathode. In a lithium-ion battery, carbon usually is used as the anode, and lithium oxide is used as the cathode.

Lead-acid batteries use sulphuric acid as an electrolyte, and li-ion batteries use lithium salt as an electrolyte. While discharging, ions flow from anode to cathode through the electrolyte, and the opposite reaction occurs while charging, as shown in Figure 12.3.

However, the internal chemical reactions of li-ion batteries generate the growth of "dendrites", metallic filaments that can perforate the separator and cause a short circuit between the anode and the cathode, leading to the phenomenon of "thermal runaway", which causes fires and explosions.

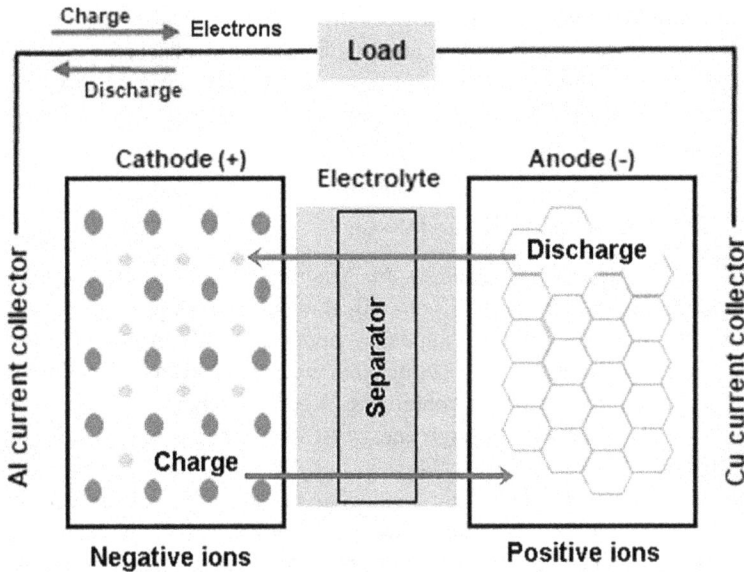

FIGURE 12.3 Lithium-ion battery charging and discharging characteristics.

A certification body has already reported that a Ex headlamp, intended for use in hazardous areas, exploded at night in the laboratory, while it was being charged for the tests next day. There have been common reports in the media of fires in mobile phones and laptops when they are not charging, indicating that internal reactions develop even when the Li-ion battery is in normal use.

Given that the internal chemical reactions of lithium-ion batteries, which generate dendrites (metallic filaments), develop silently throughout their lifetime, and that Ex certificates of conformity are issued on the basis of tests carried out only on new equipment, it is not recommended to use electronic devices equipped with Li-ion batteries, with the current construction technology in hazardous areas, as there is a risk of them acting as a source of ignition.

12.5.7 Gas Sensors

Before an industrial plant containing hazardous areas is operated for the first time, after any damage, or alterations with safety implications, its overall explosion safety must be verified, as mentioned in IEC 60079–14. The effectiveness of the explosion protection measures taken in a plant must be checked at regular intervals, and the frequency of such checks depends on the type of measure. All checks may be carried out only by competent persons, who are persons with comprehensive expertise in explosion protection as a result of their professional training, experience, and current professional activity.

As an example the performance of gas alarm systems must be verified by a competent person after installation and at suitable intervals, observing any pertinent national regulations and manufacturer's instructions. Where hybrid mixtures could arise, the detectors must be suitable for both phases and be calibrated against the possible mixtures.

12.5.8 STATIC CHARGE ON ROAD TANKERS

It is known that electrostatic charges are generated within liquids by movement during operation in processes such as liquids flowing through pipes; mixing liquids; pumping, filtering, and agitating liquids; or pouring liquids from one container to another. All moving liquids can generate static electricity, even if they flow through pipes or are contained in earthed containers. When the potential difference reaches a certain value, a spark with enough energy to ignite an atmosphere of flammable vapours that may be present is possible.

Under certain circumstances, non-conductive liquids, such as hydrocarbon-based solvents, can accumulate electrostatic charges. The generation and accumulation of static electricity can become dangerous if the liquid splashes and forms a mist inside a tank.

Manufacturers of hydrocarbon-based solvents often add antistatic agents to help dissipate electrostatic charges. This information is not usually published, and it should be verified by contacting the manufacturer.

Considering the lack of information regarding the physical characteristics of the fuels produced in Brasil, both in terms of the decay time of the accumulation of charges and the generation of electrostatic charges during transportation, a study was conducted that aimed to evaluate the conditions for the generation and accumulation of static electricity in a fuel distribution terminal, as shown in Figure 12.4.

Special attention is paid to the process of transferring flammable products within the facility, mainly during the loading and unloading operations. The products included in the study were toluene, gasoline A, aviation gasoline, metropolitan diesel, and hydrated alcohol, considering:

- Long-distance transport
- Loading and unloading operations of road tankers, with their top covers open.

The study included:

- Measurement of the static electricity accumulated in the liquids arriving at the terminal, whether by tanker trucks or pipelines, by measuring the voltages of the elements that accumulate this energy
- Assessment of the speed of dissipation of electrostatic charges between the accumulation elements and the earth by measuring the relaxation time of the charge accumulation
- Measurement of the electrical characteristics of the liquids handled at the terminal, as their Material Safety Data Sheets (MSDS) did not indicate them

FIGURE 12.4 A tank truck loading station in a fuel distribution terminal.

- Measurement of the level of static electricity accumulated during the road tanker loading process by measuring the voltage of the tank external surface
- Measurement of the speed of dissipation of the charges accumulated in the tanker truck loading process by measuring the temporal evolution of the voltage between the tank and the earth
- Study of the charge accumulated by the trucks' drivers during the filling process by measuring the voltage between the operator and the earth
- Assessment of the conditions of the earthing system of the site.

The instrument chosen to measure the voltage associated with the electrostatic charge accumulated on a surface was a portable sensor, indicated for use in the range of 0 to 20 kV, as shown in Figure 12.5.

A conductivity meter according to IEC 61620 was used to measure the physical and chemical characteristics of the products. However, this method could not be used for hydrated alcohol due to its high conductivity. Therefore, the methodology of ASTM D1169 was adopted, making it necessary to build a special cell to measure the electrical resistivity of insulating liquids, as shown in Figure 12.6.

The first measurements of the electrostatic voltages on the road tankers in relation to the earth resulted in zero values. The explanation for these values was the time that elapsed by more than an hour between the arrival of the truck at the terminal's external parking lot, and the authorization for the truck to enter for unloading. Also in the initial checks, it was possible to observe that the short distance travelled by the road tanker between the external parking lot and the unloading platform was not enough to generate appreciable voltages on its external surface.

FIGURE 12.5   Electrostatic potential meter.

FIGURE 12.6   Measurement cell under ASTM D1119 specifications.

Measurements of the electrical characteristics of the liquids handled at the terminal, especially conductivity, were performed, and the summarized data are presented in Table 12.2. This phase can be considered to be the most important of this study, given the importance of conductivity for the generation, accumulation, and dissipation of electrostatic charges. There was no reasonable indication of the expected values for conductivity. However, the measured values were compared with the value of 50 pS/m used as a reference for the characterization of the liquids as "semiconductives" or "insulated".

TABLE 12.2

Comparison between the Measured Values (Min and Max) and Tabulated by Literature, for Conductivity of Flammable Liquids

Product	Values [pS/m]		
	Min.	Max.	Tabulated
Hydrated alcohol	99×10^6	130×10^6	1.35×10^5 *
Metropolitan diesel	3.4	26.3	~ 0.1
Gasoline A	13.8	48.7	> 50
Aviation gasoline	0.38	18.3	< 50
Toluene	6.8	553	1×10^{-5}
Hexane	0.03	0.14	1×10^{-5}

Note: In literature, the value indicated for "ethyl alcohol 25°C" was found but without specifying whether anhydrous or hydrated.

To perform the measurements, the two types of loading were considered:

- "Bottom-loading": Where the fuels are injected into the tanker trucks through valves located at the bottom of the tank, without contact with the atmosphere
- "Top-loading": This is where the fuel is injected into the tanker trucks through a cap located at the top of the tank. The fuel comes from a metal piping that is open to the atmosphere inside the tank.

For all liquids under study, measurements were taken as shown in Figure 12.5, both for "top-loading" and "bottom-loading". Most of the measurements showed zero or very insignificant electrostatic voltage values. Only on the platforms with "top-loading" was it possible to measure electrostatic voltage values in the order of tens to hundreds of volts.

In the "bottom-loading" type, the electrostatic voltage was measured at several points of the tanker truck, with emphasis on the part corresponding to the middle height of the tank and near the liquid inlet valve. Zero values, or values close to zero, were found at all points. In the "top-loading" type, before the start of loading, several points of the tank without connection to the earthing cable were also checked, and later with the earthing cable connected.

In both situations, the measured values were always zero or close to zero. During the "top-loading", the electrostatic voltage was measured at several points of the tanker truck, with emphasis on the part corresponding to the middle height of the tank and near the lid located on the top.

12.5.8.1   Measured Values

The results are shown in Tables 12.2 and 12.3.

TABLE 12.3

Electrostatic Potentials Measured in Different Elements of the Fuel Distribution Terminal

Element	Measured Electrostatic Potential
Storage tanks and pipes	Always zero
Tanker trucks arriving at the terminal	Always zero
Top-loading (diesel)	Up to 13 V
Bottom-loading	Up to 4 V
Cotton clothing	Always zero
Hard hats	Up to 543 V
Respirators	Up to 302 V
Truck plastic bumpers	150 V

The resistance values of the earthing grids were between 0.23 and 12.5 Ω. The observations made during the course of the work allow us to say that the electrostatic voltage and decay time values found were insufficient to provide voltages capable of causing electrostatic discharges in the conditions of high ambient temperature and soil humidity encountered during the tests.

The reason of the low voltage on the external surfaces of the tanker trucks was pointed to the good quality of the earthing system and its good maintenance. An important point raised in the work was the very low conductivity value of hexane, which recommends carrying out studies regarding its production process to ensure greater safety in its handling, even with low stock volumes and infrequent handling. This product was not on the original list of products selected for the study.

As for hydrated alcohol, the conductivity value obtained in these tests was an order of magnitude higher than that found in literature, which highlights that this product has not yet been sufficiently precisely identified.

The analysis of the results presented shows that in "bottom-loading", the external voltage of the tank is always zero, from which it can be concluded that the danger of electrostatic discharges in this case is zero, if the safety procedures are been followed.

12.5.8.2 Safety Alert

There is the widely held myth that a tanker earthing clamp must have some mechanisms to prevent it drawing a spark from a charged object, as the tanker will arrive holding charge which it picked up on its journey. In fact, road tankers have slightly static-dissipative tires that help to prevent the build-up of charge by allowing low levels of static electricity generated during movement to flow to the earth through the road surface.

The measurements indicated that "in practice, there is little chance of the road tanker accumulating electrostatic charge on the road through the tires", and, in fact, if this were not the case, simply a person approaching an electrically charged road tanker would cause a spark. However, considering the numerous cases of road tankers that have caught fire, or exploded during loading and unloading operations, even at properly equipped terminals with trained workers and following safety procedures, it can be concluded that not only a grounded clamp but also any object with sufficient capacitance and at a lower electrical potential, which comes into contact with an electrically charged road tanker, can cause a spark to occur. This is the danger of earthing an already charged object.

12.6   CONSIDERATIONS ON EX PERSONAL CERTIFICATION

In many countries, certification is the final step in a process of recognizing a worker's professional competence. Care must be taken to define the type of certification desired, so that the proposal does not represent a factor of exclusion, causing the market to mistakenly interpret that the uncertified professional is incapable.

Among the principles of professional certification, the following stand out: Legitimacy (based on the ethical construction of the certification process), reliability (produced by a precise and suitable process), and validity (by the recognition of the value of the certification by employers and regulatory bodies).

In a country with pronounced regional characteristics, such as Brasil, it is essential to customize the certification models used by other countries, avoiding simply copying them, as the approach needs to bring together the educational and professional training systems.

Considering that in hazardous areas, the non-conformities in the execution of services can compromise the safety of the installation, workers' Ex certification has been seen as a good policy but not enough by itself.

"Competence" is defined in ISO/IEC 17024 as "ability to apply skills and knowledge to achieve intended results" on the basis of skills and knowledge obtained by training, self-study, work experience, or other appropriate means.

Among the benefits provided by professional certification, the company can get the following: Reduced need for supervisors, as the performers are proven to be able to perform the tasks perfectly; reduction of rework for fixing non-conformities; reduction of material waste; and increased productivity.

There are some Ex personnel certification schemes in the market, and one of them, launched in 2010, is IECEx, which is divided in units, one for each activity foreseen to be done in hazardous locations, including design, installation, and maintenance, as shown in Table 12.4. However, it is not enough to simply request that the candidate be Ex-certified in the job advertisement. It is necessary to know the details of the Ex certification scheme and the conditions under which those certificates are issued, especially when there have been reports of workers who were hired as "certified" but did not perform a quality work.

TABLE 12.4
IECEx Units of Competence

Reference	Title	Verify Limitation
Ex 000	Basic knowledge and awareness to enter a site that includes a classified hazardous area	NA
Ex 001	Basic principles of protection in explosive atmospheres	1
Ex 002	Perform classification of hazardous areas	3
Ex 003	Install explosion-protected equipment and wiring systems	1, 2, 3, 4
Ex 004	Maintain equipment in explosive atmospheres	1, 2, 3, 4
Ex 005	Overhaul and repair of explosion-protected equipment	1, 2, 3, 4
Ex 006	Test electrical installations in, or associated with, explosive atmospheres	1, 2, 3, 4
Ex 007	Perform visual and close inspection of electrical installations in, or associated with, explosive atmospheres	1, 3, 4
Ex 008	Perform detailed inspection of electrical installations in, or associated with, explosive atmospheres	1, 3, 4
Ex 009	Design electrical installations in, or associated with, explosive atmospheres	1, 3, 4
Ex 010	Perform audit inspection of electrical installations in, or associated with, explosive atmospheres	1, 3, 4
Ex 011	Basic knowledge of the safety of hydrogen systems	NA

Note:
Possible scope limitations are identified as:
NA—Not applicable
1. Explosion-protection technique (e.g.: Ex d);
2. Product type (e.g.: instrument, motor, etc.);
3. Equipment group (e.g.: IIA);
4. Voltage range (e.g.: 24V to 220V).

The IECEx scheme is regulated by Operational Documents (OD), and in OD 504, were the evidence guide, prerequisites, performance criteria, and scope limitations for each unit of competence are defined.

Three units of competence (Ex 000, Ex 001, and Ex 011) have no prerequisites, and so, alone, they do not provide a basis for a qualification for the person to conduct services in a hazardous area and/or on Ex equipment. Competence in all units, except Ex 000, Ex 002, Ex 005, Ex 010, and Ex 011, shall be assessed either concurrently with, or after, unit Ex 001. Competence in unit Ex 010 shall be assessed either concurrently with, or after, units Ex 002, Ex 008, and Ex 009.

It is important to note that ISO/IEC 17024 does not require persons to have "experience in a role" before they can be certified as competent. Previous experience is a prerequisite in some IECEx units of competence. However, it is surprising that this is not required for the unit Ex 002. Candidates can be certified in Ex 002 just after completing a elementary training course, which is particularly concerning.

The IECEx website shows that sales engineers, business managers, and mechanical technicians have obtained their Ex 002 certificates. However, as their formation courses did not cover industrial processes nor mathematical models for gas and dust dispersions, hiring them based solely on their Ex 002 certificates to conduct an area classification study for an industrial plant is a great risk! Therefore, before hiring an Ex professional, the following checks must be taken:

Does the scope of the certificate cover the same type of equipment, or service, for which the worker will be hired?

Was the certificate issued only upon the completion of a multiple choice test on Ex basic concepts?

Has the certification body adequate facilities to conduct practical assessments, or does it only apply written tests?

Has the certificate any limitations, such as the language that was used in the tests?

Although personal Ex certification is voluntary, there are companies that encourage their employees to obtain it, as a way of identifying the quality of the service they provide. But, such schemes certifiy on the basis of specified minimum requirements; their objective is not to reveal fantastic professionals.

Concluding, a mere curricular assessment, based only on the candidate's work history, without a practical test, may be sufficient to hire people in certain occupations, but not in the field of explosive atmospheres, as it does not identify whether the installations were correctly carried out by the worker. To avoid a situation where a worker can be hired, for having an Ex certificate, but not the required practical knowledge to correctly perform the tasks, it is recommended to apply a practical test, elaborated by a specialized consultancy, to even the Ex certified candidates.

12.7   MANAGEMENT OF CHANGE

On June 1, 1974, occurred the first major accident in the petrochemical industry, at the Nypro Factory, in north eastern England. The Flixborough plant produced caprolactam ($C_6H_{11}NO$) which was used in the production of nylon, and its explosion highlighted the importance of the Management of Change (MOC), for the safety of companies with industrial processes using flammable materials.

Summarizing what happened:

On March 27, 1974, an employee noticed a crack in the side of reactor 5 that was about 2 metres long. The maintenance supervisor suggested stopping production for seven weeks in order to inspect, repair, and test this reactor and also take the opportunity to check the other reactors that were part of the process.

Due to the financial difficulties that the company was experiencing, it was suggested that reactor 5, which had the crack, be removed for repair and that a bypass be created from reactor 4 to reactor 6, so that production would not have to be interrupted for so long. With this suggestion, the process would restart in just three days.

But the diameter required for the flanges was 28 inches, and the factory only had 20-inch pipes to replace them. The bypass was built with this reduction without prior engineering studies, and its sketch of the design was an improvised chalk drawing on the workshop floor.

On April 1, 1974, production was restarted at around 4 a.m. Almost two months later, a leak was detected in the expansion joints, and production had to be stopped for three days.

On June 1, 1974, the process resumed during the early hours of the morning. However, in the afternoon, the piping and expansion joints could not withstand the high pressure and temperature of the fluid that constantly passed between them and ruptured, releasing cyclohexane gas and forming a cloud of approximately 200 metres.

It is estimated that after 30 seconds, the gas reached the petrochemical plant's gas burner, triggering the explosion. It is believed that the cyclohexane leak was between 30 and 50 tons.

The MOC is a policy companies use to manage any health, safety, or environmental risks that arise when facilities, employees, or operations are updated, added, or otherwise modified. The details outlined in an MOC help a business unit to ensure that new or increased hazards are identified and mitigated with a given organizational change.

In the United States, as defined by OSHA 29 CFR 1910.119, the employer is required to implement written procedures that will constitute the Process Safety Information (PSI), to manage changes to process chemicals, technology, equipment, procedures and changes to facilities that impact a covered process. PSI must include information pertaining to the highly hazardous chemicals used, or produced, by the process, information pertaining to the technology of the process, and information pertaining to the equipment in the process.

The MOC procedure requires descriptions of the technical basis for the change, impact on safety and health, modifications to operating procedures, necessary time period for change, and appropriate authorizations. Any employee who will be impacted by the change must be informed and trained appropriately before the unit/ process can restart.

It is important to know the results of audits carried out by Occupational Safety and Health Administration (OSHA) in the United States, which took four years, on non-conformities in the change management system in oil refineries, installations that have considerable extents of hazardous areas. Non-conformities were found in topics such as:

- Equipment design
- Operating procedure
- Regular maintenance/repair.

And, as examples of changes in equipment design where an MOC was not utilized:

- Installing a control valve bypass
- Installing a spill guard berm under a fracturing tank along with proper earthing and bonding
- Changes to an alarm set point
- Changes to materials of construction.

If any piece of equipment is changed to equipment with different specifications from the design in the PSI (i.e. not a "replacement-inkind"), an MOC must be utilized. Likewise, changes in design, such as chemicals used, or increases/decreases in operating parameters outside their range described in the PSI also require MOC. Such changes may result in new hazards or necessitate new or additional safeguards and/or procedures.

Regarding operating procedures, OSHA 3918–08 audits found that some refineries did not utilize MOC when there were changes in them. Examples include:

- Changing procedures for the manual addition of methanol to a chloride injection tank
- Procedures for installing a new type of relief device (different from the original one).

If operating procedures are changed, a MOC is required to assess the potential hazards introduced by the change. Additionally, a MOC ensures proper training on the new operating procedures for impacted personnel, prior to start-up of the process, or the impacted part of the process.

Another recommendation is to analyse external documents before they came into operational procedures. As an example when the EPL concept was presented in Annex I of IEC 60079–14: 2007 as an "alternative method for Ex equipment selection", a controversial statement caused concern:

> It is reasonable for the owner of a remote, well secured, small pumping station to drive the pump with a "Zone 2 type" motor, even in Zone 1, if the total amount of gas available to explode is small and the risk to life and property from such an explosion can be discounted.

This was, in fact, a dangerous statement, because an installation that is carried out without respecting safety regulations cannot be considered "well secured". Furthermore, it is considered illegal in many countries to deliberately allow an explosion to occur, even in "uninhabited" places, without considering the eventual presence of inspectors and maintenance personnel.

Although the expression "small explosion" conjures up an image of something with low energy release, it can impact tanks of flammable products or other nearby equipment, leading to a large explosion via a domino effect and causing considerable material and personal damage. Therefore, the adoption of new standards must also be preceded by an MOC analysis, with the help of a specialized consultancy. And, OSHA 3918–08 audits in petroleum refineries also found that MOC was not used when there was a change in maintenance procedures. Examples include:

- Changing inspection intervals for pipes
- Changing the number of thickness measurement locations (or condition monitoring locations) on a pipe
- Changing maintenance procedures following a change (not replacement-in-kind) in process equipment.

Maintenance procedures will dictate preventive maintenance intervals and repair procedures. Similar to operating procedures, if maintenance intervals or regular repair procedures need to be changed, MOC must be utilized.

12.7.1 HUMAN FACTORS

There is a consensus that facility safety is highly dependent on human factors. Within the context of change management, there are areas for improvement, as the Process Safety Management standard requires the employer to evaluate human factors during its process hazard analysis.

Human factors can be defined as "a common term given to the widely-recognized discipline of addressing interactions in the work environment between people, a facility, and its management systems." The basic principle of assessing human factors is to determine whether the employer "fit the task and environment to the person rather than forcing the person to significantly adapt in order to perform their work".

It is important to know that during OSHA 3918–08 audits in some petroleum refineries, the following were found:

- Inadequate or unsafe accessibility to process controls during an emergency
- A lack of clear emergency exit routes
- Absence of warining signs in hazardous locations
- Inadequate, or confusing, labelling on equipment, procedures, and/or Piping and Instrument Diagrams (PID).

And, some examples were registered:

- Inadequate alarm management, which required operators to perform multiple mental calculations and be aware of an unrealistic number of simultaneous alarms in a unit
- Operation of a bypass valve that required "crawling" across a pipe rack
- Locating emergency isolation valves where operators are in harm's way during an emergency.

The OSHA 3918–08 inspections' programme also found that many refineries have over-relied on administrative controls to address human factors concerns. Over-reliance on administrative controls puts a burden on employees, invites confusion, and can increase the risk of failure-on-demand in an emergency.

Moreover, administrative controls may be inappropriate as the only protection against hazards with potentially severe consequences. It should also be realized that it may be very difficult to eliminate human errors from occurring, and therefore additional safeguards are always necessary. Personnel may not be admittedly knowledgeable in the tasks at hand, especially in stressful environments.

It is therefore useful in critical operations to design systems that might be specified as fail-safe. This applies to not only the design of the system for operation but also to maintenance activity periods, the time when most of the historically cata-strophic incidents have occurred.

A Human Reliability Analysis (HRA) can be performed to identify points that may contribute to an accidental loss. Human errors are generally related to the complexity of the equipment; human-equipment interfaces; hardware for emergency actions; and procedures for operations, testing, and training.

12.8 CLOSING REMARKS

Safe working practices encompass the procedures, guidelines, and behaviours that are implemented to ensure a safe and healthy work environment. These practices are designed to prevent accidents, injuries, and illnesses by identifying and controlling potential hazards in the workplace.

Furthermore, implementing safe working practices ensures that employees are aware of the risks and are trained to handle hazardous materials safely. This reduces the likelihood of accidents and injuries, safeguarding the health and well-being of the workforce.

The design, installation, and maintenance of bonding and earthing systems require careful attention to provide a continuous, low-resistance path between two objects, or an object and earth. All pieces of metallic or conductive equipment should be effectively earthed. This includes elements of process systems such as hoses, piping, flexible connections, or equipment supports where conductive objects may be isolated from earth by non-conductors.

Actually, many pieces of equipment may be inherently earthed through their contact with piping, building steel, or other earthed conductive equipment. The earthing of piping entering a hazardous location is a common industrial practice. If in question, the resistance to earth should be checked to determine if an external bonding or earthing connection is needed.

MOC is highly effective for ensuring the safety of facilities with hazardous areas, because it systematically evaluates and controls any changes in processes, equipment, or operating procedures. MOC also requires a thorough risk assessment of any proposed change, ensuring that all changes are documented, reviewed, and approved before their implementation.

When hiring new workers for services in hazardous locations, whether their own or outsourced, the selection process must include a practical test, even for "Ex certified workers", whose certificates should be verified about their limitations of scope. By prioritizing safe working practices, industries can create a safer and more sustainable work environment, ultimately benefiting employees and the organization as a whole.

BIBLIOGRAPHY

API RP-2214—*Recommended practice—sparks ignition properties of hand tools*. American Petroleum Institute, 2004.

API RP-505—*Recommended practice for classification of locations for electrical installations at petroleum facilities classified as class I, zone 0, zone 1, and zone 2*. American Petroleum Institute, 2018.

ATEX 114—Directive 2014/34/EU of the European Parliament and of the Council of 26 February 2014 on the harmonisation of the laws of the Member States relating to equipment and protective systems intended for use in potentially explosive atmospheres, 2014.

ATEX 137 (now ATEX 153)—Directive 99/92/EC—Directive on minimum requirements for improving the safety and health protection of workers potentially at risk from explosive atmospheres, 1999.

ASTM D1169—*Standard test method for specific resistance (resistivity) of electrical insulating liquids*. ASTM International, 2011.

Bozek, A., Martin, K. and Cole, M.—Cellular phones in class I, Division 2/Zone 2 hazardous locations. In: *IEEE PCIC—Petroleum and Chemical Industry Conference*, Philadelphia, US, 2006.

CCPS—*Guidelines for process safety fundamentals in general plant operations*. American Institute of Chemical Engineers. ISBN 978-0-8169-0564-9, 1995.

CCPS—*Guidelines for safe process operations and maintenance*. American Institute of Chemical Engineers. ISBN 978-0-8169-0627-7, 1995.

Chubb, J. N.—*Electrostatic ignition risks and tank washing operations*, 1980. Available at: www.oreco.com/wp-content/uploads/2020/03/tankwashingrisks.pdf. Accessed on Nov. 29, 2024.

Crowl, D. A. and Lovar, J. F.—*Chemical process safety: Fundamentals with applications*, 2nd edition. Prentice Hall, 2002.

Eckhoff, Rolf Kristian and Pedersen, G. H. – Can mechanical sparks ignite gases? *Chr. Michelsen Institute (CMI) report no. 02702-1*, 1988.

Eckhoff, Rolf Kristian—*Explosion hazards in the process industries*. Gulf Publishing Company, 2005.

Eckhoff, Rolf Kristian and Thomassen, Odd—Possible sources of ignition of potential explosive gas atmospheres on offshore process installations. *Journal of Loss Prevention in the Process Industries*, vol. 7, no. 4, 1994, pp. 281–294.

Floyd, H. Landis. Hidden danger: Reducing residual risk in your electrical safety program. *IEEE Industry Applications Magazine*, vol. 29, no. 4, Jul./Aug. 2023.

HSE Bulletin Field Operations Directorate (FOD) 1-2017—*Safety alert*. UK: Health and Safety Executive, 2017.

Holländer, L., Grunewald, T. and Grätz, R.—Influence of material properties of stainless steel on the ignition probability of flammable gas mixtures due to mechanical impacts. *Chemical Engineering Transactions*, vol. 48, 2016, pp. 307–312. DOI:10.3303/CET1648052.

IEC 60079-10-1—Explosive atmospheres—Part 10-1: Classification of areas—Explosive gas atmospheres. International Electrotechnical Commission, 2020.

IEC 60079-10-2—Explosive atmospheres—Part 10-2: Classification of areas—Explosive dust atmospheres. International Electrotechnical Commission, 2015.

IEC 60079-14—*Explosive atmospheres—Part 14: Electrical installation design, selection and installation of equipment, including initial inspection*. Ed. 6.0. International Electrotechnical Commission, 2024.

IEC 60079-17—*Explosive atmospheres Part 17: Electrical installations inspection and maintenance*. Ed. 6.0. International Electrotechnical Commission, 2023.

IEC 61620—*Insulating liquids—determination of the dielectric dissipation factor by measurement of the conductance and capacitance—test method*. Ed. 1.0. International Electrotechnical Commission, 1998.

IECEx OD 504—*Specification for units of competence assessment outcomes. IECEx scheme for certification of personnel competence for explosive atmospheres*. Ed. 5.0, 2022. Available at: https://www.iecex.com/dmsdocument/1556/.

ISO 7010—*Graphical symbols—safety colours and safety signs—registered safety signs*. International Organization for Standardization, 2019.

ISO/IEC 17024—*Conformity assessment—general requirements for bodies operating certification of persons*. International Organization for Standardization, 2012.

ISO/TS 20559—*Graphical symbols—safety colours and safety signs—guidance for the development and use of a safety signing system*. International Organization for Standardization, 2020.

ISGOTT—*International safety guide for oil tankers & terminals*, 6th edition. IAPH/ICS/OCIMF, 2020.

Jones, Ray A. and Jones, Jane G.—*Electrical safety in the workplace*. NFPA, 2000.

Kletz, T.—*What went wrong? Case histories of process plant disasters and how they could have been avoided*, 5th edition. Butterworth-Heinemann, 2009.

Morais, Caroline Maurieli, Correia, E., Bravim, A. and Ferreira, N.—Criteria for implementation of risk analysis recommendations in oil and gas offshore production installations. Included in: Risk, reliability and safety: Innovating theory and practice, Sep. 2016. DOI: 10.1201/9781315374987-191.

Morais, Caroline Maurieli, Patelli, E., Moura, R. and Beer, M.—Analysis and estimation of human errors from major accident investigation reports. *Journal of Risk and Uncertainty in Engineering Systems, Part B: Mechanical Engineering – American Society of Civil Engineers—American Society of Mechanical Engineers*, Oct. 2019, 6(1): 011014.

Morais, Caroline Maurieli, Yung, K. L., Patelli, E., Moura, R., Beer, M. and Johnson, K.—Identification of human errors and influencing factors: A machine learning approach. *Safety Science*, Elsevier, vol. 146, Feb. 2022.

NFPA 70—*National Electrical Code*. National Fire Protection Association, 2023.

NFPA 660—*Standard for combustible dusts and particulate solids*. National Fire Protection Association, 2024.

NFPA 921—*Guide for fire and explosion investigations*. National Fire Protection Association, 2021.

Oliveira, Aloízio Monteiro de—*Curso Básico de Segurança em Eletricidade: Manual de referência da NR-10*. ed. 1. 2007. ISBN 978-85-907318-0-1.

OSHA 29 CFR 1910.119—*Code of federal regulations—process safety management of highly hazardous chemicals*. Occupational Safety and Health Administration, 2023.

OSHA Instruction CPL 03-00-004—*Compliance directive for petroleum refinery process safety management—national emphasis program (NEP)*. Occupational Safety and Health Administration, 2007.

OSHA Instruction CPL 03-00-008—*Revised combustible dust national emphasis program (NEP)*. Occupational Safety and Health Administration, 2023.

OSHA 3918-08—*Process safety management for petroleum refineries: Lessons learned from the petroleum refinery process safety management. National emphasis program (NEP)*. Occupational Safety and Health Administration, 2017.

Portaria 115/2022—Aprova os requisitos de Avaliação da Conformidade para Equipamentos Elétricos para Atmosferas Explosivas—Consolidado. In: *INMETRO—Instituto Nacional de Metrologia, Qualidade e Tecnologia*, Brasil, 2022.

Rangel Jr., Estellito—Eletricidade estática: um risco invisível na indústria do petróleo. *Eletricidade Moderna Magazine*, Ed. Aranda, no. 386, May 2006, pp. 74–83. ISSN 0100-2104.

Rangel Jr., Estellito—Acidentes. *Section EMEx, Eletricidade Moderna Magazine*, São Paulo, ed. 468, Mar. 2013, p. 182.

Rangel Jr., Estellito—Análise de acidentes. *Section EMEx, Eletricidade Moderna Magazine*, São Paulo, ed. 549, yr. 48, Dec. 2019, p. 62.

Rangel Jr., Estellito—As normas brasileiras sobre instalações elétricas em atmosferas explosivas. In: *III ESW Brasil—Seminário Internacional de Engenharia Elétrica na Segurança do Trabalho*, Rio de Janeiro, 2007.

Rangel Jr., Estellito—A nova diretriz ATEX 2014/34. *O Setor Elétrico Magazine*, Atitude Editorial, ed. 108, Jan. 2015, pp. 68–74. Available at: http://bit.ly/3h1ebq7.

Rangel Jr., Estellito—Atualização de documentação Ex. *Section EMEx, Eletricidade Moderna Magazine*, São Paulo, ed. 457, Apr. 2012, p. 164.

Rangel Jr., Estellito—Auditorias em instalações Ex. *Section EMEx, Eletricidade Moderna Magazine*, São Paulo, ed. 395, Feb. 2007, p. 180.

Rangel Jr., Estellito—Auditorias Ex. *Section EMEx, Eletricidade Moderna Magazine*, São Paulo, ed. 479, Feb. 2014, p. 130.

Rangel Jr., Estellito—Baterias de íons de lítio (II). *Section EMEx, Eletricidade Moderna Magazine*, São Paulo, ed. 577, ano 52, May/Jun. 2024, pp. 50–51.

Rangel Jr., Estellito—Li-ion batteries: Fires and explosions. *Eletricidade Moderna Magazine*, São Paulo, ed. 549, yr. 48, Dec. 2019, pp. 42–47.

Rangel Jr., Estellito—Celulares e áreas classificadas. *Section EMEx, Eletricidade Moderna Magazine*, São Paulo, ed. 407, Feb. 2008, pp. 178, 180.

Rangel Jr., Estellito—Como escolher o treinamento Ex. *Section EMEx, Eletricidade Moderna Magazine*, São Paulo, ed. 455, Feb. 2012, p. 152.

Rangel Jr., Estellito—Conscientização Ex. *Section EMEx, Eletricidade Moderna Magazine*, São Paulo, ed. 575, yr. 52, Jan./Feb. 2024, pp. 54–55.

Rangel Jr., Estellito—Competências de profissionais Ex. *Section EMEx, Eletricidade Moderna Magazine*, São Paulo, ed. 425, Aug. 2009, p. 168.

Rangel Jr., Estellito—Computadores para áreas classificadas. *Section EMEx, Eletricidade Moderna Magazine*, São Paulo, ed. 365, Aug. 2004, pp. 196–197.

Rangel Jr., Estellito—Explosões e incêndios em baterias de lítio: como torná-las mais seguras? In: *Congresso Técnico Científico da Engenharia e da Agronomia—CONTECC*, Gramado, Brasil, Confea, 2023. Available at: https://bit.ly/45PLf8m.

Rangel Jr., Estellito—Ferramentas não-faiscantes. *Section EMEx, Eletricidade Moderna Magazine*, São Paulo, ed. 435, Jun. 2010, p. 234.

Rangel Jr., Estellito—Gerenciamento da segurança Ex. *Section EMEx, Eletricidade Moderna Magazine*, São Paulo, ed. 434, May 2010, p. 174.

Rangel Jr., Estellito—Handling changes from divisions to zones. *IEEE IAS Industry Applications Magazine*, IEEE, EUA, May/Jun. 2004, pp. 49–56. Available at: http://ieeexplore.ieee.org/document/1286611/.

Rangel Jr., Estellito—Máscaras faciais. *Section EMEx, Eletricidade Moderna Magazine*, São Paulo, ed. 560, yr. 49, Jul./Aug. 2021, p. 61.

Rangel Jr., Estellito—Motores diesel Ex. *Section EMEx, Eletricidade Moderna Magazine*, São Paulo, ed. 473, Aug. 2013, p. 128.

Rangel Jr., Estellito—O impacto das fake news na segurança industrial. *Revista Petro & Química, Editora Valete*, ed. 377, yr. 41, Dec. 2018, pp. 24–27. Available at: http://bit.ly/2BlAvCO; ISSN: 0101-5397.

Rangel Jr., Estellito—Panorama da certificação pessoal Ex. *Revista Potência, HM News Editora e Eventos*, ed. 183, yr. 16, Mar. 2021, pp. 66–79. Available at: https://bit.ly/3fWdlX3.

Rangel Jr., Estellito—Primeiro compressor de ar para zona 2 produzido no Brasil. *Eletricidade Moderna Magazine*, Ed. Aranda, ed. 534, Sep. 2018, pp. 21–23. ISSN: 0100-2104.

Rangel Jr., Estellito—Qualificação dos profissionais Ex. *Section EMEx, Eletricidade Moderna Magazine*, São Paulo, ed. 462, Sep. 2012, p. 156.

Rangel Jr., Estellito—Requisitos legais para a comercialização de materiais elétricos. *Eletricidade Moderna Magazine*, Ed. Aranda, Dec. 2001, pp. 164–171.

Rangel Jr., Estellito—Safety in electrical systems on board offshore platforms. *Macaé Offshore Magazine*, G. C. Pinto Publicidade, no. 13, Aug. 2003, p. 9.

Rangel Jr., Estellito—Treinamento Ex. *Section EMEx, Eletricidade Moderna Magazine*, São Paulo, ed. 420, Mar. 2009, p. 154.

Rangel Jr., Estellito—Treinamento Ex—II. *Section EMEx, Eletricidade Moderna Magazine*, São Paulo, ed. 535, Oct. 2018, p. 62.

Rangel Jr., Estellito and Silva, Márcio Valério.—Carregamento eletrostático em terminal de combustível. In: *V Encontro Petrobras sobre Instalações Elétricas em Atmosferas Explosivas—EPIAEx*, Rio de Janeiro, Anais, 2006.

Rangel Jr., Estellito, Motta, Glaucio, Cavalheiro, Igor, Nogueira, Leandro S., Possi, Marcus and Maurer, Rogério—*A Nova NR-10: Quem ganha e quem perde*. 1a. Rio de Janeiro: Edição, Editora Ecthos, 2020.

Safety management systems: Sharing experiences in process safety. UK: European Process Safety Centre, Institution of Chemical Engineers, 1994. ISBN 978-0-85295-356-9.

Smith, Arthur J. and Doan, Daniel R.—Sometimes overlooked safety concerns with large engineered ASD systems. In: *2017 IEEE IAS Electrical Safety Workshop*, Reno, USA, 2017.

Index

For Product Safety Concerns and Information please contact our EU
representative GPSR@taylorandfrancis.com
Taylor & Francis Verlag GmbH, Kaufingerstraße 24, 80331 München, Germany

9781032797076